QUALITY OF FRESH
AND PROCESSED FOODS

ADVANCES IN EXPERIMENTAL MEDICINE AND BIOLOGY

QUALITY OF FRESH AND PROCESSED FOODS

Edited by

Fereidoon Shahidi
Memorial University of Newfoundland
St. John's, Newfoundland, Canada

Arthur M. Spanier
Spanier Consulting
Rockville, Maryland

Chi-Tang Ho
Rutgers University
New Brunswick, New Jersey

Terry Braggins
AgResearch Ltd.
Hamilton, New Zealand

Kluwer Academic / Plenum Publishers
New York, Boston, Dordrecht, London, Moscow

ISBN 0-306-48071-9

©2004 Kluwer Academic/Plenum Publishers, New York
233 Spring Street, New York, New York 10013

http://www.wkap.nl/

10 9 8 7 6 5 4 3 2 1

A C.I.P. record for this book is available from the Library of Congress

Permissions for books published in Europe: *permissions@wkap.nl*
Permissions for books published in the United States of America: *permissions@wkap.com*

Printed in the United States of America

PREFACE

Quality is a composite term encompassing many characteristics of foods. These include color, aroma, taste, texture, general nutrition, shelf-life, stability and possible presence of undesirable constituents. Obviously deterioration of quality may lead to changes in the attributes that characterize the food in its fresh or freshly processed state. In addition, quality enhancement of products may be carried out using appropriate processing techniques. Interaction of different components present with one another could have a profound effect on sensory quality of products. Meanwhile, presence of extraneous matter such as pesticides and debris may also contribute to a compromise in the quality of foods. In addition, processing often brings about changes in many attributes of food including its nutritional value. Thus, examination of process-induced changes in food products is important. In this book, a cursory account of quality attributes of fresh and processed foods is provided. The contributors are experts in their respected fields from the international arena, to whom we are most grateful. The book is of interest to food scientists, nutritionists and biochemists in academia, government and the industry. Extensive bibliography is provided for further reading. The book may also be used as a reference in advanced undergraduate and graduate courses.

Fereidoon Shahidi
Arthur M. Spanier
Chi-Tang Ho
Terry Braggins

CONTENTS

QUALITY OF FRESH
AND PROCESSED FOODS

EFFECT OF ANIMAL PRODUCTION ON MEAT QUALITY

Morse B. Solomon[*]

1. INTRODUCTION

Until recently, improvements in the quality of meat products that reached the market place were largely the result of postharvest technologies. Extensive postharvest efforts have been implemented to improve or to control the tenderness, flavor and juiciness of muscle foods. Tenderness, flavor, and juiciness are the sensory attributes that make meat products palatable and are often the attributes which consumers consider when making their selection to purchase meat products.

Consumers have not only been interested in the quality (palatability) but have been concerned with the nutritional value, safety and wholesomeness of the meat they consume. The public has been inundated with warnings about the health risks of consuming certain types or classes of foods (in particular fat profile). Consumers became more health and weight conscious in the '80s desiring fewer calories in their diet. In fact, the '80s was considered the decade of "nutrition." However, present trends suggest that there is less concern among consumers about many of the substances previously viewed as harmful. There is also less concern about calories. Consumers appear to prefer traditional and familiar foods they have always eaten. New technologies (e.g., biotechnology) and alternative production methods appear to hold great promise for improving the quality and yield attributes of animal products.

A wide range of production practices/strategies for altering the balance between lean and adipose tissue growth and deposition in meat-producing animals are available. These include genetic selection and management (production) strategies. More recently, the confirmation of the growth-promoting and nutrient repartitioning effects of somatotropin, somatomedin, ß-adrenergic agonists, immunization of animals against target circulating hormones or releasing factors, myostatin mutations polar overdominance (callipyge mutation) and gene manipulation techniques have given rise to a technological revolution for altering growth and development in meat producing animals.

---

[*]Mention of trade names or companies does not constitute an implied warranty or endorsement by the USDA or the author. USDA, ARS, BARC-East, Beltsville, MD 20705

Quality of Fresh and Processed Foods, edited by Shahidi et al.
Kluwer Academic/Plenum Publishers, 2004.

1

1.1. Genetic Selection and Management Strategies

The main genetic alteration during the past 40 years has been to decrease carcass fatness and increase lean tissue deposition. These alterations have been via growth rate and mature size of meat animals particularly through genotypic and sex manipulation. There have been numerous papers on the effects of breed and sex condition on carcass composition and meat quality and therefore, this will not be reviewed in this paper.

1.2. Growth-Promoting Agents

Growth-promoting agents are substances that enhance growth rate of animals without being used to provide nutrients for growth such as nutrient partitioning agents. Growth-promoting agents are anabolic; that is, they produce more body tissues and thereby result in more rapid animal growth. Growth-promoting agents cause changes in carcass composition, mature weight, and efficiency of growth. Many substances qualify as "growth-promoting agents" despite their varied origin and chemical nature. Examples of growth promoting agents are the following: antibacterial agents (e.g., antibiotics), rumen modifiers (e.g., monensin, lasalocid), steroids (e.g., testosterone, estrogen), ß-adrenergic agonists (e.g., clenbuterol) and somatotropin (growth hormone). Growth-promoting agents influence growth in three ways: 1) by stimulating feed intake (appetite stimulants) and thereby increasing the supply of nutrients available for growth, 2) by altering the efficiency of the digestive process resulting in improved supply and/or balance of nutrients derived per unit of feed consumed (antibacterial agents, rumen additives, i.e., ionophores), 3) by altering the manner in which the animals utilize or partition absorbed nutrients for specific growth processes.

The effects of various growth-promoting agents on improving efficiency of growth, changing nutrient requirements thereby increasing lean meat yield and concomitantly decreasing fat have been thoroughly reviewed in books edited by (Hanrahan, 1987; Sejrsen et al., 1989; van der Wal, P. et al., 1989; Heap et al., 1989; National Research Council, 1994; Campion et al., 1989; Schreibman et al., 1993; Pearson and Dutson, 1991; Wood and Fisher, 1990). It is not the intent of this chapter to review all the plethora of literature on growth and development but rather discuss the implications these interventions have on muscle and meat quality.

1.3. Appetite Stimulants

Many growth-promoting agents influence voluntary feed intake; however, responses in voluntary intake are generally small and inconsistent. Stimulating appetite (intake) could result in improved growth rate and the production of edible protein (lean tissue). Scientists are investigating the use of drugs to control appetite. If successful substances are developed, it may be possible to enhance meat production in the future by increasing or decreasing an animal's appetite.

1.4. Antibacterial Growth Promoters

A wide variety of antibiotics (antibacterial agents) enhance growth in primarily nonruminant animals. Some common antibiotics that are used to remedy clinical infections are successful at promoting growth. Included in these common antibiotics are: penicillins, tetracyclines, bacitracin, avoparcin and virginiamycin. These antibiotics are active against gram-positive bacteria and, in most instances, are not retained in the tissues of the animal. As much as a 20% improvement in growth rate and feed efficiency have been observed with the administration of antibacterial agents to nonruminants.

1.5. Rumen Modifiers

A special class of antibiotics whose principal site of action is the reticulorumen of ruminants has been investigated. These include: monensin (Rumensin, Romensin) and lasalocid (Bovatec) which are classified as ionophores. Rumen additives are generally administered to growing ruminant animals. The type of diet fed to the animal receiving rumen modifiers has an effect on the outcome. Voluntary intake is depressed when rumen modifiers are included in concentrate diets, whereas little depression in intake is observed when included in forage-based diets. Improvements in growth rate and productivity are generally less than 10%. The mode of action of rumen modifiers is poorly understood but is thought to be a result of altering the metabolism (digestive process) of rumen microflora. They act against gram-positive bacteria in the rumen causing a shift in patterns of volatile fatty acid production, improved digestive efficiency, reduced bloat and production of methane and hydrogen. In a study using lambs the effects of energy source and ionophore (lasalocid) supplementation on carcass characteristics, lipid composition and meat sensory properties reported that ionophore supplementation had no significant effect on either carcass/meat quality or yield characteristics (Solomon *et al.*, 1996; Fluharty *et al.*, 1999). Diet (forage vs concentrate) offered more opportunities for manipulating carcass composition without jeopardizing meat quality than the use of ionophore supplementation.

1.6. Anabolic Steroids and Related Substances

The importance of steroids such as androgens and estrogens in regulating growth is evident from distinct differences between male and females in growth and composition. Castration of males, which removes their primary source of androgens (testosterone), is the oldest method of manipulating growth by non-nutritional mechanisms. A number of steroid-related growth-promoting agents that are either available for commercial use or have been extensively researched exist. These are listed in Table 1. These agents are most effective in castrated males or females (especially in ruminants). Naturally occurring and synthetic estrogens and androgens have been used to improve efficiency of growth and carcass composition of meat animal for more than 50 years. Steroid-related growth-promoting agents generally increase live weight gain and improve carcass lean:fat ratios in ruminants but are not as effective in swine. In fact many of the growth-promoting agents are not approved for use in growing swine in the United States. Their mode of action is unclear but many consider a direct effect on muscle cells. A review of the biosynthesis and metabolism of the naturally occurring estrogens and androgens has been published (Hancock *et al.*, 1991). The literature on growth-performance responses to anabolic steroids indicates great variability, ranging

Table 1. Steroid-related growth-promoting agents

Class	Active Agent(s)	Trade Name
Natural estrogens	Estradiol-17B	Compudose®
Estrogen analogues	Diethylstilbesterol	DES
	Hexestrol	
	Zeranol (resorcyclic acid lactone)	Ralgro®
Natural androgens	Testosterone	
Androgen analogues	Trenbolone acetate	
Natural progestagens	Progesterone	Finaplix®
Progestagen analogues	Melengestrol acetate	
Combined products	Trenbolone acetate + estradiol	MGA
	Testosterone + estradiol	Revalor®
	Testosterone propionate + estradiol	Implix BF®
	benzoate	Synovex-H®
	Progesterone + estradiol	Implix BF®
	Progesterone + estradiol benzoate	Synovex-S®
	Zeranol + trenbolone acetate	Forplix®

from no response in feed lot bulls (Richards *et al.*, 1986) to a 70% increase in average daily gain in heifers (Bouffault and Willemart, 1983). The efficacy of anabolic steroid implants is summarized in several reviews (Hancock *et al.*, 1991; Galbraith and Topps, 1981; Schanbacher, 1984; Muir, 1985; Beermann, 1989).

In a recent study (Fritsche *et al.*, 1999) looking at the effects of different growth-promoting implants on muscle morphology in finishing steers, these researchers found Revalor implants to be most effective followed by Ralgro implants in increasing lean mass through muscle fiber hypertrophy. Synovex implant had the least pronounce capacity for muscle growth enhancement. Other studies (see exogenous somatotropin section) have compared combining anabolic steroids with somatotropin administration in cattle.

1.7. Endogenous Somatotropin

A group of peptide hormones, i.e., somatotropin (ST), growth hormone-releasing factor (GRF), somatostatin, insulin-like growth factor-I (IGF-I), insulin, and thyrotrophic hormone, work in harmony to regulate and coordinate the metabolic pathways responsible for tissue formation and development. Even though relationships between growth and circulating levels of some of these peptide hormones have often produced conflicting results, the majority of data indicates that the genetic capacity for growth is related to increased circulating levels of somatotropin and IGF-I in livestock.

The anterior pituitary secretes three hormones (ST, prolactin and thyroid stimulating hormone) that influence growth and carcass composition. Somatotropin, often called growth hormone (GH), is the most notable with commercial growth-promoting potential. Somatotropin is a small, single chain polypeptide, made up of 191 amino acids, secreted by the pars distalis of the pituitary's adenohypophysis. The structure of somatotropin varies between species. Release of ST is stimulated by growth hormone releasing factor (GRF or GHRH) produced in the hypothalamus (Hardy, 1981). Understanding how to control the production of these hormones within the meat animal is a long-term goal of scientists.

1.8. Exogenous Somatotropin

1.8.1. Porcine Somatotropin

There is a growing database supporting the use of pituitary derived porcine somatotropin (pST) as an agent to improve efficiency of growth and carcass composition in swine. Turman and Andrews (1955) and Machlin (1972) were the first to demonstrate that daily exogenous administration (injection) of highly purified pST dramatically altered nutrient use resulting in improved growth rate and feed conversion of growing-finishing pigs. Pigs injected with pST had less (35%) fat and more (8%) protein. However, their original observations were of little practical significance because purification of porcine ST from pituitary glands was not economical. A single dose required 25-100 pituitary glands. More recently, the development of recombinant deoxyribonucleic acid (DNA) technology has provided a mechanism for large scale production of somatotropin. The gene for ST protein is inserted into a laboratory strain of *Escherichia coli* which can be grown on a large scale and from which ST can be purified and concentrated for use (Harlander, 1981). There is also a growing database supporting the use of recombinantly derived pST (rpST). No significant differences in the effectiveness of pST as compared to rpST have been observed.

With greater emphasis on lean tissue deposition and less lipid, the optimal genetic potential for protein deposition of an animal is a very important concept in that this potential, or ceiling, defines the protein requirement of the animal. In defining the optimal/genetic potential for protein deposition, ST is used as a tool to maximize genetic potential for protein accretion. Administration of pST to growing pigs elicits a pleiotropic response that results in altered nutrient partitioning. In studies with growing pigs, significant improvements of 40% in average daily gain and 30% in feed conversions can be achieved by administration of pST. Research has also shown that the effect of pST is enhanced by good management and nutritional practices (Campbell *et al.*, 1988; Campbell *et al.*, 1990; Caperna *et al.*, 1990). Furthermore, a 60% reduction in carcass fat and a 70% increase in carcass protein content can be attained. The magnitude of response has varied in the various studies performed since the initial classical studies of Turman and Andrews (1955) and Machlin (1972). Different interpretations in response have been attributed to differences in experimental designs. These include initial and final weight of pigs, length of study, genotype, sex, dose of ST, nutritional conditions, and time of injection. However, despite these differences in design, it is quite apparent that pST or rpST increases average daily gain by as much as 40%, decreases carcass fat deposition by as much as 60%, and concomitantly increases carcass protein (lean) accretion by 70%.

Daily injections of pST have been the method of choice for pST administration. Porcine ST must be administered by injection because it is a protein and would be inactivated by digestive enzymes if given orally. The response from pST administration is not related to the site of injection nor to the depth of injection. Recently, researchers (Evans *et al.*, 1991; Evock-Clover *et al.*, 1992) investigated the effects of daily pST administration versus injecting larger doses of pST over an extended period of time. Frequency of administration influenced the magnitude of responsiveness to pST treatment. They indicated that optimal benefit would be realized by a delivery system which mimicked a daily surge of pST.

Mechanisms of pST action have been reviewed and discussed in numerous reviews (several listed above). Clearly, pST affects many metabolic pathways that influence the flow of nutrients among various tissues of the body. Many have concluded that the mechanism by which pST decreases fat content in pigs is via the inhibition of lipogenesis). These changes in metabolism and cell proliferation lead to the alteration of carcass composition via the reduction of nutrients normally destined to be deposited as lipid to other tissues. Studies have demonstrated that regardless of the stage of growth, pigs respond to exogenous pST at all ages. However, rapidly growing animals do not seem to benefit from exogenous pST as much as do finishing animals with respect to fat alterations. This is not surprising since pigs that are growing rapidly and are more efficient are not producing much fat.

1.8.1a. Meat Tenderness. Administration of pST represents a technology with a promise not only for improving production efficiency but also as a means for packers and retailers to offer leaner pork products. A multitude of pST studies have been performed in evaluating the flexibility of pST technology in conjunction with other variables, e.g., management and environmental conditions, to alter carcass and muscle characteristics of morphology as well as meat palatability. These studies typically found that administration of pST to barrows reduced tenderness (increased shear force) by as much as 39% in the longissimus (LM) muscle and 15% in the semimembranosus (SM) when compared to controls. Although the reason for the increase in shear values representing reduced tenderness remains unclear, Solomon et al. (1988) proposed that it may involve an alteration of the muscle composition and/or the cold shortening phenomenon of muscle. Solomon et al. (1990) found that pST administration to boars (Table 2) and gilts lowered shear force of the LM (13.9 and 17.1%, respectively), thus improving tenderness muscle. Klindt et al. (1995) reported that extended administration of pST for 18 weeks to barrows resulted in reduced tenderness and juiciness, but was not observed with a 6 week pST administration period. It is difficult to determine whether these tenderness differences found in barrows will be perceived by the consumer; however, it should be noted that most of the shear force values reported were within the shear values associated with normal pork products. Research is needed to determine whether consumers and/or trained sensory panels can detect tenderness differences or other possible problems with pST-treated pork.

In a study (Solomon et al., 1994), time postmortem of sampling muscle from pST-treated barrows for subsequent shear force analysis had a significant effect on tenderness. Differences in shear force tenderness between pST and control pigs were virtually eliminated when loin chops were removed from the carcass and frozen within 1.5 hours postmortem compared to controls (frozen 5-days postmortem). Some of the inconsistencies reported in the literature for shear force and tenderness as a result of pST administration may be a result of inconsistencies in the time that the meat sample is removed and subsequently frozen.

Minimal observable differences in processing yields, color retention or composition of products from control and pST-treated pigs have been observed. From the wealth of literature, it appears that pST affects carcass composition and not quality (other than possibly tenderness). However, pale, soft, exudative (PSE) muscle have been observed in a couple of pST studies (discussed in pST Muscle Morphology section).

Table 2. Comparison[a] of total carcass and lean tissue lipid and cholesterol content, longissimus muscle area and shear force for transgenic, pST treated fed pigs

Item	T-Control	T-bST	T-oST	pST	pST-Control	SEM	Significance T	Significance pST
Carcass								
Total lipid, g/100g	27.00	4.49	4.82	18.64	25.18	1.7	*	*
Cholesterol, mg/100g	68.71	77.18	67.87	68.48	70.72	2.5	*	NS[c]
Lean								
Total lipid, g/100g	2.89	1.38	.96	2.33	3.21	.5	*	*
Cholesterol, mg/100g	48.64	55.58	49.33	50.13	45.38	1.5	NS	NS
Longissimus muscle								
10th rib back fat, mm	24.8	2.2	2.4	13.2	18.7	1.5	*	*
Area, cm²	33.91	32.37	33.17	42.61	33.29	3.1	NS	*
Shear force, kg/1.3 cm	3.32	3.46	3.88	4.83	5.61	.5	NS	*
Fiber type, %								
SO	12.4	7.2	13.3	14.0	12.5	3.9	*	NS
FOG	20.0	24.2	22.9	24.5	21.0	2.4	*	NS
FG	67.6	68.6	63.8	61.5	66.5	2.8	NS	NS
Fiber area, µm²								
SO	3053	2694	3166	3713	3311	226	*	*
FOG	3669	1979	2180	3121	2785	225	*	*
FG	4359	2749	4356	4795	4344	407	*	*
Giant fiber								
Number	0	0	0	3.3	.3		*	*
Area, µm²				7147	7248			

[a]Wet weight basis. [b]T-control = control boars for transgenics; T-bST = transgenics (boars) with bovine somatotropin gene; T-oST = transgenics (boars) with ovine somatotropin gene; pST = exogenously treated boars with porcine somatotropin.
[c]NS indicates not significant (P>.05); *=P<.05.

1.8.1b. Muscle Morphology. The consensus is that pST exerts a hypertrophic response on carcass muscles, which can be seen at both the cellular level (fiber types) and with the naked eye (carcass conformation and loin-eye area). Most of the research demonstrates that muscle fiber area increased in size with the use of pST. However, a rate limiting factor in the hypertrophic response of muscles to pST administration was the level of dietary protein used (Solomon *et al.*, 1994) in combination with pST administration. This confirms that the beneficial effects of pST are dependent on good management and nutritional practices. Porcine ST administration has little effect on the (percentage) distribution of muscle fiber types (Table 2). Solomon *et al.* (1986) and Solomon (1988) reported that alterations in muscle fiber type populations are often associated with differences in physiological maturation rates of the animals studied.

Sorensen *et al.* (1996) observed that genotypes with relatively large muscle fibers are less responsive to pST treatment than genotypes with relatively small muscle fibers. Others (Solomon *et al.*, 1988; 1989; 1990; 1991; 1994) observed hypertrophied (giant) fibers in muscles from pST-treated pigs (Table 2). Whether the giant fiber anomalies occurred through increased activity associated with compensatory (flux) adaptations or from fibers undergoing degenerative changes has yet to be determined. The occurrence of giant muscle fibers has been associated with stress-susceptible pigs, which exhibit pale, soft, exudative muscle (PSE). In two pST studies (Solomon *et al.*, 1990; 1991) pST-treated pigs exhibited PSE muscle (30 and 62% incidence, respectively) which is much higher than the 12% that is reported as normal occurrence by packers. One explanation offered for the increased incidence of PSE in these studies was a seasonal effect. The experiments were conducted during the summer with average temperatures ranging above 35°C for the duration of the experiments. Perhaps pST administration in conjunction with elevated environmental temperatures may induce the PSE syndrome. However, one can not discount the possibility that the occurrence of PSE in these two studies and the absence of PSE in other studies could be due to genetic differences between pigs used in the different studies. Aalhus *et al.* (1997) did not observe an increase in the proportion of giant fibers or an interactive effect with the halothane genotype. They indicated that the effects of pST appear to be muscle and gender specific, which may be due to differences in maturity and rates of growth at the time of pST administration. Although they observed minor reductions in the muscle color (paler) and increased drip as well as higher shear values, these quality differences were not always significant nor did they result in PSE meat.

Ono *et al.* (1995) reported on the effects of pST administration to barrows on muscles located within different regions of the body (i.e., locomotive vs postural muscles). Based on the analysis of changes of the percentage of muscle fiber types from 20 to 90 kg body weight, they suggested that the fiber type transformation from small slow-twitch oxidative (SO) to large fast-twitch glycolytic (FOG) fibers plays an important role in muscle size enlargement as animals grow. Treatment with pST showed differential effects on muscle fiber growth among the different fiber types and different locations of muscles in the body, suggesting some relationship between pST and muscle function and metabolism and/or muscle maturation. Ji *et al.* (1998) found that the enhanced muscle growth achieved by pST was not associated with altered expression of the p94 or α-actin gene (key genes relative to muscle growth), or an increase in the abundance of any calpastatin transcription product.

1.8.1c. Carcass Composition. Lipid composition studies have demonstrated that the lipid content from pST-treated pigs was as much as 27% less in the lean tissue and 23% less in the subcutaneous fat compared to controls. Carcasses from pST-treated pigs contained 22% less saturated fatty acids (SFA), 26% less monounsaturated fatty acids (MUFA) and no difference in polyunsaturated fatty acids (PUFA) compared to controls (Solomon et al., 1990). The administration of pST resulted in lean tissue containing as much as 40% less SFA, 37% less MUFA and no difference in PUFA compared to controls. The subcutaneous fat from pST-treated pigs contained 33% less SFA, 24% less MUFA and 9% less PUFA than controls. Cholesterol content in the subcutaneous fat from pST pigs was 12% higher than from control pigs. Cholesterol content of intramuscular fat was similar. These results indicate that significant reductions in total lipid and all three classes of fatty acids can be achieved using pST. This represents a favorable change in regard to human dietary guidelines. Few differences in fatty acid profiles of the intramuscular fat extracted from cooked pork rib chops as a result of pST administration have been observed (Prusa et al., 1989). Cholesterol content of cooked chops from pigs receiving pST in their study were greater than controls. Differences in cholesterol values may have resulted from less concentration of the cholesterol during cooking (heating).

Lonergan et al. (1992) reported that pST treatment did not alter the overall fatty acid saturation in subcutaneous or perirenal fat depots, but resulted in a greater reduction of the more saturated middle and inner layers of subcutaneous fat at the 10th rib location. As a result, the more unsaturated outer layer of subcutaneous fat was present in a greater proportion in pST-treated pigs. Oksbjerg et al. (1995) found a change to more polyunsaturated fatty acids and less saturated fatty acids in the backfat of pST treated gilts.

The decrease in total SFA and MUFA, and virtual no change in PUFA, support the conclusion that the mechanism by which pST decreases carcass fat content in pigs is by the inhibition of lipogenesis. A decrease in fat synthesis would lead to a decrease in the production of SFA and MUFA with little effect (change) in the amount of PUFA. The majority of PUFA in pig tissues are the result of dietary fatty acids linoleic and linolenic, and are not synthesized. However, the possibility of an increased turnover of storage lipids, at the level of triacylglycerol synthesis or hydrolysis, exists (Clark et al., 1992).

Use of pST is similar to cattle implants and anabolic agents used to enhance growth in that all cause increases in growth and more efficient utilization of feed. However, they are different in chemical structure. Porcine ST is a protein that is not active orally and is readily digested like any protein. Since pST is a protein and is broken down in the gastrointestinal tract, human ingestion of pST would present no dangers because digestive processes would inactivate the protein and provide no residues. Digestion would break the protein down into its component amino acids, making it available for normal metabolic processes. This has been one of the concerns for the acceptance of the use of pST in pigs. Caperna et al. (1994) reported that daily pST treatment to growing barrows enhanced collagen deposition in the skin, head and viscera, whereas non-collagen protein deposition and collagen maturation were enhanced in the carcass tissues.

1.8.2. Bovine and Ovine Somatotropin

There is evidence that exogenous bovine and ovine ST improves efficiency of growth and lean-to-fat ratio in ruminant animals. The database is much less extensive than it is for swine. Bovine somatotropin has been found to increase gain and protein deposition in feedlot steers, but reports in the literature suggest the greatest effect on steers is in the noncarcass fraction (Early et al., 1990; Moseley et al., 1992). An additive effect of estrogen and somatotropin treatment on weight gain and protein deposition of steers has been suggested (Preston *et al.*, 1995; Enright *et al.*, 1990). Rumsey *et al.* (1996) found that rbST (Somavubove) and the estrogenic growth promoter Synovex-S independently increased growth and protein deposition in young beef steers. The combined treatments consistently demonstrated an additive effect on growth rate, carcass growth, protein deposition, and the energy efficiency of protein deposition. The combination of these two growth promoting agents was also effective in some muscles beyond the response obtained with either treatment alone (Ono *et al.*, 1996; Elsasser *et al.*, 1998). The overall muscle growth affects were greater when the two growth-promoting agents were combined than when they were administered singularly. Similar results were observed (Hughes *et al.*, 1998) using rbST (Posilac) and Revalor-S (a trenbolone acetate and estrogen implant).

Vann *et al.* (1998) observed that rbST alone administered to creep-fed beef calves increased muscle mass but did not affect satellite cell number or concentration of myosin light chain-1f mRNA. The increased muscling appeared to be the result of a greater distribution of FG fibers which possess larger cross-sectional areas than the other fibers (Vann *et al.*, 2001).

1.9. Growth Hormone Releasing Factor

Somatocrinin, often called growth hormone releasing factor (GRF) or growth hormone releasing hormone (GHRH), is a peptide hormone belonging to the glucagon family of the gut. Effects on growth of the administration or manipulation of GHRH are likely to be due to direct effects on the secretion of ST. Exogenous GRF administration increases concentrations of ST in serum of meat animals (Moseley *et al.*, 1985). Long-term administration of GRF has been shown to stimulate growth in the rat and in man by increasing both the secretion and synthesis of pituitary ST (Clark *et al.*, 1986).

2.0. Somatostatin

Somatostatin, another hormone produced by the hypothalamus acts directly on the adenohypophysis of the pituitary gland to inhibit ST release (Hardy, 1981). Endogenous ST secretion could be enhanced by GRF agonists or somatostatin antagonists. Enkephalins, also hypothalamic peptides, stimulate ST release and it is likely that enkephalin agonists could also be used to enhance endogenous secretion (Convey, 1988).

Although somatostatin was originally purified from the hypothalamus, it is now recognized that many cells and tissues throughout the body secrete this peptide. One of these networks releases somatostatin to affect ST release (Tannenbaum, 1985). Several studies (Spencer, 1986) used somatostatin antibodies to attempt to neutralize circulating levels of somatostatin in blood and promote growth by increasing blood ST concentrations. Although this technique appeared to be an attractive possibility for manipulating growth, immunization against somatostatin did not consistently stimulate whole body growth. Lack of a response may be because of the multitude of sources (cells and tissues) that secrete somatostatin. Immunization against somatostatin has led to other strategies based on immunization techniques which may neutralize or amplify hormonal signals via receptors for hormones.

Passive (Vale *et al.*, 1977) and active (Varner *et al.*, 1980) immunization against somatostatin markedly increased serum concentrations of ST. As a result of the increased concentration of ST, researchers set out to determine if immunization against somatostatin would improve growth on carcass lean:fat ratio. As much as a 22% improvement in ratio of gain and 13% improvement in efficiency of gain in lambs was reported (Spencer *et al.*, 1983). However, no significant effect on carcass composition was realized. This would be expected if the effect was mediated via ST or ST receptors (Spencer *et al.*, 1983).

2.1. Beta-Adrenergic Agonists

The discovery of ß-adrenergic agonists, which are chemical analogs of epinephrine, norepinephrine and catecholamines, is a promising development in growth promotion applications of animals. Included in this class of compounds are clenbuterol, cimeterol, ractopamine, L-644,969 and L-640,033 and isoproterenol. ß-adrenergic agonists are orally active and thus may be administered in the feed. Many are chemically stable and extremely potent, making successful development of implants for cattle, sheep and swine likely. ß-agonists improved live weight gain (15%) and feed conversion efficiency (15%) in meat-producing animals in research trials. Carcass protein (muscle) content is substantially increased (25%) while carcass fat is decreased (30%). These changes were observed in intact and castrated males and females. Unlike anabolic steroids, the effects of ß-agonists were not dependent on sex of the animal. They represent an altered pattern of metabolism such that nutrients are directed or partitioned away from adipose (fat) tissue and directed towards lean (muscle) tissue. For this reason, the term "nutrient repartitioning agents" is commonly applied to adrenergic agonists. Mechanisms (Table 3) by which ß-agonists influence growth have been reviewed (Hanrahan *et al.*, 1986; Williams 1987; Moloney *et al.*, 1991).

Table 3. Biological effects of β-adrenergic agonists on adipose and skeletal muscle tissue.

Tissue	Effect	Physiological Process Affected
Adipose	↓[a]	Glucose uptake
		Glucose oxidation unaffected
	↓	Lipid synthesis
	↓	Lipogenic enzyme activity
	?[b]	Insulin stimulation of glucose metabolism
	↑[c]	Basal lipolysis
	↑	Catecholamine-stimulated lipolysis
	↑	Insulin inhibits stimulation of lipolysis
		Insulin binding unaffected
		Somatotropin binding unaffected
Skeletal muscle (growth)	↓	Protein degradation
	↑	Protein synthesis
	↑	Satellite cell proliferation

[a] ↓ = decreases
[b] ? = questionable
[c] ↑ = increases.

Changes in carcass lean content is primarily due to hypertrophy (increased cell size), rather than hyperplasia (increased cell number). ß-agonists appear to exert their effects on skeletal muscle by reducing degradation rate without altering the rate of protein synthesis. Muscle tissue is continually being degraded and resynthesized in animals. Nitrogen retention in skeletal muscle is increased by ß-agonists yet decreased in skin/hide and visceral organs. This decrease in nitrogen retention of non-carcass components, together with reduced fat deposition, accounts for the increased carcass yield in animals treated with ß-agonists. Hypertrophy of cross-sectional area of skeletal muscle fiber types is typical of all red-meat-producing animals treated with ß-agonists. Stimulating increased rates of lipolysis (fat breakdown) and decreased rates of lipogenesis (fat synthesis) reflect decreased carcass fat deposition (Muir, 1988). Additionally, the supply of energy available for fat synthesis may be reduced in treated animals, both because an increased proportion of dietary energy is used for protein synthesis and because the ß-agonists elicit a general increase in metabolic rate.

Growth-promoting beta-agonists were found to reduce plasma levels of insulin and somatomedin-C but did not elevate plasma ST levels. This suggests the concept that growth-promoting ß-agonists work directly through skeletal muscle cell receptors and not indirectly through the elevation of plasma ST or insulin concentrations (Muir, 1988). A critical factor in the usefulness of ß-agonists is likely to be the degree to which their effects are retained after withdrawal, since a recommended withdrawal period is likely to be required. Growth rates decline following withdrawal. Carcass effects are not affected as rapidly but the advantages are not sustained after prolonged periods of withdrawal.

2.1.1. Porcine

In the majority of studies, daily gain was not significantly increased by feeding of ß-agonists. In fact, some studies showed a depression in growth rate. The repartitioning effects of ß-agonists appear to increase with dose rate in pigs and result in a decrease in carcass fat. Minimal differences in response to sex (gilt versus barrow versus boar) have been observed. Minor increases in meat toughness have been reported for pigs treated with ß-agonists. In a recent review (Warriss, 1989), it was concluded that ß-agonists fed to pigs do not promote pale, soft, exudative (PSE) meat but might lead to a greater propensity for dark, firm, dry meat. ß-agonists appear to have no effect on water-holding capacity and ultimate muscle pH. Ruminants appear to be more responsive than swine to β-agonists.

2.1.2. Bovine and Ovine

Bulls, steers and heifers as well as rams, wethers and ewes have been studied in different trials as to their response to treatment with ß-agonists. Responses have ranged from no response to an improvement of 48% in growth rate and feed conversion. Beef carcasses treated with ß-agonists during the finishing phase contain less fat and more lean than controls. ß-agonists are very effective in repartitioning utilizable energy away from fat deposition and toward protein accretion in rams, wethers and ewes. Ruminants seem to be quite susceptible to meat toughening when treated with ß-agonists.

Hamby et al. (1986) were among the first to report that toughening of meat (lamb) occurs in ß-agonist-treated animals. Shear-force in the longissimus muscle increased 114% in treated lambs. Although treated lambs had less carcass fat, Henby *et al.* (1986) concluded that some factors other than cold-shortening were involved in the 114% increase in toughness as a result of a low r^2 value for linear regression of shear-force on neutral lipid content. The hypothesis that reduced tenderness is due to reduced protein degradation has been considered (Moloney et al., 1991). Collagen and its respective cross-linking are a major determinant in the texture (tenderness) of cooked meat (Bailey, 1987). If ß-agonists cause an increase in growth by reducing protein degradation, this would allow the collagen molecules more time to cross-link (Bailey, 1987). The increase in cross-linking would enhance the toughness independently of post-mortem proteolytic degradation. Beermann et al. (1989) demonstrated a significant, as much as 70%, reduction of μM calcium dependent proteinase (CDP) activity in skeletal muscle of lambs fed a ß-agonist for 3 or 6 weeks. This reduction was synonymous with a dose-dependent increase in shear-force values of treated lambs.

3.0. Emerging Metabolic Modifiers

In a recent paper by Caperna et al. (2000), emerging metabolic modifiers for use in livestock production practices were cited and reviewed. Included in their list were ractopamine, conjugated linoleic acid (CLA), betaine and leptine. These authors pointed out that while a major challenge of today's livestock industries is to produce leaner, healthier meat products, today's producers are expected to meet the demands on a decrease in land mass. Environmental concerns, such as ground water contamination and solid waste accumulation must be addressed by these producers.

Ractopamine, recently FDA approved for use by the swine industry, is a member of the beta-adrenergic agonist family of compounds which was described earlier. Conjugated linoleic acid (CLA) is a mixture of positional and geometric isomers of linoleic acid with unique chemo-protective properties of preventing/reducing cancer (Pariza and Hargraves, 1985), heart disease (Nicolosi et al., 1997), obesity (Park et al., 1999), diabetes mellitus (Park et al., 1997), and improving muscle (lean) mass (Houseknecht et al., 1998; Dugan et al., 1997), bone growth (Dunshea et al., 1998), and enhancing the immune system (Li and Watkins, 1998). Several experiments have been conducted using swine and dairy cattle fed CLA. Carcass fat was significantly reduced with a concomitant increase in carcass lean when CLA was included in the diet (Cook et al., 1993; Ostrowska et al., 1999). Betaine is an amino acid (trimethyl-lycine) formed by oxidation of choline (Kidd et al., 1997). The use of betaine as an ingredient in livestock diets was initially introduced by the feed industry as a replacement for methionine and choline in poultry diets (Caperna et al., 2000). Studies suggest that dietary betaine reduces and alters carcass fat in broilers (Saunderson and Mackinlay, 1990) and pigs (Cadogan et al., 1993; Virtanen and Campbell, 1994). There are however, mixed reports on the effects of betaine on loin eye area and backfat thickness in pigs. Betaine appears to be more consistent as a metabolic modifier that alters carcass composition in poultry than in pigs. Energy balance regulation is a complex process involving several controlling systems. Control of the metabolic processes that regulate feed intake and the efficiency and rate of growth is an area of being investigated by researchers. Leptin, a 16 kD protein is involved in energy metabolism and the role of leptin in regulating feed intake and appetite control is being determined in order to promote more efficient animal production practices. Leptin treatment of rats has been shown to cause a dose-dependent decrease in food intake, loss of body weight and loss of fat depots suggesting a specific lipoatrophic activity of leptin (Chen et al., 1996b).

4.0. Transgenic Animals

4.0. Transgenic Animals

Development of recombinant DNA technology enabled scientists to isolate single genes, analyze and modify their nucleotide structures, make copies of these isolated genes, and transfer copies into the genomes of livestock species. Such direct manipulation of genetic composition is referred to as "genetic engineering" and the term "transgenic animal" denotes an animal whose genome contains recombinant DNA.

The dramatic achievements in molecular biology during the past decade and the development of micromanipulation for early-stage embryos provided the combined capabilities for introducing cloned genes into the mouse genome in 1980 (Brinster *et al.*, 1985; Palmiter and Brinster, 1982). The transfer of genes was immediately recognized as an important scientific achievement. However, the subsequent creation of the "super" mouse by the transfer of a rat somatotropin gene provided the convincing evidence that demonstrated the potential offered by gene transfer (Palmiter *et al.*, 1982). At the Ohio University (Athens) transgenic mice carrying a modified version of the bovine somatotropin gene that originally created the "super" mouse was found to produce "mini" mice (approximately half the size as the controls). Modifying a somatotropin gene and incorporating it into mouse DNA that in turn prevents stimulating growth lends itself as a powerful tool for probing the hormone's function. This suggests that somatotropin does more than promote growth. The first U.S. patent for a transgenic animal, a mouse expressing a foreign oncogene created by Harvard University researchers, was issued in 1988. The issue of that patent triggered intense criticism from animal rights activists, ethicists and environmentalists. As a result, the government did not issue any further patents. Environmentalists and others argue that there are dangers in releasing transgenic animals into the environment, as well as ethical issues that have not been fully explored.

Scientists had been struggling with practical problems involved in transferring genes from one animal to another. Scientists inserting genes from a variety of different species, for example the pig, have encountered significant problems. One major difficulty was the inability to insert genes precisely into animals' DNA, the building block of genetics. The gene is inserted into the nucleus of a cell with the hope that the gene lands in an appropriate location. In addition, many pigs develop severe health problems, e.g., ulcers, pneumonia, arthritis, cardiomegaly, dermatitis and renal disease. The successful use of genetic engineering to enhance carcass composition, and efficiency of meat production in livestock depends on many factors. These include identification, isolation and modification of useful genes or groups of genes that influence meat quality and quantity. Control of the time and level of expression of the inserted genes in transgenic animals so that their health status is either improved or not diminished affects the successful insertion of these genes into the genome. The progress and methods of transferring genes in farm animals has been published (Engel *et al.*, 1995). Wells (2000) published a review of the current techniques and applications for genome modification as it relates to meat science. Recently, Pursel *et al.* (1999) reported on the successful expression of insulin-like growth factor-I (IGF-I) in skeletal muscle of transgenic swine. The general health of the IGF-I transgenic and control pigs did not differ in physical appearance, behavior, ability to tolerate summertime heat stress or reproductive capacity. These types of problems were observed for the transgenic pigs expressing bovine growth hormone (Pursel *et al.*, 1989).

In April 1997 a Scottish scientist Ian Wilmut successfully cloned a fully grown sheep. Named "Dolly," the sheep represented the first time DNA from a mammal had been used to produce an exact genetic replica. The Roslin Institute research team in Edinburgh took DNA from single mammary cells from a 6-year-old ewe, inserted it into fertilized eggs, then implanted the eggs in 13 ewes, one of which became pregnant and gave birth to the cloned lamb, Dolly. This method proved that mature cells other than reproductive cells contain DNA capable of programming regeneration of an entire animal. In August of 1997, a Wisconsin based company, Deforest successfully cloned a Holstein bull calf. In June of 1999 researchers at the University of Connecticut announced the birth of a cloned calf from an adult farm animal using cells from the ear of the adult, not the reproductive organs. Similarly, scientists in Hawaii announced the successful cloning of mice using tail cells from adult males, again the successful exhibition of cloning without using reproductive organs/cells. The first cloned pigs (Polejaeva et al., 2000), from PPL Therapeutics Inc. of Blacksburg, Virginia and Midlothian, Scotland were born March 2000. A new approach was used in cloning these piglets. The new approach involved two nuclear transfers. The researchers first injected the donor nucleus, from an adult cell, into an unfertilized egg. Then, once the nucleus expanded to form a pronucleus, they transferred it to a fertilized egg. The researchers hypothesized that factors present in the pronucleus and not needing to artificially activate an unfertilized egg contributed to the success of cloning the piglets. However, cloning still remains less than efficient with five piglets resulting from 401 implanted fertilized ova. With regards to the cloning of farm animals most of the research focuses on state-of-the-art science and cutting-edge methodologies, technical improvements, and current progress towards producing transgenic animals for medical and agricultural applications. To date the effects of cloning farm animals on carcass and meat composition and quality have not been investigated.

A tremendous amount of variation in carcass components, such as muscle development, fat content distribution, tenderness and flavor, exists among and within breeds of each species. Animal breeders have successfully utilized selection from this genetic variation to improve farm animals for many years. Unfortunately, the quantitative genetic approach has yielded few clues regarding the fundamental genetic changes that accompanied the selection of animals for superior carcass attributes. Few single genes have been identified that have major effects on carcass composition. A national effort to map the genes of meat animals is underway. In cattle, the double-muscle gene is responsible for both muscle hyperplasia and hypertrophy and enhanced lean tissue deposition. In sheep, the callipyge mutated gene is responsible for muscle hypertrophy and enhanced lean tissue accretion. In pigs, the halothane (Hal) sensitivity gene is associated with increased yield of lean meat and porcine stress syndrome. Pigs homozygous for "Hal" are susceptible to stress and have a high incidence of pale-soft-exudative (PSE) meat. These genes offer considerable potential for investigation of carcass composition in meat producing animals. However, except for the Hal gene, which has been identified as a single mutation in the ryanidine receptor gene, the specific product of each gene remains to be identified.

4.0.1. Swine with Growth-Related Transgenes

A number of transgenic pigs containing various ST transgenes have been raised (98). Production of excess ST in transgenic animals caused multiple physiological affects, but did not result in "giantism" as was expected based on the earlier production of "super" mice as described (Palmiter et al., 1982). However, transgenic pigs that have excess ST levels exhibited numerous unique carcass traits. Reduced carcass fat, alteration of muscle fiber profiles, thickening of the skin, enlargement of bones, and redistribution of major carcass components occurred in transgenic pigs. Some of these effects are similar to those observed after daily injection of pST, while others are considerably different. Possibly, these differences are the consequence of continual presence of excess ST in the transgenics while injections of pST provide a daily pulse of excess ST.

4.0.1a. Carcass Composition and Meat Tenderness. The first transgenic animals for which carcass and meat quality was evaluated were the T-bST pigs at the USDA-ARS-Beltsville, MD research facility. Carcass fat was dramatically reduced (Figure 1) in transgenic pigs that expressed a bST transgene at five different live weights (Solomon *et al.*, 1994). This difference in fat became greater among transgenic and non-transgenic littermates as the pigs approached market eight. Total cholesterol content of ground carcass tissue of bST transgenic pigs was not different from sibling control pigs at any of the designated weights. However, as body weight increased from 14 to 92 kg, the cholesterol content decreased for both groups of pigs. Analysis of fatty acids showed that carcasses of bST transgenic pigs consistently contained less total (expressed as g/100 g tissue)SFA than sibling control pigs at each body weight. These differences in SFA were primarily a result of reductions in palmitic, stearic and myristic acids. Both myristic and palmitic acids have been reported to be hyperlipidemic and hypercholesterolemic in humans. Consumption of hypercholesteremic fatty acids by humans has come under attack by health professionals in the US. Carcasses from bST transgenic pigs contained less total MUFA and PUFA fatty acids than sibling control pigs. Similar observations were made for T-oST (transgenic pigs with an ovine somatotropin gene). Carcasses and lean tissue from transgenic pigs had near the optimum ratio of 1:1:1 for SFA:MUFA:PUFA as recommended (National Research Council, 1988).

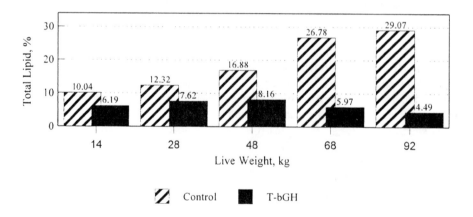

Figure 1. Carcass lipid accretion in control and transgenic pigs from 14 to 92 kg.

When carcasses were separated into the four primal (pork) cuts, the hams of the bST transgenic pigs were significantly larger and the loins were significantly smaller than those of the sibling control pigs (Solomon and Pursel, 1994). The intramuscular fat for each primal cut (lean portion only) showed large differences between bST transgenic pigs and the controls. In spite of these dramatic reductions of fat throughout all primal cuts, evaluation of tenderness by shear-force determination indicated there were no significant differences between the two groups of pigs for the longissimus (loin) muscle (Table 2).

Although somatotropin is considered the primary growth-promoting hormone in mammals, many of its effects are thought to be mediated by insulin-like growth factor-I (IGF-I). Transgenic pigs have been produced using a skeletal α-actin regulatory sequence to direct expression of an IGF-I gene specifically in skeletal muscle (Pursel et al., 1999). The underlying rationale was to initiate a paracrine response with IGF-I to enhance muscle development without altering the general physiology that might occur from an endocrine response. Founder T-pigs were mated to non-T pigs to produce G1 transgenic and sibling control progeny. In comparison to sibling controls, T-females and intact males had less carcass fat (9.9 and 8.1%, respectively), and more muscle (8.6 and 3.6%, respectively) (Pursel et al., 1999). In a follow up study, using barrows and gilts (Eastridge et al., 1999), IGF-I transgene pigs had larger (34%) loin eye areas and heavier (range 9-24%) muscle weights of five major muscles of the carcass (Eastridge et al., 1999). Neither average daily gain nor feed efficiency differed for T and control pigs. T and control pigs did not differ in general appearance, and no gross abnormalities, pathologies, or health-related problems were encountered as was observed for transgenic pigs with somatotropin genes. Thus, enhancing IGF-I specifically in skeletal muscle had a positive effect on carcass composition of swine. Shear force values for control pigs was 6.52 kg and 6.42 kg for IGF-I T-pigs (Eastridge et al., 1999).

4.0.1b. Muscle Morphology. Morphological evaluation (Table 2) of bST transgenic-pig skeletal muscles revealed bST transgenic pigs had fewer SO fibers and more FOG fibers than control pigs. The population of FG was similar between the transgenic and controls; however, the classical porcine fiber arrangement with SO fibers grouped in clusters surrounded by FOG and FG fibers was less evident in the transgenic muscle (Solomon et al., 1991). Morphological fiber profiles for T-bST pigs resembled that of bovine muscle rather than porcine muscle. Hypertrophied (giant) fibers, that were identified in pST-injected pigs, were not observed in bST transgenic pigs (Table 2). The shift in the percentage of SO fibers to FOG fibers in the bST transgenic pigs has not been identified in pigs that have received daily injections of pST.

Muscle fiber growth patterns in bST and oST transgenic pigs differ markedly from that seen in muscle of pigs injected daily with pST. All three fiber types are enlarged in pST-treated pigs whereas in bST transgenic pigs, only SO fibers appear to hypertrophy during growth compared to controls. In the T-oST pigs both the SO and FG fibers hypertrophy similar to controls during growth, whereas the FOG fibers remain much smaller than controls (Table 2). No giant fibers were observed in muscle tissue from bST transgenic pigs. Even though bST transgenic pigs were highly stress sensitive, there were no signs of pale, soft, exudative meat.

The IGF-I transgene pigs from the Pursel and co-workers (1999) study exhibited an increase in FG fibers and a decrease in FOG fibers (Bee et al., 1997). All fibers increased in size with the hypertrophic response being greatest for the SO fibers followed by the FOG and FG fibers. The IGF-I transgene pigs from the (Eastridge *et al.*, 1999) study showed that there was no difference in fiber type percentages between the T-pigs and controls but that the increase in muscle mass was due to an increase in muscle fiber area (hypertrophy) for all three fiber types. Bee *et al.* (1997) also reported that IGF-I transgene expression altered the distribution of slow and fast isomyosin forms.

4.0.2. Bovine and Ovine with Growth-Related Transgenes

To date, transgenic cattle produced by microinjection of DNA into pronuclei is inefficient and extremely costly, in large part due to the cost of maintaining numerous pregnancies to term. Numerous pregnancies result in non-transgenic progeny. The success rate in both bovine and ovine is significantly less than that for swine. No carcass data are available for transgenic bovine or ovine. The cloning technology described by Wilmut in 1997 has introduced the successful cloning of cattle and sheep, however, these research programs have not looked at carcass data at this point.

5. CONCLUSIONS

Potential for manipulation of growth and composition of farm animals has never been greater than at present due to the wide array of strategies for altering the balance between lean and fat. Recent discoveries of repartitioning effects of somatotropin, select ß-adrenergic agonists, as well as the variety of growth-promoting agents, and gene manipulation techniques offer a wide range of strategies. Although progress is being made, much more needs to be accomplished. Eating quality and safety must not be sacrificed as leaner animals are developed. We are still a long way from fully understanding the integrated mechanisms resulting from manipulation of growth and carcass composition and possible effects on meat quality (either positive or negative) as a result of the techniques described in this chapter.

6. REFERENCES

Aalhus, J. L., Best, D. R., Costello, F., and Schaefer, A. L., 1997, The effects of porcine somatotropin on muscle fibre morphology and meat quality of pigs of known stress susceptibility, *Meat Sci.*, **45**:283-295.
Bailey, A. J., 1987, Connective tissue and meat quality. Proceedings of 33rd International Congress of Meat Sci and Technology, p. 152.
Bee, G., Solomon, M. B., and Pursel, V. G., 1997, Expression of an IGF-I transgene on skeletal muscle morphology in swine, *FASEB J.*, **11**:A415, #2405.
Beermann, D. H., 1989, Status of current strategies for growth regulation, in: *Animal Growth Regulation*, D. R. Campion, G. J. Hausman, R. J. Martin, eds., Plenum, New York, pp. 377-400.
Beermann, D. H., Wang, S. Y., Armbruster, G., Dickson, H. W., Rickes, E. L., and Larson, J. G., 1989, Influences of beta-agonist L-665,871 and electrical stimulation on postmortem muscle metabolism and tenderness in lambs. Proceedings of 42nd Annual Reciprocal Meat Conf. **42**:54.
Bouffault, J. C., and Willemart, J. P., 1983, Anabolic activity of trenbolone acetate alone or in association with estrogens, in: *Anabolics in Animal Production*, E. Meissonnier, ed., Office International des Epizooties, Paris.
Brinster, R. L., Chen, H. Y., Trumbauer, M. E., Yagle, M. K., and Palmiter, R. D., 1985, Factors affecting the efficiency of introducing foreign DNA into mice by microinjecting eggs, *Proceedings of Natl. Acad. Sci. USA* **82**:4438-4444.

Cadogan, D. J., Campbell, R. G., Harrison, D., and Edwards, A. C., 1993, The effects of betaine on the growth performance and carcass characteristics of female pigs, in: *Manipulating Pig Production*, E. S. Batterham, ed., Australasian Pig Sci. Assoc., Attwood, Victoria , p. 219.

Campbell, R. G., Johnson, R. J., King, R. H., Taverner, M. R., and Meisinger, D. J., 1990, Interaction of dietary protein content and exogenous porcine growth hormone administration on protein and lipid accretion rates in growing pigs, *J. Anim. Sci.*, **68**:3217-3225.

Campbell, R. G., Steele, N. C., Caperna, T. J., McMurtry, J. P., Solomon, M. B., and Mitchell, A. D., 1988, Interrelationships between energy intake and exogenous porcine growth hormone administration on the performance, body composition and protein and energy metabolism of growing pigs weighing 25 to 55 kilograms body weight, *J. Anim. Sci.*, **66**:1643-1655.

Campion, D. R., Hausman, G. J., and Martin, R. J., 1989, *Animal Growth Regulation*, Plenum, New York.

Caperna, T. J., Figares, I. F., Steele, N. C., and Campbell, R. G., 2000, Emerging metabolic modifiers. Proceedings of 53rd Annual Reciprocal Meat Conf., pp 57-62.

Caperna, T. J., Gavelek, D., and Vossoughi, J., 1994, Somatotropin alters collagen metabolism in growing pigs, *J. Nutr.*, **124**:770-778.

Caperna, T. J., Steele, N. C., Komarek, D. R., McMurtry, J. P., Rosebrough, R. W., Solomon, M. B., and Mitchell, A. D., 1990, Influence of dietary protein and recombinant porcine somatotropin administration in young pigs' growth, body composition and hormone status, *J. Anim. Sci.*, **68**:4243-4252.

Chen, G., Koyama, K., Yuan, X., Lee, Y., Zhou, Y. T., O'Doherty, R., Newgard, C. B., and Unger, R. H., 1996b, Disappearance of body fat in normal rats induced by adenovirus-mediated leptin gene therapy, *Proc. Natl. Acad. Sci. USA* **93**:14795-14799.

Clark, R. G., Chambers, G., Lewin, J., and Robinson, I. C. A. F., 1986, Automated repetitive microsampling of blood: growth hormone profiles in conscious male rats, *J. Endocrinology*, **111**:27-35.

Clark, S. L., Wander, R. C., and Hu, C. Y., 1992, The effect of porcine somatotropin supplementation in pigs on the lipid profile of subcutaneous and intermuscular adipose tissue and longissimus muscle, *J. Anim. Sci.*, **70**:3435-3441.

Convey, E. M., 1988, Strategies to increase meat yield and reduce fat/cholesterol, in: Proceedings of 6th Biennial Symp. American Academy Vet. Pharm. and Therapeutics on Animal Drugs and Food Safety, pp. 27-32.

Cook, M. E., Miller, C. C., Park, Y., and Pariza, M. W., 1993, Immune modulation by altered nutrient metabolism: nutritional control of immune - induced growth depression, *Poultry Sci.*, **72**:1301-1305.

Dugan, M. E. R., Aalhus, J. L., Schaeffer, A. L., and Kramer, J. K. G., 1997, The effect of conjugated linoleic acid on fat to lean repartitioning and feed conversion in pigs, *Can. J. Anim. Sci.*, **77**:723-725.

Dunshea, F. R., Ostrowka, E., Muralitharan, M., Cross, R., Bauman, D. E., Pariza, M. W., and Skarie, C., 1998, Dietary conjugated linoleic acid decreases backfat in finisher gilts, *J. Anim. Sci.*, **76**(Suppl. 1): No. 508.

Early, R. J., McBride, B. W., and Ball, R. O., 1990, Growth and metabolism in somatotropin-treated steers: I. Growth, serum chemistry and carcass weights, *J. Anim. Sci.*, **68**:4134.

Eastridge, J. S., Solomon, M. B., Pursel, V. G., and Mitchell, A. D., 1999, Response of porcine skeletal muscle enhanced by an IGF-I transgene. 52nd Annual Reciprocal Meat Conference, p 125.

Elsasser, T. H., Rumsey, T. S., Kahl, S., Czerwinski, S. M., Moseley, W. M., Ono, Y., Solomon, M. B., Harris, F., and Fagan, J. M., 1998, Effects of Synovex-S® and recombinant bovine growth hormone (Somavubove®) on growth responses of steers: III. Muscle growth and protein responses, *J. Anim. Sci.*, **76**:2346-2353.

Engel, K. H., Takeoka, G. R., and Teranishi, R., 1995, Genetically Modified Foods: Safety Issues, ACS Symp. Series 605, American Chemical Society, Washington, D.C., 1995.

Enright, W. J., Quirke, J. F., Gluckman, P. D., Breier, B. H., Kennedy, L. G., Hart, I. C., Roche, J. F., Coert, A., and Allen, P., 1990, Effects of long-term administration of pituitary-derived bovine growth hormone and estradiol on growth in steers, *J. Anim. Sci.*, **68**:2345.

Evans, F. D., Osborne, V. R., Evans, N. M., Morris, J. R., and Hacker, R. R., 1991, Effect of different patterns of administration of recombinant porcine somatotropin on growth performance and economic returns of pigs in the starter-grower vs. finisher phases of production, *Can. J. Anim. Sci.*, **71**:355-360.

Evock-Clover, C. M., Steele, N. C., Caperna, T. J., and Solomon, M. B., 1992, The effect of frequency of recombinant porcine somatotropin (rpST) administration on growth performance, tissue accretion rates, and hormone and metabolic concentration in pigs, *J. Anim. Sci.*, **70**:3709-3720.

Fluharty, F. L., McClure, K. E., Solomon, M. B., Clevenger, D. D., and Lowe, G. D., 1999, Energy source and ionophore supplementation effects on lamb growth, carcass characteristics, visceral organ mass, diet digestibility, and nitrogen metabolism, *J. Anim. Sci.*, **77**:816-823.

Fritsche, S., Solomon, M. B., and Rumsey, T. S., 1999, Influence of growth promoting implants on morphology of longissimus muscle in growing steers, *FASEB J.*, **13**(4):A53.

Galbraith, H., and Topps, J. H., 1981, Effects of hormones on the growth and body composition of animals, *Nutr. Abstr. Rev. Ser., B* **52**:521.

Hamby, P. L., Stouffer, J. R., and Smith, S. B., 1986, Muscle metabolism and real-time ultrasound measurement of muscle and subcutaneous adipose tissue growth in lambs fed diets containing a ß-agonist, *J. Anim. Sci.,* **63**:1410-1421.

Hancock, D. L., Wagner, J. F., and Anderson, D. B., 1991, Effects of estrogen and androgens on animal growth, in: *Growth Regulation in Farm Animals: Advances in Meat Research*, Vol. 7, A. M. Pearson and T. R. Dutson, eds., Elsevier Appl. Sci., London, pp. 255-297.

Hanrahan, J. P., 1987, *Beta-agonists and their Effects on Animal Growth and Carcass Quality*, Elsevier Appl. Sci., London.

Hanrahan, J. P., Quirke, J. F., Bomann, W., Allen, P., McEwan, J. C., Fitzsimons, J. M., Kotzian, J., and Roche, J. F., 1986, β-agonists and their effects on growth and carcass quality, in: *Recent Advances in Animal Nutrition*, W. Haresign and D. J. A. Cole, ed., Butterworths, London, pp. 125-138.

Hardy, R. N., 1981, Endocrine Physiology, Edward Arnold (Publishers) Ltd, pp. 80-89.

Harlander, S. K., 1981, Marketing and consumer acceptance of bST, in: *Proc. Bovine Somatotropin Vet. Learning Systems, Inc.*, pp 25-30.

Heap, R. B., Prosser, C. G., and Lamming, G.E., eds., 1989, Biotechnology in Growth Regulation. Proceedings of an international symposium held at AFRC Inst of Anim Physiology and Genetics Res, Babraham, Cambridge, United Kingdom.

Houseknecht, K. L., Vanden Heuvel, J. P., Moya Camarena, S. Y., Portocarrero, C. P., Peck, L. W., Nickel, K. P., and Belury, M., 1998, Dietary conjugated linoleic acid normalizes impaired glucose tolerance in the Zucker diabetic fatty fa/fa rat, *Bioch. Biophy. Res., Com.* **244**:678.

Hughes, N. J., Schelling, G. T., Garber, M. J., Eastridge, J. S., Solomon, M. B., and Roeder, R. A., 1998, Skeletal muscle morphology alterations due to Posilac® and Revalor®-S treatments, alone or in combination in feedlot steers. Proceedings of Western Section, *Amer. Soc. Anim. Sci.,* **49**:88-92.

Ji, S. Q., Frank, G. R., Cornelius, S. G., Willis, G. M., and Spurlock, M. E., 1998, Porcine somatotropin improves growth in finishing pigs without altering calpain 3 (p94) or α-Actin mRNA abundance and has a differential effect on calpastatin transcription products, *J. Anim. Sci.,* **76**:1389-1395.

Kidd, M. T., Ferket, P.R., and Garlich, J. D., 1997, Nutritional and osmoregulatory functions of betaine, *World's Poult. Sci. J.,* **53**:125-139.

Klindt, J., Buonomo, F. C., Yen, J. T., and Baile, C. A., 1995, Growth performance, carcass characteristics, and sensory attributes of boars administered porcine somatotropin by sustained-release implant for different lengths of time, *J. Anim. Sci.,* **73**:3585-3595.

Li, Y., and Watkins, B. A., 1998, Conjugated linoleic acids alter bone fatty acid composition and reduce *ex vivo* bone PGE2 biosynthesis in rats fed n-6 or n-3 fatty acids, *Lipids,* **33**:417-425.

Lonergan, S. M., Sebranek, J. G., Prusa, K. J., and Miller, L. F., 1992, Porcine somatotropin (pST) administration to growing pigs: effects on adipose tissue composition and processed product characteristics, *J. Food Sci.,* **57**:312-318.

Machlin, L. J., 1972, Effect of porcine growth hormone on growth and carcass composition of the pig, *J. Anim. Sci.,* **35**:794-800.

Moloney, A., Allen, P., Joseph, R., and Tarrant, V., 1991, Influence of beta-adrenergic agonists and similar compounds on growth, in: *Growth Regulation in Farm Animals, Adv. in Meat Research*, Vol. 7, A. M. Pearson and T. R. Dutson, ed., Elsevier Appl. Sci., New York, pp. 455-513.

Moseley, W. M., Krabill, L. F., Friedman, A.R., and Olsen, R. F., 1985, Administration of synthetic human pancreatic growth hormone-releasing factor for five days sustains raised serum concentrations of growth hormone in steers, *J. Endocrinology,* **104**:433-439.

Moseley, W. M., Paulissen, J. B., Goodwin, M. C., Alaniz, G. R., and Claflin. W. H., 1992, Recombinant bovine somatotropin improves growth performance in finishing beef steers, *J. Anim. Sci.,* **70**:412.

Muir, L. A., 1985, Mode of action of exogenous substances on animal growth–an overview, *J. Anim. Sci.,* **61**(Suppl. 2):154.

Muir, L.A., 1988, Effects of beta-adrenergic agonists on growth and carcass characteristics of animals, in: *Designing Foods*, National Res. Council, Washington, DC, pp. 184-193.

National Research Council, 1988, Designing Foods, Animal Product Options in the Market Place, Natl. Academy Press, Washington, DC.

National Research Council, 1994, *Metabolic Modifiers: Effects on the Nutrient Requirements of Food-producing Animals*, National Academy Press, Washington, DC.

Nicolosi, R. J., Rogers, E. J., Kritchevsky, D., Scimeca, J. A., and Huth, P. J., 1997, Dietary conjugated linoleic acid reduces plasma lipoproteins and early aortic atherosclerosis in hypercholesterolemic hamsters, *Artery*, **22**:266-277.

Oksbjerg, N., Petersen, J. S., Sorensen, M. T., Henckel, P., and Agergaard, N., 1995, The influence of porcine growth hormone on muscle fibre characteristics, metabolic potential and meat quality, *Meat Sci.*, **39**:375-385.

Ono, Y., Solomon, M. B., Elsasser, T. H., Rumsey, T. S., and Moseley, W. M., 1996, Effects of Synovex-S® and recombinant bovine growth hormone (Somavubove®) on growth responses of steers: II. Muscle morphology and proximate composition of muscles, *J. Anim. Sci.*, **74**:2929-2934.

Ono, Y., Solomon, M. B., Evock-Clover, C. M., Steele, N. C., and Maruyama, K., 1995, Effects of porcine somatotropin administration on porcine muscles located within different regions of the body, *J. Anim. Sci.*, **73**:2282-2288.

Ostrówska, E. M., Muralitharan, M., Cross, R. F., Bauman, D. E., and Dunshea, F. R., 1999, Dietary conjugated linoleic acids increase lean tissue and decrease fat deposition in growing pigs, *J. Nutr.*, **129**:2037-2042.

Palmiter, R. D., Brinster, R. L., 1986, Germ-line transformation of mice, *Annual Rev. Genet.*, **20**:465-478.

Palmiter, R. D., Brinster, R. L., Hammer, R. E., Trumbauer, M. E., Rosenfeld, M. G., Birnberg, N. C., and Evans, R. M., 1982, Dramatic growth of mice that develop from eggs microinjected with metallothionein-growth hormone fusion genes, *Nature*, **300**:611.

Pariza, M. W., and Hargraves, W. A., 1985, A beef-derived mutagenesis modulator inhibits initiation of mouse epidermal tumors by 7, 12-dimethylbenz[a]anthracene, *Carcinogenesis*, **6**:591-593.

Park, Y., Albright, K. J., Liu, W., Storkson, J. M., Cook, M. E., and Pariza, M. W., 1997, Effect of conjugated linoleic acid on body composition in mice, *Lipids*, **32**:853-858.

Park, Y., Albright, K. J., Storkson, J. M., Liu, W., Cook, M. E., and Pariza, M. W., 1999, Changes in body composition in mice during feeding and withdrawal of conjugated linoleic acid, *Lipids* **33**:243-248.

Pearson, A. M., and Dutson, T. R., 1991, Growth Regulations in Farm Animals, Vol. 17: Advances in Meat Research, Elsevier, Essex, United Kingdom.

Polejaeva, I. A., Chen, S-H., Vaught, T. D., Page, R. L., Mullins, J., Ball, S., Dai, Y., Boone, J., Walker, S., Ayares, D. L., Colman, A., and Campbell, K. H. S., 2000, Cloned pigs produced by nuclear transfer from adult somatic cells, *Nature*, **407**:86-90.

Preston, R. L., Bartle, S. J., Kasser, T. R., Day, J. W., Veenhuizen, J. J., and Baile, C. A., 1995, Comparative effectiveness of somatotropin and anabolic steroids in feedlot steers, *J. Anim. Sci.*, **73**:1038.

Prusa, K. J., Love, J. A., and Miller, L. F., 1989, Composition, water-holding capacity and pH of muscles from pigs supplemented with porcine somatotropin, *J. Food Qual.*, **12**:467-473.

Pursel, V. G., Pinkert, C. A., Miller, K. F., Bolt, D. J., Campbell, R. G., Palmiter, R. D., Brinster, R. L., and Hammer, R. E., 1989, Genetic engineering of livestock, *Science*, (Washington, DC) **244**:1281-1288.

Pursel, V. G., and Rexroad, C. E., Jr, 1993, Status of research with transgenic farm animals, *J. Anim. Sci.*, **71**(Suppl. 3):10-19.

Pursel, V. G., Wall, R. J., Mitchell, A. D., Elsasser, T. H., Solomon, M. B., Coleman, M. E., DeMayo, F., and Schwartz, R.J., 1999, Expression of insulin-like growth factor-I in skeletal muscle of transgenic swine, in: *Transgenic Animals in Agriculture*, J. D. Murray, G. B. Anderson, A. M. Oberbauer and M. M. McGloughlin, ed., CAB International, pp. 131-144.

Richards, J. E., Mowat, D. N., and Wilton, J. W., 1986, Ralgro implants for intact male calves, *Can. J. Anim. Sci.*, **66**:441.

Rumsey, T. S., Elsasser, T. H., Kahl, S., Moseley, W. M., and Solomon, M. B., 1996, Effects of Synovex-S® and recombinant bovine growth hormone (Somavubove®) on growth responses of steers: I. Performance and composition of gain, *J. Anim. Sci.*, **74**:2917-2928.

Saunderson, C. L. and Mackinlay, J., 1990, Changes in body-weight, composition and hepatic enzyme activities in response to dietary methionine, betaine and choline levels in growing chicks, *Br. J. Nutr.*, **63**:339:349.

Schanbacher, B. D., 1984, Manipulation of endogenous and exogenous hormones for red meat production, *J. Anim. Sci.* **59**:1621.

Schreibman, M. P., Scanes, C. G., and Pong, P. K. T., 1993, *The Endocrinology of Growth, Development, and Metabolism in Vertebrates*, Academic, San Diego, CA.

Sejrsen, K., Vestergaard, M., and Neimann-Sorensen, A., 1989, *Use of Somatotropin in Livestock Production*, Elsevier Appl. Sci., New York, NY.

Solomon, M. B., 1988, Relationship of muscle fiber types to the development of physiological maturity in meat animals. ASPC/NC-111 Symposium on Methods to Improve the Lean Yield of Lambs, pp 82-88.

Solomon, M. B., Berry, B. W., Fluharty, F. L., and McClure, K. E., 1996, Effects of energy source and ionophore supplementation on lipid composition and sensory properties of lamb, *J. Anim. Sci.*, **74**(Suppl. 1):162-170.

Solomon, M. B., Campbell, R. G., and Steele, N. C., 1990, Effect of sex and exogenous porcine somatotropin on longissimus muscle fiber characteristics of growing pigs, *J. Anim. Sci.*, **68**:1176-1182.

Solomon, M. B., Campbell, R. G., Steele, N. C., and Caperna, T. J., 1989, Occurrence of giant fibers and pale, soft, exudative (PSE) muscle in pigs treated with porcine somatotropin (pST). Proceedings of 35th International Congress of Meat Sci. and Technology, Vol. 3, pp. 1077-1085.

Solomon, M. B., Campbell, R. G., Steele, N. C., and Caperna, T. J., 1991, Effects of exogenous porcine somatotropin administration between 30 and 60 kilograms on longissimus muscle fiber morphology and meat tenderness of pigs grown to 90 kilograms, *J. Anim. Sci.*, **69**:641-645.

Solomon, M. B., Campbell, R. G., Steele, N. C., Caperna, T. J., and McMurtry, J. P., Effect of feed intake and exogenous porcine somatotropin on carcass traits and longissimus muscle fiber characteristics of pigs weighing 55 kg live weight, *J. Anim. Sci.*, **66**:3279-3284.

Solomon, M. B., Caperna, T. J., Mroz, R.J., and Steele, N.C., 1994, Influence of dietary protein and recombinant porcine somatotropin administration in young pigs: III. Muscle fiber morphology and meat tenderness, *J. Anim. Sci.*, **72**:615-621.

Solomon, M. B., Caperna, T. J., and Steele, N. C., 1990, Lipid composition of muscle and adipose tissue from pigs treated with exogenous porcine somatotropin, *J. Anim. Sci.*, **68**(Suppl. 1):217.

Solomon, M. B., and Pursel, V. G., 1994, Partitioning of carcass components of transgenic pigs. Proceedings of 40th International Congress of Meat Sci. and Technology, S-VII.01, pp. 11-17.

Solomon, M. B., Pursel, V. G., Paroczay, E. W., Bolt, D. J., Brinster, R. L., and Palmiter, R. D., 1994, Lipid composition of carcass tissue from transgenic pigs expressing a bovine growth hormone gene, *J. Anim. Sci.*, **72**:1242-1246.

Solomon, M. B., Steele, N. C., Caperna, T. J., and Pursel, V. G., 1991, A further look at the effects of growth hormone on morphological muscle characteristics in pigs. Proceedings of 37th Intl. Congress of Meat Sci. and Technology (1), pp. 497-501.

Solomon, M. B., West, R. L., and Hentges, J. F., Jr., 1986, Growth and muscle development characteristics of purebred Angus and Brahman bulls, *Growth*, **50**:51-67.

Sorensen, M. T., Oksbjerg, N., Agergaard, N., and Petersen, J. S., 1996, Tissue deposition rates in relation to muscle fibre and fat cell characteristics in lean female pigs (Sus scrofa) following treatment with porcine growth hormone (pGH), Comp. Biochem Physiol Vol. 113A, No. 2, pp. 91-96.

Spencer, G. S. G., 1986, Hormonal manipulation of animal production by immunoneutralization, in: *Control and Manipulation of Animal Growth*, P. J. Buttery, D. B. Lindsay and N. B. Haynes, ed., Butterworths, London, pp. 279-291.

Spencer, G. S. C., Garssen, G. J., and Bergstrom, P. L., 1983, A novel approach to growth promotion using auto-immunization against somatostatin. II. Effects on appetite, carcass composition and food utilization in lambs, *Livestock Production Sci.*, **10**:469-477.

Spencer, G. S. C., Garssen, G. J., and Hart, I. E., 1983, A novel approach to growth promotion using auto-immunization against somatostatin. I. Effects on growth and hormone levels in lambs, *Livestock Production Sci.*, **10**:25-37.

Tannenbaum, G. S., 1985, Physiological role of somatostatin in regulation of pulsatile growth hormone secretion, in: *Advances in Experimental Medicine and Biology*, Vol. 188, Somatostatin, Y. C. Patel and G. S. Tannenbaum, ed., Plenum Publishing Corp., New York, pp. 229-259.

Turman, E. J., and Andrews, F. M., 1955, Some effects of purified anterior pituitary growth hormone on swine, *J. Anim. Sci.*, **14**:7-15.

Vale, W., Riveir, C., and Brown, M., 1977, Regulatory peptides of the hypothalamus, *Ann. Rev. Physiol.*, **39**:473-527.

van der Wal, P., Nieuwhof, G. J., and Politiek, R. D., eds., 1989, Biotechnology for Control of Growth and Product Quality in Swine, Implications and Acceptability. Proceedings of an international symposium organized by the Wageningen Agric. Univ., Wageningen, Netherlands.

Vann, R. C., Althen, T. G., Smith, W. K., Veenhuizen, J. J., and Smith, S. B., 1998, Recombinant bovine somatotropin (rbST) administration to creep-fed beef calves increases muscle mass but does not affect satellite cell number or concentration of myosin light chain-1f mRNA, *J. Anim. Sci.*, **76**:1371-1379.

Vann, R. C., Althen, T. G., Solomon, M. B., Eastridge, J. S., Paroczay, E. W., and Veenhuizen, J. J., 2001, Recombinant bovine somatotropin (rbST) increases size and proportion of fast-glycolytic muscle fibers in semitendinosus muscle of creep-fed steers, *J. Anim. Sci.*, **79**:108-114.

Varner, M. A., Davis, S. L., and Reeves, J. J., 1980, Temporal serum concentrations of growth hormone, thyrotropin, insulin, and glucagon in sheep immunized against somatostatin, *Endocrinology*, **106**:1027-1032.

Virtanen, E. and Campbell, R., 1994, Reduction of backfat thickness through betaine supplementation of diets for fattening pigs, Handbuch der Tierische Veredlung **19**:145-150. Verlag H. Kamlage, Osnabruek, Deutchland.

Warriss, P. D., 1989, The influence of ß-adrenergic agonists and exogenous growth hormone on lean meat quality, Proceedings Brit. Soc. Anim. Prod. winter meeting, p. 52.

Wells, K. D., 2000, Genome modification for Meat Science: Techniques and Applications. Proceedings of 53rd Annual Reciprocal Meat Conf. pp. 87-93.

Williams, P. E. V., 1987, The use of ß-agonists as a means of altering body composition in livestock species, *Nutr. Abstr. Reviews,* **57B**:453-464.

Wood, J. D., and Fisher, A. V., eds., 1990, *Reducing Fat in Meat Animals*, Elsevier Appl. Sci., Bristol, UK.

QUALITY ASPECTS OF PORK MEAT AND ITS NUTRITIONAL IMPACT

Fidel Toldrá, Miguel A. Rubio, José L. Navarro and Lucio Cabrerizo [1]

1. INTRODUCTION

Pork meat production in Europe has followed a steady increase over the last decade, with about 19,000,000 tons produced in 1,999, and is expected to remain fairly stable or even experiment a small increase over the next few years. Spain constitutes the second largest producer of pig meat in the European Union. The production has risen to about 2,900,000 tons in 1,999. Pork meat is recognised as an important part of the European diet. However, consumers perception in recent years is not so good because they believe pork meat contains a high amount of visible fat with a high content in saturated fatty acids and cholesterol.

Modern pig breeding and feeding techniques are leading to a leaner meat with an increased proportion of unsaturated fatty acids in its lipids. In fact, the composition in fatty acids mainly depends on the genetic origin, age, weight at slaughter, feed composition and husbandry systems (Morgan et al., 1992; Toldrá et al., 1996). In some studies, the effect of genetics on intramuscular fat is broken down into free fatty acids, triacylglycerols and phospholipids (Gandemer et al., 1992; Armero et al., 2002). These authors found a strong effect of genetic type on the fatty acid composition of phospholipids but only a slight effect on the composition of triacylglycerols. Other studies have evaluated the effect of different pig sire types and sex on carcass traits and meat quality (Armero et al., 1999a) as well as on endogenous enzymatic activity such as cathepsins and exopeptidases (Armero et al., 1999b, 1999c). Fatty acid composition is also affected by the anatomical location of the muscle, especially the phospholipid fraction which contributes to the muscle variability (Leseigneur-Meynier and Gandemer, 1991; Hernández et al., 1998). In the case of feed, the change of the fatty acid composition in a monogastric animal such as pork is relatively easy.

Different pork meats, resulting from different crossbreeds and feeds, have been assayed in our laboratory during the last years for quality (pH, color, drip loss and fat

[1] Fidel Toldrá and José L. Navarro, Instituto de Agroquímica y Tecnología de Alimentos (CSIC), PO Box 73, 46100 Burjassot (Valencia), Spain. Miguel A. Rubio and Lucio Cabrerizo, Endocrinología y Nutrición, Hospital Clínico San Carlos, Madrid, Spain

Quality of Fresh and Processed Foods, edited by Shahidi et al.
Kluwer Academic/Plenum Publishers, 2004.

content) and composition in fatty acids of polar and non-polar lipids. This work presents the results obtained with the crossbreed and feed which gave the best results in both quality and proportion in unsaturated fatty acids. The selected crossbreed was based on a Landrace x Large White pig and the feed consisted in a mixture of barley, wheat, corn and soy meal. This meat, which is branded as Porcidiet® and commercialized by Vaquero Meat Industries, was also chosen for studying its nutritional impact in a controlled diet with <10% saturated fat and < 300 mg/d of cholesterol.

2. TECHNOLOGICAL QUALITY OF THE SELECTED CROSSBREED

Pork quality may experience large variations in quality depending on the susceptibility to stress. So, two extremes of quality like pale, soft and exudative (PSE) and dark, firm and dry (DFD) are usually found (Kauffman *et al.*, 1993). The classification is usually based on pH, measured at 2 and 24 hours postmortem in the muscle *Semimembranosus*, color parameter (L) and drip loss (DL). Thus, the carcasses from the selected crossbreed were classified according to the following values (Kauffman *et al.*, 1993; Flores *et al.*, 1999):

Pale, soft, exudative (PSE)	$pH_{2h}<5.8$, L>50, DL>6%
Red, firm, non-exudative (RFN)	$pH_{2h}>5.8$, $pH_{24h}<6.0$, L=44-50, DL<6%
Dark, firm, dry (DFD)	$pH_{24h}>6.0$, L<44, DL>3%

Based on these parameters, 82% of the carcasses were classified as normal (RFN), 15% as PSE and only 3% DFD. This postmortem quality has also an effect on the most important eating quality properties, namely tenderness and flavor. In the case of tenderness, it appears that DFD meat is more tender while PSE remains controversial (Tornberg, 1966; Flores *et al.*, 1999). In the case of flavor, DFD meat also shows a good quality with a low sour and salty tastes and high sweet taste (Flores *et al.*, 1999). In any case, PSE meat presents a high percentage of dripping that causes unsatisfaction to consumers.

The selected crossbreed showed a low intramuscular fat content and a relatively low content in cholesterol (see Table 1). These results are thus in accordance to the dietary recommendations for a reduced intake of fat and cholesterol with an adequate fatty acid composition.

Table 1. Characteristics, expressed as means (X) and standard deviations (SD), of the muscle *Longissimus dorsi.*

	X	SD
Moisture (g/100g)	75.6	0.51
Protein (g/100g)	21.8	2.1
Total lipid (g/100g)	2.86	0.35
Phospholipids (g/100g)	0.529	0.042
Free fatty acids (g/100g)	0.025	0.005
Triglycerides (g/100g)	2.31	0.221
Cholesterol (mg/100g)	46.1	6.1

3. LIPIDS COMPOSITION

3.1 Intramuscular lipids

The fatty acid composition of the muscle *Longissimus dorsi* is shown in Table 2. These lipids can be separated into polar and non-polar lipids. The neutral fraction containing non-polar lipids averages for more than 80% of the total lipids while the polar lipids are mostly phospholipids (see Table 1). The fatty acid composition of both fractions is shown in Table 2. Oleic acid, which is considered as a cholesterol lowering fatty acid, constitutes the major fatty acid (36.6%). This fatty acid is especially present in the neutral fraction (42.4%). When considering the saturated fatty acids and cardiovascular diseases, they contribute to the increase in LDL cholesterol but the individual fatty acids do not contribute to the same extent. For instance, myristic and palmitic acids (1.5% and 25.1%, respectively) raise LDL cholesterol while stearic acid (12.7%) is considered to have no net effect. So, even though total SFA amounts for 39.3%, only 26.6% are really contributing to the increase in LDL cholesterol. The ratios M/S and P/S are 1.01 and 0.53, respectively. The last one is above the recommended ratio 0.45 (Enser *et al.,* 1996).

Linolenic acis is the main source of ω-3 acids while linoleic and arachidonic acids constitute the major ω-6. Other ω-3 fatty acids like eicosapentaenoic (EPA) and docosahexaenoic (DHA) acids are present only in a very low percentage (data not shown). The ω-6/ω-3 ratio is proposed to be between 4 and 8 (Jakobsen, 1999).

Muscle tissue contains several endogenous antioxidant systems responsible for the oxidative stability of the lipids. Some of these antioxidants include histidine-containing dipeptides carnosine and anserine. Their concentrations in the muscle were in the range 300-320 mg/100g and 14-16 mg/100g, respectively (Aristoy and Toldrá, 1998). However, these values experience a high drop in the case of oxidative muscles. It is recommended to add vitamin E in the feed, at least 100-200 ppm for several weeks, for further protection against oxidation.

3.2 Adipose tissue lipids

The fatty acid composition of the adipose tissue is shown in Table 3. This composition tends to reflect the dietary fatty acids during fattening. The major fatty acid is oleic acid (31.64%) followed by linoleic acid (25.39%). The M/S and P/S ratios are 0.91 and 0.79, respectively. The P/S ratio is also quite above the recommended value 0.45 and approaching to 1.0 as the ultimate goal. Furthermore, the $\Sigma\omega$-6/$\Sigma\omega$-3 is approaching the recommended range of 4-8.

Table 2. Fatty acid composition (expressed as % of total fatty acids) of muscle *Longissimus dorsi* as well as into its neutral and polar fractions. Results expressed as means (X) and standard deviations (SD)

Fatty acid	Muscle		Neutral		Polar	
	X	SD	X	SD	X	SD
C 14:0	1.55	0.32	1.97	0.33	0.32	0.06
C 16:0	25.10	0.86	26.19	1.53	22.10	1.41
C 18:0	12.62	0.44	11.91	0.82	14.49	0.61
C 16:1	2.79	0.36	3.49	0.55	0.69	0.11
C 18:1	36.47	2.85	42.35	2.53	11.45	0.77
C 20:1	0.47	0.06	0.52	0.07	0.15	0.06
C 18:2	16.49	2.87	11.38	3.08	37.37	1.38
C 18:3	1.14	0.28	1.17	0.31	0.97	0.13
C 20:2	0.49	0.08	0.43	0.09	0.66	0.12
C 20:3	0.30	0.06	0.10	0.03	1.04	0.16
C 20:4	2.18	0.38	0.25	0.07	9.83	1.06
C 22:4	0.25	0.04	0.08	0.02	0.84	0.17
Total SFA	39.42	0.91	40.23	1.42	37.03	1.24
Total MUFA	39.74	3.15	46.36	2.93	12.26	0.85
Total PUFA	20.84	3.58	13.41	3.56	50.70	1.50
Ratio M/S	1.01		1.15		0.33	
Ratio P/S	0.53		0.33		1.37	
$\Sigma\omega$-3	1.14		1.17		0.97	
$\Sigma\omega$-6	18.97		11.73		48.2	
$\Sigma\omega$-6/$\Sigma\omega$-3	16.64		10.0		49.7	

4. NUTRITIONAL IMPACT

The current nutritional recommendations for the prevention and treatment of the hyperlipidaemia requires a controlled intake of saturated fat at levels below 10% of the total energy and of the dietary cholesterol lower than 300 mg/day (NCEP,1994). Basically, these recommendations imply the reduction in the ingestion of whole dairy products and red meats. Although there are some data that suggest a bigger freedom in the consumption of those red meats with a higher content in stearic acid, as the beef tallow (Bonanome *et al.*, 1988), for their neutral effect on the cholesterol concentrations, the scientific societies have not pronounced in favor of the free consumption of this type of red meats.

The studies that have analyzed the effects of the habitual intake of red meats on the lipid profile have had little impact in the current nutritional recommendations for the prevention of atherosclerosis. The intake of red meats, including pork meat, in comparison to the consumption of poultry meat or fish has not demonstrated that it negatively alters the lipid pattern in subjects with normolipemia (O'Brian *et al.*, 1980 Flynn *et al.*, 1982) nor in hyperlipidaemic patients (Gascon *et al.*, 1996, Watts *et al.*, 1988, Davison *et al.*, 1999). Whenever the general diet is adjusted to a controlled intake of saturated fat and cholesterol and include different portions of red meats low in total fat,

Table 3. Fatty acid composition (expressed as % of total fatty acids) of adipose tissue. Results expressed as means (X) and standard deviations (SD).

Fatty acid	Adipose tissue	
	X	SD
C 14:0	1.40	0.17
C 16:0	23.78	1.90
C 18:0	11.67	1.17
C 16:1	1.71	0.30
C 18:1	31.64	3.57
C 20:1	0.45	0.07
C 18:2	25.39	5.05
C 18:3	2.64	0.51
C 20:2	0.78	0.09
C 20:3	0.10	0.03
C 20:4	0.19	0.04
C 22:4	0.07	0.02
Total SFA	37.02	2.79
Total MUFA	33.81	3.85
Total PUFA	29.17	5.68
Ratio M/S	0.91	
Ratio P/S	0.79	
$\Sigma\omega$-3	2.64	
$\Sigma\omega$-6	25.68	
$\Sigma\omega$-6/$\Sigma\omega$-3	9.73	

all the refereed studies have not pointed out significant modifications in the lipid composition. Subjects with hyperlipidaemia, having a typically western diet, were allocated to another diet restricted in saturated fat and cholesterol, but incorporating the intake of lean red meats. The results showed a significant decrease in total cholesterol (- 8.6%) and c-LDL (- 11.1%) (Watts et al., 1988). Those data suggest that this type of lean meat can be freely incorporated in diets for the treatment of different hyperlipidemia. In fact, these results have also been confirmed in open studies, under conditions of free-living and long-term follow-up, as the one described recently (Davison et al., 1999) that included 191 patients with hypercholesterolemia that were assigned a diet with red meats (beef or pork) versus white meats (poultry or fish) over a 36 week period. This study did not show significant differences on the lipid profile among the two intervention groups.

The meat consumption in Spain is 53.4 kg / per capita/ year, poultry meat being most consumed (13.7 kg), followed by pork (10.4 kg), beef (7.1 kg), and other types of meat (9.3 kg) (MAPA, 1999). The consumption of processed pork meat (cooked and dry-cured products) is also quite important (12.9 kg) and follows a long tradition. However, dietary recommendations for the prevention of the hyperlipidaemia usually restrict the global consumption of pork meat (either lean or those processed meats). In these cases, poultry meats, lean cuts of veal and fish constitute an important part of the routine diets for the control of the hyperlipidemias.

Keeping in mind this situation, we conducted a comparative study among the intake of lean veal meat and lean pork meat (Porcidiet® with less than 5% content in total fat), in healthy volunteers to analyze their influence on the lipid profile. A controlled study, crossover and randomized was designed. During 12 weeks, 44 healthy subjects (22 men and 22 women) used to typical occidental diet rich in saturated fat and cholesterol, under free-living conditions, consumed during this period a diet according to the recommendations of the step 1 of the NCEP (NCEP, 1994) (30% total fat, <10% saturated fat and <300 mg/day of dietary cholesterol) that includes a daily portion of lean veal meat (6 weeks) or lean pork meat (6 weeks). The preliminary results of this study shows a reduction in plasma lipids. Both types of lean meats decreased significantly the content of total cholesterol (around 5%) and triacylglycerols (around 6%), without significant differences among both types of evaluated meats.

These data corroborate the results of other previously published studies and suggest that the intake of red lean meats, like pork meat, can be included in standard diets and even as a part of the therapeutic diets for the hyperlipidaemic patients.

5. REFERENCES

Armero, E., Flores, M., Toldrá, F., Barbosa, J-A., Olivet, J., Pla, M., and Baselga, M., 1999a., Effects of pig sire types and sex on carcass traits, meat quality and sensory quality of dry-cured ham, *J. Sci. Food Agric.,* **79**:1147-1154.

Armero, E., Barbosa, J.A., Toldrá, F., Baselga, M., and Pla, M., 1999b, Effects of the terminal sire type and sex on pork muscle cathepsins (B, B+L and H), cysteine proteinase inhibitors and lipolytic enzyme activities, *Meat Sci.,* **51**:185-189.

Armero, E., Baselga, M., Aristoy, M-C., and Toldrá, F., 1999c, Effects of sire types and sex on pork muscle exopeptidase activity and the content of natural dipeptides and free amino acids, *J. Sci. Food Agric.,* **79**:1280-1284.

Armero, E., Navarro, J. L., Nadal, M. I., Baselga, M., and Toldrá, F., 2002, Lipid composition of pork muscle as affected by sire genetic type, *J. Food Biochem.,* **26**:91-102.

Bonanome, A., Grundy, S. M., 1988, Effect of dietary stearic acid on plasma cholesterol and lipoprotein levels, *N. Engl. J. Med.,* **318**:1244-1248.

Davison, M. H., Hunninghake, D., Maki, K. C, Kwiterovich , P. O., and Kafonek S., 1999, Comparison of the effects of lean red meat vs lean white meat on serum lipid levels among free-living persons with hypercholesterolemia: a long-term, randomized clinical trial. *Arch. Inter. Med.,* **159**:1331-1338.

Flores, M., Armero, E., Aristoy, M-C., and Toldrá, F., 1999, Sensory characteristics of cooked pork loin as affected by nucleotide content and post-mortem meat quality, *Meat Sci.,* **51**:53-59.

Flynn , M. A., Naumann, H. D., Nolph ,G. B., Krause, G., and Ellersieck, M., 1982, Dietary meats and serum lipids, *Am. J. Clin. Nutr.,* **35**:935-942.

Gandemer, G., Viau, M., Caritez, J. C., and Legault, C., 1992, Lipid composition of adipose tissue and muscle in pigs with an increasing proportion of Meishan genes, *Meat Sci.,* **32**:105-121.

Gascon, A., Jacques, H., Moorjani, S., Deshaies, Y., Brun, L.D., and Julien , P , 1996, Plasma lipoprotein and lipolitic activities in response to the substitution of lean white fish for other animal protein sources in premenopausal women. *Am. J. Clin. Nutr.,* **63**:315-321.

Hernández, P., Navarro, J-L., and Toldrá, F., 1998, Lipid composition and lipolytic enzyme activities in porcine skeletal muscles with different oxidative pattern, *Meat Sci.,* **49**:1-10.

Hernández, P., Navarro, J-L., and Toldrá, F., 1999, Effect of frozen storage on lipids and lipolytic activities in the *Longissimus dorsi* muscle of the pig, *Z. Lebensm. Unters. Forchs.,* A, **208**:110-115

Kauffman, R. G., Sybesma, W., Smulders, F. J. M., Eikelenboom, G., Engel, B., Van Laack, R. L. J. M., Hoving- Bolink, A. H., Sterrenburg, P., Nordheim, E. V., Walstra, P., and Van der Waal, P. G., 1993, The effectiveness of examining early post-mortem musculature to predict ultimate pork quality, *Meat Sci.,* **34**:283-300.

Leseigneur-Meynier, A., and Gandemer, G., 1991, Lipid composition of pork muscle as related to metabolic type of fibres, *Meat Sci.,* **29**:229-241.

Ministerio de Agricultura, Pesca y Alimentación (MAPA), 1999, La alimentación en España. Madrid: Ministerio de Agricultura, Pesca y Alimentación.

Morgan, C. A., Noble, R. C., Cocchi, M., and McCartney, R., 1992, Manipulation of the fatty acid composition of pig meat lipids by dietary means, *J. Sci. Food Agric.*, **58**:357-368.

National Cholesterol Education Program (NCEP), 1994, Second report of the expert Panel on Detection, Evaluation and Treatment of high blood cholesterol in adults (Adult Treatment Panel II), *Circulation*, **89**:1329-1445.

O'Brien B. C., and Reiser, R., 1980, Human plasma lipid responses to red meat, poultry, fish, and eggs, *Am. J. Clin. Nutr.*, **33**:2573-2580.

Toldrá, F., Reig, M., Hernández, P., and Navarro, J-L., 1996, Lipids from pork meat as related to a healthy diet, *Recent Res. Devel. Nutr.*, **1**:79-86.

Toldrá, F., and Flores, M., 2000, The use of muscle enzymes as predictors of pork meat quality, *Food Chem.*, **69**:387-395.

Tornberg, E., 1996, Biophysical aspects of meat tenderness, *Meat Sci.*, **43**:S175-S191.

Watts, G.F., Ahmed ,W., Quiney, J., Houlston, R., Jackson, P., Illes, C., 1988, Effective lipid lowering diets including lean meat. *BMJ*, **296**:235-237.

6.   ACKNOWLEDGEMENTS

Grants 1FD97-1864 from the Comisión Interministerial de Ciencia y Tecnología (CICYT-FEDER, Madrid, Spain) and FIS 0279/98 from Ministerio de Sanidad y Consumo (Madrid, Spain) are acknowledged. Financial support from Vaquero Foundation for R+D on Pork Meat and Vaquero Meat Industries (Madrid, Spain) is also fully acknowledged. Porcidiet® is a trade mark by Vaquero Meat Industries.

MEAT FLAVOR: CONTRIBUTION OF PROTEINS AND PEPTIDES TO THE FLAVOR OF BEEF.[1]

A. M. Spanier [a], M. Flores [b], F. Toldrá [b], M-C. Aristoy [b], Karen L. Bett [c], P. Bystricky [d], and J. M. Bland [c]

1. INTRODUCTION

The flavor of a food is one of the principle factors involved in a consumer's purchase decision; meat is no exception. This makes it necessary for food technologists to have a thorough understanding of flavors and changes in flavors if they are to prepare products that consumers will purchase repeatedly. Such knowledge is particularly important in meat and meat products, since the development and deterioration of meat flavor is a continual process (Spanier *et al.,* 1988, 1990, 1992a, 1993; St. Angelo *et al.,* 1988) that involves both the generation and loss of desirable flavor components (St. Angelo *et al.,* 1988; Drumm and Spanier 1991; Spanier and Drumm-Boylston 1994) and the formation of off-flavor compounds (Bailey 1988; Spanier *et al.,* 1992b; St. Angelo *et al.,* 1992; Timms and Watts 1958). The development of many of these flavor producing components is associated with the process of postmortem aging (Spanier *et al.,* 1997), with the end-point cooking temperature (Spanier *et al.,* 1996), and with the process of lipid oxidation (St. Angelo 1992).

According to Hedrick *et al.* (1994) and Hamm (1986), meat is comprised of 76% water, 18.5% protein, 3% lipid, 1.5% non-protein nitrogen and 1% minerals. The carbohydrates, lipids, and proteins of meat all contribute in one way or another to its final flavor. While the literature is filled with references to the contribution of lipids and carbohydrates to meat

---

[1]   Mention of brand or firm names does not constitute an endorsement by the U.S. Department of Agriculture over others of a similar nature not mentioned.

[a]   U.S.D.A., Agricultural Research Service, Beltsville Area, Animal and Natural Resources Institute, Food Technology and Safety Laboratory, Bldg. 201, BARC-East, Beltsville, MD 20705 U.S.A.

[b]   Instituto de Agroquimica y Technologia de Alimentos, C.S.I.C., Department of Food Science, P.O. Box 73, 46100 Burjassot, Valencia, Spain

[c]   U.S.D.A., Agricultural Research Service, Southern Regional Research Center, Food Processing and Sensory Quality, 1100 Robert E. Lee Blvd., New Orleans, LA 70124 U.S.A.

[d]   University of Veterinary Medicine, Department of Meat Hygiene and Technology, Komenskeho 73, 041 81 Kosice, Slovakia

Quality of Fresh and Processed Foods, edited by Shahidi et al.
Kluwer Academic/Plenum Publishers, 2004.

flavor, little attention has been paid to the overall contribution of meat protein and peptides to meat flavor. On the other hand, the contribution of meat protein to human nutrition is well known, conferring nearly half of the protein nutrient intake in the US Diet (USDA 1984). Bodwell and Anderson (1986) indicate that meat contains a high proportion of highly digestible and essential amino acids such as tyrosine and cysteine. Generally protein from meat and meat products are 95 - 100% digestible while plant protein may be as low as 65-75% digestible. Other than the interaction of amino acids and peptides with reducing sugars to form meaty flavor notes via Strecker degradation, the importance and impact of meat peptides and amino acids to the taste-component of meat flavor has been neglected. This chapter deals with the contribution and impact of these compounds of protein origin to meat flavor.

2. DISCUSSION

2.1. Flavor

Several antemortem factors, e.g., age, breed, sex, nutritional status of the animal, muscle type (Hopkins, 1981), final end-point cooking temperature (Spanier and Miller, 1996), manner (moist, dry, convection, microwave, etc.) of cooking, and storage (type and duration), contribute to meat flavor quality (Carmack *et al.*, 1995; Imafidon and Spanier, 1994; Smulders *et al.*, 1992) (Figure 1). The flavor (Spanier *et al.*, 1997) and textural (Koohmaraie, 1994; Koohmaraie *et al.*, 1988) changes that occur during the postmortem aging period and during subsequent cooking and storage (Spanier and Miller, 1996) are important contributors to final meat flavor.

Factors Affecting Meat Flavor

ANTEMORTEM:	COOKING:
Age, Breed	Dry (roast, broil, bake, bar-b-que) vs.
Sex, Fodder quality	Wet (boil, stew)
Feeding regimen	Oven vs. Microwave
Fat level and compostion	Rate of heating
	high temp. / short time
POSTMORTEM:	low temp. / long time
Manner of slaughter	End-point temperature
Manner of storage after slaughter	Fresh cooked vs. Cooked/rewarmed
Length of aging time (PMA)	Fat level and composition
Animals condition at slaughter	POST-COOKING STORAGE:
(stressed vs. unstressed).	Refrigerated vs. frozen
Primal cut	Aerobic (O_2) vs. Anaerobic (CO_2, N_2)

Figure 1. *A list of some factors affecting meat flavor*

Meat shows a significant alteration in the level of numerous chemical components (sugars, organic acids, peptides, and free amino acids, and metabolites of adenine nucleotide metabolism such as ATP) during the postmortem aging period. Many of these changes are due to hydrolytic activity (Spanier *et al.,* 1990; Koohmaraie, 1994; Dransfield, 1994). These compounds, continuously created during the postmortem aging process, serve as a pool of primary precursors, reactive flavor chemicals and intermediates (Figure 2) that interact during cooking to form meat flavor characteristics (Spanier and Miller, 1996). It is apparent, therefore, that the development of meat flavor precursors is a highly dynamic process requiring further discussion and investigation. While we know many of the "words" which constitute flavor quality in meat (Spanier *et al.,* 1997; ASTM, 1996; St. Angelo *et al.,* 1994) (Figure 3) we do not fully understand the "syntax" of the words that comprise the complete flavor sentence i.e., give meat its overall flavor. We must learn not just what these words are, but how they work together to offer the final flavor perceived by the consumer.

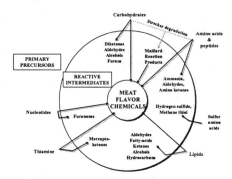

Figure 2. Meat flavor chemicals.

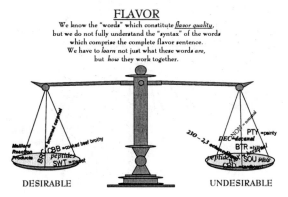

Figure 3. Flavor balance beam.

2.2. Postmortem Aging Changes

2.2.1a. *Morphological and biochemical events during the conversion of muscle to meat.*
After slaughter, muscle tissue is in a state of anoxia (no or low blood flow) and is in a
situation where the blood no longer brings new metabolic energy sources to the muscle nor
does it remove metabolites. Along with the anoxia is a drop in the intracellular levels of ATP
(Spanier *et al.,* 1985, 1997) as well as a visible increase in the number of lysosomes
(Imafidon and Spanier, 1994; Spanier and Weglicki, 1982; Spanier *et al.,* 1985) (Figure 4).
The lysosome is an important organelle in muscle since it contains many hydrolytic enzymes
(Bird, 1975) that are capable of hydrolyzing the muscle proteins to its constituent amino
acids (Bird, 1975; Bird *et al.,* 1978, 1980). Unlike the lysosomes of non-muscle cells,
muscle lysosomes of muscle are part of the sarcoplasmic reticular system (Bird, 1975, Bird
et al., 1978) (Figure 5). These lysosomes are intimately involved in the production of meat
peptide and amino acid flavor principles and precursors. They eventually lose their stability
and release many of their hydrolytic enzymes into the sarcoplasm , i.e., into the cytoplasm
of muscle. This occurs because tissue ATP levels are depleted due to inactivation of electron
transport system and because tissue lactate levels begin to increase. This leads to the release
of various enzymes (see "2" below) and proteins and peptides (see "3 and 4" below) which
have the capacity to (i) generate more flavor components, and (ii) generate additional
substance that can react and interact separately or together to form flavor compounds with
continued aging and/or during cooking and subsequent storage after cooking.

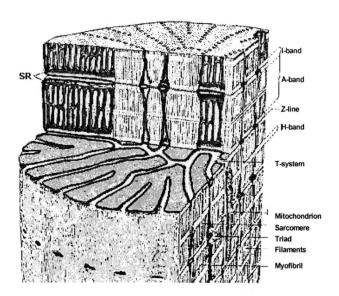

Figure 4.   Schematic representation of mammalian skeletal muscle.

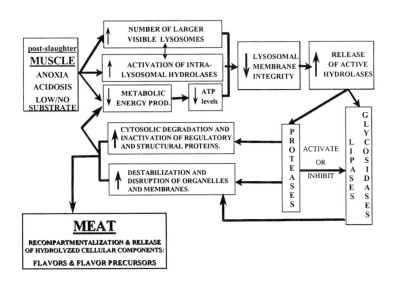

Figure 5.   Flow diagram for event as muscle is converted to meat.

2.2.1b. Enzymes. We have previously reported (Spanier *et al.*, 1990) on the changes in the activity of several key marker enzymes in aging meat; these included total protein, EGTA-ATPase, Ca^{++}-ATPase, –acetyl-β-glucosaminidase, cathepsin D, 5'-nucleotidase and creatine kinase. The enzyme specific-activity reached a maximum value by 3.5 hours postslaughter and was followed by a rapid drop in activity of all enzymes by 26 hours. Thereafter, a gradual decline in activity was observed through 360 hours (15 days). While total enzyme activity was declining, at 360 hours it was still higher than the activity immediately post slaughter. The high level of activity of membrane markers such as 5'-nucleotidase (sarcolemmal membranes), EGTA (myofibrillar fractions) and Ca^{2+}-ATPase (sarcoplasmic reticulum) suggest that these compartments are 'leaky' and thus more permeable to ions and other compounds during the PMA process.

2.2.1c. Proteins and peptides (gels and CE, in vivo). The leaky membranes and redistribution of hydrolytic enzymes such as glycosidase, proteases, lipases, etc. (Spanier *et al.*, 1990), make contact between hydrolytic enzymes and their substrates more likely. Furthermore, many of these hydrolytic enzymes show optimal activity at pH less than neutral, a condition that exists in aging meat (Spanier *et al.*, 1997). Sodium dodecyl sulfate-polyacrylamide gel electrophoresis (Spanier *et al.*, 1997) and capillary electrophoresis have shown both a creation (Figure 6; peaks #4, 6, 7, and 11) and destruction (Figure 6; peak #3) of peptides with continuous postmortem aging. It has been suggested that the newly appearing peaks are degradation products of the disappearing peak (Spanier *et al.*, 1997). This data was correlated with human sensory perception and other instrumental methods of flavor assessment (see below).

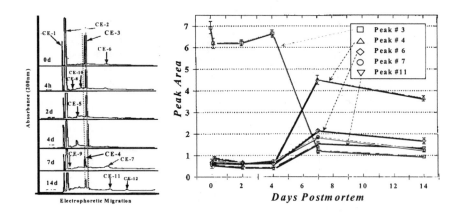

Figure 6. Electropherogram of capillary electrophoresis (CE) of meat extract stored from 0 to 14 days (left) and plot of area under peak for several peptide fragments (right).

2.2.1d. Peptides (in vitro). Because muscle contains many classes of proteinases (Bird *et al.,* 1977; 1978, 1980) each having its own inherent optimal pH for activity against various substrates, we examined the effect of pH on the creation of peptide products as measured by release of TCA (trichloroacetic acid)-soluble material. Figure 7 shows that the creation of TCA-soluble material may be prevented by addition of specific proteolytic inhibitors (10μM leupeptin and 2 μM pepstatin) at all pH levels examined. However, optimal production of TCA soluble material (peptides and amino acids) is seen between pH 5.0 and 6.0 the range that is optimal for most thiol-dependent proteinase activity such as cathepsins B, H, and L, and also the pH that most beef products reach during aging (Shahidi *et al.,* 1986).

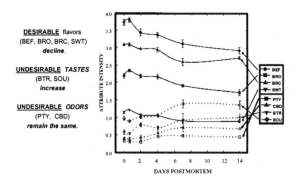

Figure 7. Descriptive sensory analysis of Brangus beef at times postharvest.

2.2.1e. Descriptive sensory analysis. During postmortem aging there is a change in intensity of several sensory attributes as a function of storage (Figure 8). There is a graded decline in desirable descriptive notes such as beefy (BEF), brothy (BRO), browned caramel (BRC) and sweet (SWT). The flavors (defined by intensity units, IU) showed a decline of 0.13, 0.05, 0.10, and 0.05 IU/day, respectively. On the other hand, the undesirable odor descriptors painty (PTY) and cardboard (CBD) and the taste descriptors bitter (BTR) and sour (SOU) show a moderate rate of increase in intensity of 0.04, 0.07, 0.06 and 0.80 IU/day, respectively. Correlation coefficients determined for both linear regression analysis of the rates and from ANOVA analysis of unlike groups were larger than 0.9 in all cases.

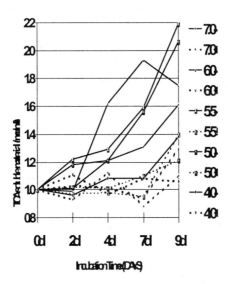

Figure 8. Effect of pH and storage on the production of peptides (TCA soluble material) in mat. Solid lines represent meat without the protease inhibitors leupeptin and pepstatin and dashed lines represent meat containing the protease inhibitors.

A correlation matrix (not shown) compared the sensory attributes over the range of treatments (days postmortem). Statistical evaluation indicated a very strong negative correlation between the bitter (BTR) taste and the more desirable flavor beefy (BEF), brothy (BRO), browned caramel (BRC), and sweet (SWT). On the other hand, the taste of sour (SOU) gave strong negative correlation only to browned-caramel (BRC). The undesirable aroma note, painty (PTY), had a strong negative correlation only with beefy (BEF), while the undesirable aroma note, cardboard (CBD), had a negative correlation with both beefy (BEF) and browned-caramel (BRC). All four desirable flavor descriptors, i.e. BEF, BRO, BRC, and SWT, had a strong positive correlation with each other. All of the sensory descriptors seem to have a sudden 'hump' or change between 4 and 7 days postmortem (Spanier *et al.,* 1997).

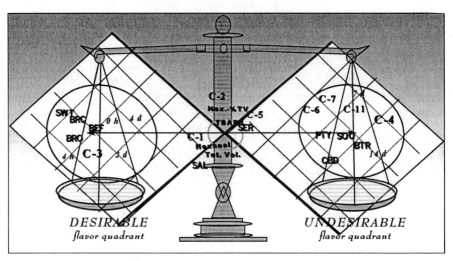

Figure 9.   Multivariate principal component analysis of meat flavor descriptors from descriptive sensory analysis, and peaks obtained by capillary electrophoreisis (C) and by GC . The bivariate plot is rotated 45° to permit visualizaton of the data over a balance beam.

A bivariate plot is created from multivariate principal components solution to compare treatments, sensory, instrumental and chemical data (Figure 9). This type of statistical analysis provides an empirical summary of the pattern of intercorrelations among the variables. In that these results segregate in opposite regions of the grid, the plot is rotated 90° and superimposed over a diagram of a balance beam to show the weighting of the opposing factors (Figure 9). For example, the data for the desirable flavors SWT, BRC, BEF and BRO group on the left of the rotated plot while the undesirable flavors such as CBD, PTY, BTR and SOU group on the right side of the plot. Chemical attributes, such as thiobarbituric acid reactive substances (TBARS), hexanal, and total volatiles, cluster near the origin of the plot (or fulcrum of the balance) indicating they have no weight (no effect or correlation) with the sensory attributes under these experimental conditions. Analysis of the multivariate factors reveals that meat from freshly slaughtered animals (0h) and meat aged for 4 h, and 2 and 4 days (4h, 2d, 4d) cluster in the same region (*left*) as the desirable flavors BEF, BRO, BRC, and SWT. On the other hand, meat aged 7 to 14 days (7d, 14d) group (*right*) with the undesirable flavors PTY, CBD, SOU, BTR. While these data reveal that a trained descriptive sensory panel can accurately assess the flavor of meat (Figure 8), it is unlikely that a typical consumer could clearly distinguish these subtle, yet significant, changes. On the other hand, these changes, no matter how subtle, are important and should be examined for a better understanding of the impact of the different flavor-precursor compounds generated during the postmortem period.

2.3 Dry-curing

Cured products represent a large portion of processed meats usually defined as wet- or dry-cured (Flores and Toldra, 1993). In wet cured products the curing ingredients are dissolved in water to form a pickle or brine which is introduced or injected into the meat.

On the other hand, dry-curing is a traditional process where the curing ingredients are rubbed onto the surface of the meat. In general, the dry-curing process consists of several stages: salting (about 9 - 11 days at 2-4°C), washing, post-salting (20-40 days at 2-4°C) for salt equalization, and ripening-drying (7-12 months at 12-22°C). These dry-cured products are subjected to extensive ripening or aging periods where the generation of dry-cured flavor takes place through the action of biochemical reactions of a proteolytic and lipolytic nature. The main curing ingredients are salt, nitrate and/or nitrite. The primary role of salt in curing is to act as a bacteriostatic agent, but it also affects the flavor of the product and controls the muscle enzyme systems (Toldrá et al., 1997).

Dry-cured hams are products typically from the Mediterranean area and include products such as Spanish Serrano, Italian Parma, and French Bayonne hams. The drying or ripening process has variations depending on the traditions of the country in which it is processed. Hams in the U.S. and many other countries are subject to a final smoking stage.

The high quality of dry-cured ham depends on its unique flavor. However the increased production cost of long-term dry-curing makes the product less competitive in the market. Several studies have attempted to reduce the processing time (Marriot et al., 1987, 1992), but the length of the ripening-drying stage is necessary in order to have complete cured color formation and dry-cured flavor development (Toldrá et al., 1997).

An increase in the concentration of free amino acids has been reported both in postmortem muscle during meat aging (Nishimura et al., 1988, 1990; Okitani et al., 1981) and in meat products such as dry-cured and cooked ham (Aristoy and Toldrá, 1991; Buscailhon et al., 1994; Toldrá et al., 1995). This increase in free amino acids has been attributed to the action of muscle amino peptidases active at neutral pH (Nishimura et al., 1990; Aristoy and Troldrá, 1991). The contribution of peptides and amino acids to the improvement of meat taste have been reported in meat prepared at different cooking temperatures (Spanier et al., 1988; Spanier and Miller, 1993, 1996). Also the contribution of free amino acids accumulated during processing is relevant for the development of the specific dry-cured flavor (Flores et al., 1997). Accumulation of amino acids is of great importance not only because of their specific taste (Kato et al., 1989; Nishimura and Kato, 1988), but also because of their involvement in further Strecker and Maillard reactions generating volatile flavor compounds (Nishimura and Kato, 1988). This is the case with dry-cured ham where many volatile compounds are generated through lipid oxidation (Barbieri et al., 1992; Berdaque et al., 1991), and other compounds, such as sulfide compounds, pyrazines, and methyl branched aldehydes and alcohols are generated through Strecker degradation of amino acids (Flores et al., 1997).

2.4 Cooking Temperature

The flavor compounds and precursors developed during the postmortem aging period are further altered during cooking and heating yielding a product with a new pleasing flavor. Heating of meat to high temperature has two effects (1) it destroys most microbial pathogens and spoilage organisms thereby enhancing safety and (2) it changes the food's flavor by altering the precursors and final flavor components. Heating of meat, particularly above 50°C, alters the activity of various hydrolytic enzymes (Spanier *et al.*, 1990). Heating also causes structural and functional changes in meat proteins that create additional peptides (Figure 10) and amino acids that may have their own inherent flavor (Nishimura and Kato 1988) and are also free to react with reducing sugars of meat to form new taste and odor compounds via Maillard reaction and Strecker degradation (Bailey, 1988; Shahidi *et al.*, 1986).

Different primal cuts have been shown to exhibit different flavor profiles (St. Angelo 1992; Carmack *et al.*, 1995). These cuts react uniquely to heating because of the distinct fiber types (Hedrick *et al.*, 1994), the differences in relative fat disposition (Slover *et al.*, 1987), and the different intracellular localization of various cellular components (Spanier and Miller 1996). Indeed, it has been shown that protein of rib eye (*longissimus* muscle) are more sensitive to heating than those of top round (*semimembranosus* muscle), permitting more peptides and amino acids to be formed and localized in the soluble fraction (Spanier and Miller, 1996).

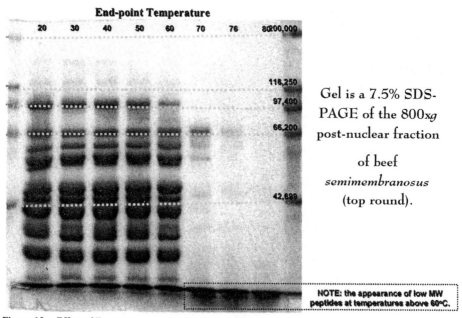

Gel is a 7.5% SDS-PAGE of the 800x*g* post-nuclear fraction

of beef
semimembranosus
(top round).

NOTE: the appearance of low MW peptides at temperatures above 60°C.

Figure 10.   Effect of Temperature on Meat Proteins and Peptides.

2.5 Cooking and Storage after Cooking

2.5.1a. *Effect on meat peptides*. Precooked beef products, often referred to as "convenience" and "institutional" foods, comprise 35% of the beef sold and consumed in America today. In 1994, this represented almost $14 billion in consumer expenditures on meat. Therefore, a thorough understanding of the flavor of meat and those factors that affect the meat flavor is critical to continued sales in this large consumer market.

Re-warming of cooked meat causes changes in the structure and function of meat proteins (Spanier *et al.*, 1988). Even after heating beyond 60°C, thiol dependent proteases retain significant activity that have the capability of producing more flavor compounds and flavor reactive material during storage (Spanier *et al.*, 1990). Post-storage warming of meat cooked to different end-point temperatures had been shown by SDS-PAGE to produce new flavor peptides and amino acids (Figure 11) as well as flavor reactive components isolated by chromatographic methods (Spanier *et al.*, 1988). Meat extracts examined sequentially by size exclusion chromatography followed by HPLC (Spanier *et al.*, 1988) showed that a potential desirable flavored peptide disappeared after 2 days of post-cooking refrigerated storage (i.e., a hydrophilic peak), but showed no change in the peak area of the peptide material associated with an undesirable flavor (hydrophobic peak). While the breakdown of the hydrophobic peptides may be a result of fragmentation of proteinaceous material by free radical mechanisms associated with lipid oxidation (Spanier *et al.*, 1992b) it may also be due to the residual proteolytic enzyme activity in precooked meat (Spanier *et al.*, 1990; Spanier and Miller, 1996).

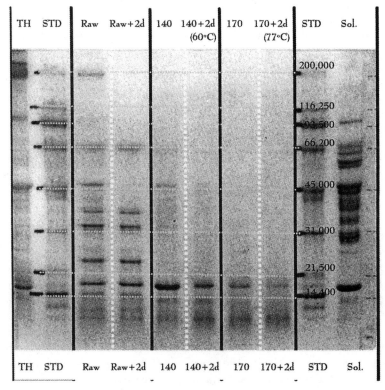

Figure 11. Effect of temperature and storage on the protein and peptide content of roasted semi-membranosus mini-cubes. Gel was a 7.5% - 21.0% SDS-page gel gradient.

Along with the change in protein caused by the heating-associated with cooking of meat is a change in the activity of many meat enzymes, particularly the proteases (Spanier *et al.,* 1990; Spanier and Miller, 1996). Meat thiol dependent proteases such as cathepsin B and L maintain 30% of their precooking activity after heating above 60°C and through approximately 80°C. Thus, these enzymes are available to produce new flavor compounds during storage and during the re-warming after storage of precooked meat.

2.5.1b. Effect on meat protein (SDS-PAGE). The effect of storage on the formation of low molecular weight (MW) proteinaceous material (i.e., peptides and amino acids) in precooked meat may also be demonstrated by the traditional technique of gel electrophoresis (Figure 11). Not only is there the generation of new and more low MW material as cooking temperature is increased (compare column 3, 5, and 7 of Figure 11 and all of Figure 10), but this pattern of peptides expands further when examining storage followed by re-warming (columns 4, 6, and 8 of Figure 11). The production of proteinaceous flavor materials is not only affected to various degrees by the cooking-temperature, but is also affected by storage. A complete understanding of the heating- and storage-induced formation of these proteinaceous products in the various primal cuts is crucial to our being able to determine the optimal conditions for producing and formulating meat products of the highest flavor quality.

2.6 SAVORY TASTE-ENHANCING PEPTIDE (STEP): A specific meat peptide

Thus far we have discussed, to varying degrees, the general changes in meat protein, peptides and amino acids as a function of postmortem aging, dry-curing, primal cut, cooking/heating, and post-cooking storage. A specific meat protein [given the acronym STEP for savory taste-enhancing peptide] should also receive attention. STEP was originally named 'delicious peptide' by Yamasaki and Maekawa (1978) who found this material in papain digests of beef. The majority of laboratories have reported this octapeptide to have a 'delicious', *umami,* and savory flavor (Spanier *et al.,* 1997; Kuramitsu *et al.,* 1997; Nakata *et al.,* 1995; Tamura *et al.,* 1989) although one laboratory recently reported that there was little or no detectable flavor or flavor enhancing properties (Hau *et al.,* 1997). However, the careful examination of the latter paper has revealed many experimental flaws (Hau *et al.,* 1997). Based on sensory evaluation of synthetic peptides analogs and the octapeptide itself, a mechanism for the ligand's response at the taste receptor was presented (Spanier *et al.,* 1997); the theory gained independent support from the sensory work of Nakata *et al.* (1995), Wang *et al.* (1996) and others while the mechanistic portion of the theory is supported by the 2 dimensional NMR spectroscopy and quenched molecular dynamics research approach taken and reported by Cutts and colleagues (1996). Characteristics of STEP may be found in Figure 12.

As a result of the development of a working theory for STEP interaction at the taste receptor, it was hypothesized that a dimer of STEP should be a more effective flavor effector than the monomer (Spanier *et al.,* 1997). The dimer of STEP was given the acronym of NExT STEP where NExT represented 'new extra tasty'; alternatively, the dimer may be written as $STEP_2$. Sensory evaluation of the materials revealed that the observed flavor was quite different if MSG, the monomer, or the dimer were tasted in (i) aqueous solution, (ii) in

Some STEP Characteristics {<u>S</u>avory <u>T</u>aste <u>E</u>nhancing <u>P</u>eptide}

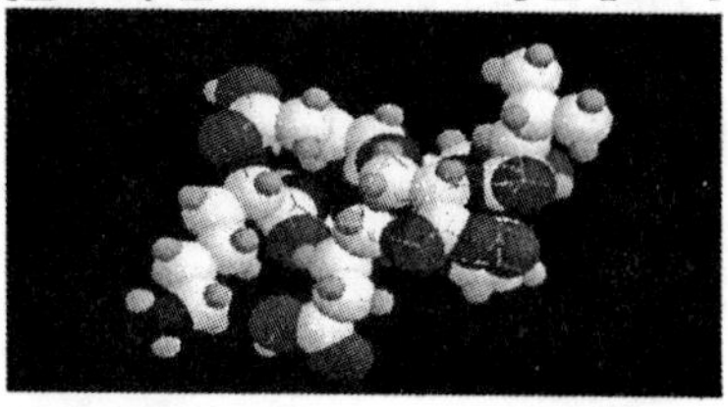

Found: Beef $\underline{MW} = 848$

Sequence: lys-gly-asp-glu-glu-ser-leu-ala

Physical Characteristic: heat stable; β - sheet

Taste in Water: full-mouth, rounded-out flavor (*umami*, savory)

Taste with other compounds: variable: depending on mixture it may enhance, modulate, or potentiate.

Produced by: proteolysis of a larger parent protein (not *actin* or *myosin*) by thiol dependent proteases of muscle.

Figure 12. Some properties of STEP (Savory Taste Enhancing Peptide).

simple mixtures such as a sucrose solution or in (iii) complex mixtures such as gravies. In aqueous solutions, MSG, STEP and NExT STEP gave a full-mouth, rounded-out sensation usually associated with MSG and savory compounds. Furthermore, NExT STEP in aqueous solution gave an unexpected effect of lingering from 5 to 60 min with a mean of 16 min in panel evaluation. Simple solutions such as a sucrose gave variable results depending upon both the concentration of the sucrose solution and upon the "question" asked of the panelists (Spanier *et al.*, 1997). Each of the three solutions seemed, for the most part, to be equally effective in stimulating the taste response in simple solution. Evaluation of the materials in more complex mixtures such as beef and chicken broths indicated that there was indeed a difference in the response of the flavor enhancers most likely due to the interaction of the material with other broth components such as nucleotides (Spanier *et al.*, 1997). Data recently obtained in our laboratory have indicated that STEP and NExT STEP both seem to have an effect on the flavor volatiles as well as on taste components; similar responses were shown by Maga and Lorenz (1972) for MSG in beef gravy. These apparently conflicting results raised some interesting questions which will be dealt with in subsequent publications. However, for now it is important to point out that the concept of taste potentiation is based on misinterpreting the meaning of the word, "enhance" Scott Hegenbart, editor of Food Product design (Feb., 1996), Sara Kemp of Givaudan-Roure in Clifton, NJ and Gary Beauchamp, Director of the Monell Chemical Senses Center in Philadelphia, PA have proposed the following limited definitions for enhance. They propose that *flavor potentiator* should be used for substances that increase the perceived intensity of the flavor of another substance by a mechanism other than simply adding to it. A *flavor modulator* would be the term for substances that appear to potentiate taste, but are actually suppressing other tastes rather than giving true potentiation. Finally, a *flavor enhancer*, would then be restricted to hedonic improvement i.e., a substance that directly increases the pleasantness of the flavor of another substance. Experimentations from our laboratory have indicated that STEP and NEXT STEP seem to function in all three manners depending upon the food they are added to.

What is the *NExT STEP* in flavor enhancement?

Use of biotechnology will permit

the engineering of microbes (*bacteria, fungi, yeast*) capable of expressing and producing STEP and/or NExT STEP for use as an ***external flavor additive***.

genetic insertion of the flavor compound into various foods **to directly enhance the flavor quality of the product** AND **any food it is used with**

e.g., tomato => sauce, paste, & juice
or grape => jam & wine, etc.

Figure 13. Possible future directions.

2.6.1 Where do we go from here? The potential to enhance the quality of our muscle foods by utilizing the knowledge gained from research on STEP is tremendous. If the parent protein or protein of STEP origin can be identified, such information can assist farmers and animal breeders in the development or selection of those breeds of cattle or other meat animals that will have more of the parent protein in their tissues thus promoting more highly flavored meat products. This knowledge should also help us to develop better post-slaughter management methods that will promote the production of STEP during the post mortem aging period. Additionally, such compounds have a potential for use in non-muscle foods (Figure 13). The material may be used directly as an added flavoring to enhance the flavor quality of the food on which it is sprinkled in a manner similar to salt and pepper, but yet without the negative effect that high salt has on the cardiovascular system. Alternatively, such material can be engineered into other commodities yielding a product not only with enhanced flavor but a product that may have similar enhanced flavor when made into other secondary products.

3. SUMMARY

While the majority of meat flavor is lipid in origin, the contribution of peptides and amino acids to overall meat flavor should not be overlooked. Amino acids and peptide levels have been shown to change with postmortem aging in muscle and with dry-curing, a process similar to PMA. Variation in protein, peptide, and amino acid composition have also been shown to occur with heating and with post-heating storage of meat. This makes a large pool of reactive components that may directly affect flavor or indirectly affect flavor by reacting with reducing sugars to form Maillard reaction products and Strecker degradation products that impact meat flavor. Further research in this area should continue with particular emphasis on natural peptide flavor enhancers, modulators, and potentiators.

4. REFERENCES

Aristoy, M. C. and Toldrá, F., 1991, Deproteinization techniques for HPLC amino acid analysis in fresh pork muscle and dry-cured ham, *J. Agric. Food Chem.*, **39**:1792-1795.

ASTM, in, *Aroma and Flavor Lexicon for Sensory Evaluation: Terms, Definitions, References and Examples*, DS66, G. V. Civille and B. G. Lyons, eds., ASTM, West Conshohocken, PA, 1996

Bailey, M. E., 1988, Inhibition of warmed-over flavor, with emphasis on Maillard reaction products, *Food Technol.*, **42**:123-126.

Barbieri, G., Bolzoni, L., Parolari, G., Virgili, R. Buttini, R., Careri, M., and Mangia, A., 1992, Flavor compounds of dry cured ham, *J. Agric. Food Chem.*, **40**:2389-2394.

Berdaque, J. L., Denoyer, C., LaQuéré, J. L., and Semon, E., 1991, Volatile compounds of dry-cured ham., *J. Agric. Food Chem.*, **39**:1257-1261.

Bird, J. W. C., Schwartz W. N., and Spanier, A. M., 1977, Degradation of myofibrillar proteins by cathepsin B and D., *Acta. Biol. Med. Germ.*, **36**:1587-1604.

Bird, J. W. C., Carter, J., Triemer, R. E., Brooks, R. M. and Spanier, A. M., 1980, Proteinases in cardiac and skeletal muscle, *Fed. Proc.*, **39**:20-25.

Bird, J. W. C., Spanier, A. M. and Schwartz, W. N., Cathepsin B and D: Proteolytic activity and Ultrastructural Localization in Skeletal Muscle, in: *Protein Turnover and Lysosome Function*, ed., H. L. Segal and D. J. Doyle, Academic Press, New York, pp. 589-604.

Bird, J. W. C. in, Vol. 4., *Lysosomes in Biology and Pathology*. ed. J.T. Dingle, and R.T. Dean, American Elsevier, New York, 1975, pp. 75-102.

Bodwell, C. E. and Anderson, B. A., 1986, Nutritional composition and value of meat and meat products, in: *Muscle as Food*, P. J. Bechtel, P. J. Academic Press, Inc., Orlando, FL, pp. 321-369.

Buscailhon, S, Monin, S., Cornet, G. and Bousset, J., 1994, Time-related changes in nitrogen fractions and free amino acids of lean tissue of french dry-cured ham, *Meat Science*, **37**:449-456.

Carmack, F. F., Kastner, C. L., Dikeman, M.E., Schwenke, J.R., and Garcia Zepeda, C. M., 1995, Sensory evaluation of beef-flavor-intensity, tenderness, and juiciness among major muscles, *Meat Sci.*, **39**:143-147.

Cutts, R. J., Howlin, B. J., Mulholland, F., and Webb, G. A., 1996, Low-energy conformations of delicious peptide, a food flavor. Study by quenched molecular dynamics and NMR., *J. Agric. Food Chem.*, **44**:1409-1415.

Dransfield, E. 1994, Optimization of tenderization, ageing and tenderness, *Meat Sci.*, **36**:105-121.

Drumm, T. D. and Spanier, A. M., 1991, Changes in the content of lipid autoxidation and sulfur-containing compounds in cooked beef during storage, *J. Agric. Food Chem.*, **39**:336-343.

Flores, M. and Toldrá, F., Curing: Process and applications, in: *Encyclopedia of Food Science, Food Technology and Nutrition*, R. Macrae, R. Robinson, M. Sadler, and G. Fullerlove, ed., Academic Press, London, 1993, pp. 1277-1281.

Flores, M., Aristoy, M. C., Spanier, A. M., and Toldrá, F., 1997, Correlations of non-volatile components and sensory properties of Spanish Serrano" dry-cured ham, Effect of processing time, *J. Food Sci.*, **62**:1235-1242.

Flores, M., Grimm, C. C., Toldrá, F., and Spanier, A. M., 1997, Correlations of sensory and volatile compounds of Spanish "Serrano" dry-cured ham as a function of two processing times, *J. Agric. Food Chem.*, **45**:2178-2186.

Hamm, R., 1986, Functional properties of the myofibrillar system and their measurements, in: *Muscle as Food*, P. J. Bechtel ed., Academic Press, Inc., Orlando, FL., pp. 135-199.

Hau, J., Cazes, D., and Fay, L. B., 1997, Comprehensive study of the Beefy Meaty Peptide, *J. Agric. Food Chem.*, **45**:1351-1355.

Hedrick, H. B., Aberle, E. D., Forrest, J. C., Judge, M. D., and Merkel, R. A., 1994, *Principles of Meat Science*, Kendall/Hunt Publishing Co., Dubuque, IA., 3rd edition

Hopkins, D. T., 1981, *Protein Quality in Humans: Assessment and In Vitro Estimation.*, C. E. Bodwell, J. S. Atkins, and D. J. Hopkins, AVI Publishing Co., Westport, CT. 1981, pp. 169-194.

Imafidon, G. I., and Spanier, A. M., 1994, Unraveling the Secret of Meat Flavor, *Trends in Food Science & Technology*, **5**:315-321.

Kato, H., Rhue, M. R., and Nishimura, T., 1989, Role of free amino acids and peptides in food taste, *Flavor Chemistry. Trends and Developments*, R. Teranishi, R. G. Buttery, and F. Shahidi, ed., ACS Symposium series 388, ACS Books, Inc., Washington, DC., pp. 158-174.

Koohmaraie, M., Babiker, A. S., Merkel, R. A., and Dutson, T. R., 1988, Role of Ca`·`- dependent proteases and lysosomal enzymes in post-mortem changes in bovine skeletal muscle, *J. Food Sci.*, **53**:1253-1257.

Koohmaraie, M., 1994, Muscle proteinases and meat aging, *Meat Science,* **36**:93-104.

Kuramitsu, R., Nakata, T., Kamasaka, K., Fukama, T., Nakamura, K., Tamura, M., and Okai, H., 1997, Role of ionic Tastes in Food Materials, in: *Chemistry of Novel Foods*., A. M. Spanier, M. Tamura, H. Okaim and O. Mills, Allured Publishing Corp., Carol Stream, Il, pp. 19-32.

Maga, J. A., and Lorenz, K., 1972, The effect of flavor enhancers on direct headspace gas-liquid chromatography profiles of beef broth, *J. Food Sci.*, **37**:963-964.

Marriot, N. G., Graham, P. P., Shaffer, C. K., and Phelps, S. K., 1987, Accelerated production of dry-cured hams, *Meat Sci.,* **19**:53-64.

Marriot, N. G., Graham, P. P., and. Claus, J. R., 1992, Accelerated dry-curing of pork legs (hams): A review, *J. Muscle Foods,* **3**:159-168.

Nakata, T., Takahashi, M., Nakatani, M., Kuramitsu, R., Tamura, M., and Okai, H., 1995, Role of basic and acidic fragments in delicious peptides (lys-gly-asp-glu-glu-ser-leu-ala) and the taste behavior of sodium and potassium salts in acidic oligopeptides, *Biosci. Biotech. Biochem.*, **59**:689-693.

Nishimura, T. and Kato, H., 1988, Taste of free amino acids and peptides, *Food Rev. Int.*, **4**:175-194.

Nishimura, T. Okitani, A., Rhue. M. R. and Kato, H., 1990, Survey of neutral aminopeptidase in bovine, porcine and chicken skeletal muscle, *Agric. Biol. Chem.*, **54**:2769-2775.

Nishimura, T., Rhue, M. R., Okitani, A. and Kato, H., 1988, Components contributing to the improvement of meat taste during storage, *Agric. Biol. Chem.,* **52**:2323-2330.

Okitani, A., Otsuka, Y., Katakai, R., Kondo, Y. and Kato, H., 1981, Survey of rabbit skeletal muscle peptideases active at neutral pH regions., *J. Food Sci.*, **46**:47-51.

Shahidi, F., Rubin, L.J. and D'Souza, L. A., 1986, Meat flavor volatiles: A review of the composition, techniques of analysis, and sensory evaluation, *CRC Crit. Rev. Food Sci. Nutr.*, **24**:141-243.

Slover, H. T., Lanza, E., Thompson, R. H. Jr., Davis, C. S. and Merola, G. V., 1987, Lipids in raw and cooked beef, *J. Food Comp. Anal.*, **1**:26-37.

Smulders, J. M., Toldrá, F., Flores, J. and Prieto, M., 1992, *New Technologies for Meat and Meat Products*, J. M. Smulders, F. Toldrá, and M. Prieto, eds., Audet Tijdschriften bv., The Netherlands, pp.389.

Spanier, A. M., and Drumm-Boylston, T., 1994, Effect of temperature on the analysis of beef flavor volatiles: A look at carbonyl and sulfur-containing compounds, *Food Chem.*, **50**:251-259.

Spanier, A. M., Flores, M., McMillin, K. W., and Bidner, T. D., 1997, The effect of postmortem aging on meat flavor quality. Correlation of treatments, sensory, instrumental, and chemical descriptors, *Food Chem.*, **59**: 531-538.

Spanier, A. M., and Weglicki, W. B., 1982, Calcium-tolerant adult canine myocytes: Preparation and response to anoxia/acidosis, *Am. J. Physiol.*, **243**:H448-H455.

Spanier, A. M., Vercellotti, J. R., and James, C. Jr., 1992a, Correlation of sensory, instrumental, and chemical attributes of beef as influenced by structure and oxygen exclusion, *J. Food Sci.,* **57**:10-15.

Spanier, A. M., Miller, J. A., and Bland, J. M., 1992b, Lipid oxidation: Effect on meat proteins, *Lipid Oxidation in Foods*, American Chemical Soc. Books, Inc., Washington, DC, pp. 104-119.

Spanier, A. M., Edwards, J. V., and Dupuy, H. P., 1988, The W.O.F. process in beef: A study of meat proteins and polypeptides, *Food Tech.*, **42**:110-118.

Spanier, A. M., and Miller, J. A. Jr., 1996, Effect of Temperature on the quality of muscle food, *J. Muscle Foods*, **7**:355-375.

Spanier, A. M., Dickens, B. F., and Weglicki, W. B., 1985, Response of canine cardiocyte lysosomes to ATP, *Am. J. Physiol.*, **249** *(Heart Circ. Physiol.* **18**):H20-H28.

Spanier, A. M., McMillin, K. W., and Miller, J. A., 1990, Enzyme activity levels in beef: Effect of postmortem aging and end-point temperature, *J. Food Sci.*, **55**:318-322, 326.

Spanier, A. M., and Miller, J. A., 1993, Role of proteins and peptides in meat flavor, in: *Food Flavor & Safety*, A. M. Spanier, H. Okai, and M. Tamura, American Chemical Society, Washington, DC, pp. 78-99.

St. Angelo, A. J., Vinyard, B. T., and. Bett, K. L., 1994, Sensory and Statistical Analyses in Meat Flavour Research, in: *Flavor of Meat And Meat Products*, F. Shahidi, Blackie Academic & Professional, London, pp. 266-290.

St. Angelo, A. J., Vercellotti, J. R., Dupuy, H. P., and Spanier, A. M., 1988, Assessment of beef quality: a multidisciplinary approach, *Food Technol.*, **42**:133-138.

St. Angelo, A. J., Spanier, A. M., and Bett, K. L., 1992, in: *Lipid Oxidation in Foods*, American Chemical Society Books, Inc., Washington, DC, pp. 140-160.

St. Angelo., A. J., 1992, in: *Lipid Oxidation in Food*, ACS Symposium Series No. 500, American Chemical Society, Washington, DC.

Tamura, M., Nakatsuka, T., Tada, Kawasaki, Y., Kikuchi, E., and Okai, H., 1989, The relationship between taste and primary structure of "delicious peptide" (lys-gly-asp-glu-glu-ser-leu-ala) from beef soup, *Agric. Biol. Chem.*, **53**:319-325.

Timms, M. J. and Watts, B. M., 1958, Protection of cooked meats with phosphates, *Food Tech.*, **12**:240-243.

Toldrá, F., Flores, M., and Aristoy, M. C., 1995, Enzyme generation of free amino acids and its nutritional significance in processed pork meats, in: *Recent developments in food science and Nutrition*, G. Charalambous, ed., G. Elsevier Sci., B.V. Amsterdam, The Netherlands, pp. 1303-1322.

Toldrá, F., Flores, M., and Sanz, Y., 1997, Dry-cured ham flavour: enzymatic generation and process influence. *Food Chem.*, **59**:523-530.

Toldrá, F., Flores, M., Navarro, J. L., Aristoy, M. C., and Flores, J., 1997, New developments in dry-cured ham, in: *Chemistry of Novel Foods*, A. M. Spanier, M. Tamura, H. Okai, and O. Mills, eds., Allured Pub. Co. Carol Stream, IL, pp. 259-272.

USDA in, *USDA nationwide food consumption survey, 1977-1978, nutrient intakes: individuals in 48 states.* U.S.D.A., Washington, DC, 1984, Report No. 1-2.

Wang, K., Maga, J. A., and Bechtel, P. J., 1996, Taste properties and synergism of beefy meaty peptide, *J. Food Sci.*, **61**:837-839.

Yamasaki, Y., and Maekawa, K., 1978, A peptide with delicious taste, *Agric. Biol. Chem.*, **42**:1761-1765.

THE EFFECTS OF EXTENDED CHILLED STORAGE ON THE ODOR AND FLAVOR OF SHEEPMEAT

Terry J. Braggins[*], Michael P. Agnew, Deborah A. Frost, Colleen Podmore, Tracey L. Cummings, and Owen A. Young

1. INTRODUCTION

New Zealand's geographic position remote from Northern Hemisphere markets is an extra challenge that the meat industry has had to address to remain internationally competitive. Traditionally, New Zealand exported most of its sheepmeat frozen. However, in a drive to market table-ready cuts of beef and lamb, a carbon dioxide controlled atmosphere packaging system was developed so raw product could be shipped chilled ($-1.5°C$) and yet maintain a hygienic storage life of at least 16 weeks. One advantage of this type of storage is that the meat has sufficient time to tenderize during transport. This process, known as 'ageing', is also important for the development of meat flavor (Bouton et al., 1958).

Coppock and MacLeod (1977) found that total organic volatiles increased with ageing period and odor descriptors indicated a trend from weak, bland and unappetising to strong, savoury, appetising and roasted, after 4 weeks storage of vacuum-packed beef at 0.6-$3.3°C$.

Gill (1988) noted that the use of a carbon dioxide controlled atmosphere for chilled meat storage 'stripped' ovine flavor from sheepmeat and 'gamey' flavor from venison. He also commented that with prolonged storage, red meat develops certain flavors due to peptide formation during the hydrolysis of proteins.

Few, if any, comprehensive studies have been published on the effects extended storage of chilled fresh meat packaged in a CO_2 atmosphere has on meat proteins and lipids and subsequent cooked flavor and odor attributes. This lack of information is probably because vacuum-packed meat, which has a much shorter storage life, has, until recently, been the preferred method of storing chilled meat, particularly beef. Also, in Northern Hemisphere markets long-term storage of meat is not normally required because of meat processors' proximity to market.

The most comprehensive study of the effects of extended storage of meat on chemical changes and flavour development is that of Toldrá et al. (1997) and Flores et al. (1998) who studied dry-cured ham. Dry-curing consists of several stages of salting, washing and finally ripening-drying-ageing for 7 to 12 months at 12 to 22°C. Generation

*AgResearch Ltd., MIRINZ Centre, Private Bag 3123, Hamilton, New Zealand

Quality of Fresh and Processed Foods, edited by Shahidi et al.
Kluwer Academic/Plenum Publishers, 2004.

of characteristic dry-cured flavour takes place through the action of proteolysis and lipolysis.

Another issue is that of meat pH. The pH of chilled meat, which is ideally around 5.6 reportedly increases on chilled storage in beef (Parrish *et al.*, 1969, Boakye and Mittal, 1993) and lamb (Moore and Gill, 1987). Since higher pH can affect flavour and potentially promote microbial growth, pH during long-term storage could be an important issue.

This report summarizes three studies done under contract for Meat New Zealand at the AgResearch Ltd.'s MIRINZ Centre, formerly MIRINZ Food Technology & Research Ltd. These studies have contributed to the understanding of changes taking place in sheepmeat stored chilled for extended periods in CO_2 atmosphere and vacuum packs. The studies were as follows:

1.1. Flavor, pH, Amino Acid and Fatty Acid Changes in Sheepmeat During Extended Chilled Storage in a CO_2 Atmosphere

Changes in the odor and flavor of lamb legs during 16 weeks chilled storage in carbon dioxide were monitored and compared with control legs stored frozen under vacuum. Changes in soluble proteins, free fatty acids and free amino acids were measured, as these degradation products could be responsible for the development of 'livery/offaly' flavor and could also explain the increase meat pH during prolonged chilled meat storage.

1.2. Effect of Rate of Rigor Onset and Different Packing Methods

This study compared the effect of different rates of rigor onset (fast *vs* slow) on the odor and flavor of chilled ($-1.5°C$) lamb legs stored for 12 weeks in either a CO_2 atmosphere or a vacuum pack. Measurements included those of glycolytic metabolites, drip loss and pH, meat pH, and total free amino acid content. The aim of these measurements was to investigate pH increases observed during chilled storage.

1.3. Effect of Rate of Rigor Attainment and Early Oxygen Exclusion on Warmed Over Flavor

The effects of fast and slow rigor attainment, early product temperature abuse and early oxygen exposure were examined with respect to subsequent generation of rancid flavor in cooked sheepmeat. Various metabolites were also measured.

2. METHODOLOGY

2.1. Carcass Treatments and Sample Preparation

Unless stated otherwise, Coopworth lambs of one flock for each experiment, were electrically stunned and conventionally slaughtered by throat cut. Carcasses were immediately dressed without electrical stimulation and held at $5 \pm 1°C$ for 6 hours, then at $3°C$ for 22 hours.

2.1.1. Flavor, pH, Amino Acid and Fatty Acid Changes in Sheepmeat During Extended Chilled Storage in a CO_2 Atmosphere

Both hind legs of each of 50 carcasses were removed and the ultimate pH of each *semimembranosus* muscle was measured using a pH probe inserted into a cut. Pairs of legs were sorted into five groups comprising 10 legs each, with equal numbers of contralateral legs in each of two subgroups.

One subgroup of lamb legs was packed with 1.5 litres of CO_2 (containing less than 500 ppb oxygen) per kg of meat in aluminium foil-lined gas-impermeable polypropylene bags and stored at $-1.5°C$ (chilled treatment). The other subgroup of legs was placed in the same type of bag, vacuum packed, and stored at $-35°C$ (frozen control). Treatments were stored for 0, 4, 8, 12 and 16 weeks at $-1.5°C$ or $-35°C$. At these times the frozen controls were thawed overnight in a $10°C$ room.

Immediately after microbial swabbing (a 5-cm² area of surface was swabbed for aerobic plate count), the volume of any drip contained within each bag was measured. The pH of the *semimembranosus* muscle was remeasured, close to the original measurement site. Each leg was then deboned, the muscle and fat diced, and then minced together by two passes through a 3 mm plate. The pH of each mince was measured by homogenising duplicate 1-g samples of mince in 10 mL of distilled water. A subsample of mince was set aside for a repeat homogenisation and measurement two hours after mincing, to allow time for any dissolved CO_2 – which could affect pH – to evolve from the meat.

2.1.2. Effect of Rate of Rigor Onset and Different Packing Methods

Ninety-six lambs were randomly allocated to two rigor treatment groups. For one rigor group (slow rigor process), the fresh carcasses were electrically stimulated (low voltage) for 60 seconds, then chilled rapidly ($0°C$ at 5 m sec^{-1} airflow for 24 hours). For the fast rigor group, the carcasses were high-voltage stimulated for 120 seconds, followed by slow chilling rate ($10°C$ for 15 hours then $4°C$ to 24 hours).

The ultimate pH of each *semimembranosus* muscle was measured 24 hours after slaughter in duplicate by inserting the pH probe 1.5 cm into the muscle. The legs were then allocated to one of three packaging treatments (chilled CO_2, chilled vacuum, frozen vacuum) and to one of four storage times (3, 6, 9, 12 weeks) in a way designed to minimize the effect of animal variation.

For chilled storage, legs were either vacuum packed in clear Cryovac vacuum bags, which have a low but measureable O_2 transmission (vacuum treatment) or evacuated then packed with 1.5 litres of CO_2 per kg of meat in aluminium foil-lined bags (CO_2 chilled treatment). Oxygen transmission through these bags is below detection. Both treatments were stored at −1.5°C. For frozen storage, legs were placed in foil-lined bags, vacuum packed, and stored at −35°C (frozen control). Groups of legs were stored for 3, 6, 9 and 12 weeks.

At the end of each storage period, the bags were warmed to about 10°C, opened, and the surfaces swabbed for aerobic plate count. The pH of the muscle was remeasured at a site within 1 cm of the first pH measurement. The *semimembranosus* and *biceps femoris* muscles were then removed from each leg. The former muscle was transversely sectioned and the pH measured at a freshly exposed cut surface. The pH of the *biceps femoris* muscle was also measured.

Two 1-g samples of *semimembranosus* muscle were then excised from the freshly exposed surface close to the interior pH site. The samples were used for homogenate pH measurement and metabolite assays.

The remainder of each leg was stored at −1.5°C overnight. Legs were deboned, the remaining muscle was minced, a 1-g sample of mince was taken for homogenate pH measurement, and the remaining mince was subjected to sensory evaluation.

2.1.3. Effect of Rate of Rigor Attainment and Early Oxygen Exclusion on Warmed Over Flavor

Twenty-seven lambs were electrically stunned and slaughtered and dressed, but not subjected to electrical stimulation.

2.1.3a   Rigor attainment and rancidity experiments. For experiments contrasting rigor attainment conditions, the hind legs were removed and alternate sides assigned to two rigor treatment groups. For one treatment (air exposure, high-temperature (fast) rigor attainment legs were placed in foil-laminate bags. However, the bags were not sealed at that time, so that much of the surface remained in contact with air. These bags were placed at 30°C for 6 hours before an additional 30 hours in a blast chiller at 4°C. A vacuum was then applied and the bags stored at −1.5°C for 0, 5 or 10 weeks.

For the oxygen-exclusion, slow rigor treatment, the legs were placed in foil bags and immediately vacuum packed. The bags were then immersed in water maintained at 10 ± 0.1°C for 10 hours before final storage at −1.5°C.

At the end of each storage period, bags were opened, and meat surfaces swabbed for aerobic plate count. The pH of the *semimembranosus* muscle after storage was remeasured 1 cm into the interior of the exposed muscle surface. An oblong piece of lean meat (about 1 cm x 1 cm x 4 cm deep into the muscle) was also dissected, and cut into cubes at 0.5, 1, 2 and 3 cm along the length starting at the surface. Each cube was homogenized in distilled water and the pH of the resulting slurry measured. The pH of the *semimembranosus* muscle was again measured directly by inserting the pH probe deep inside the muscle near where the 3 cm measurement was taken. A 1-g sample of *semimembranosus* muscle was excised from the same site for lactate, pyruvate, glucose, glycogen, and glucose-6-phosphate determinations.

The remainder of each leg was then deboned and minced twice through a 3 mm plate. Three 1-g samples of mince were taken for homogenate pH measurement (post-storage mince pH), total free amino acid and malondialdehyde analyses. The remaining mince was subjected to sensory evaluation.

2.1.3b   Early exposure to oxygen experiment. Striploins were excised from carcasses immediately after legs were removed. The loins were wrapped in polyethylene kitchen film, which is highly permeable to oxygen, then either immediately vacuum packed in foil-laminate bags or left exposed to air, the early anoxic and early oxic treatments, respectively. The oxic striploins were mounted on a grid allowing air contact at nearly all surfaces. Both treatments were held at 10°C for 24 hours, when the oxic samples were vacuum packed and all 54 bags were cooled to −1.5°C and held for 0, 5 or 10 weeks.

Two days after slaughter, the 0-week samples were tested as follows: bags were opened and swabbed for aerobic plate count. pH was measured by probe inserted 1 cm into the muscle. A 25 mm steak was cut from one end of the loin and the freshly cut surface was wrapped in polyethylene film and allowed to bloom at room temperature for 2 hours before colour measurements were made.

After the steak was cut, each loin was diced and minced twice through a 4 mm plate. Immediately after mincing the meat was vacuum packed and held for 15 minutes in a refrigerator until cooked for sensory panel assessment.

2.2. Sensory Analysis

For each of the experiments sensory analysis followed a general theme. Minced meat for evaluation was cooked to an internal temperature of 75°C in covered stainless steel beakers in a water bath at 100°C. Minces were stirred regularly to ensure even cooking. The cooking time for each mince was between 20 to 30 minutes.

In several experiments, the rendered fat and broth from each cooked sample was poured into glass beakers. About 2-g of the separated fat was removed for instrumental analysis and the remaining fat and broth was returned to the cooked mince. The mixture was reheated for about one minute before presentation to panelists. (A sample was also removed for malondialdehyde analysis.)

The rendered fat samples were centrifuged at nominally 60°C and to yield clean dry fat. The clear supernatants were transferred to glass vials and packed to ensure complete protection from oxygen. Vials were stored at −80°C until instrumental analysis.

For rancidity work, cooked minces were stored under kitchen film for 1, 4 and 7 days at 5°C. They were thus exposed to oxygen but did not dry out. Minces were then reheated to 75°C, and a sample of the rendered fat was recovered and treated as above. Remaining fatty broth was returned to the cooked mince destined for sensory evaluation and malondialdehyde analysis.

In any one sensory session, 15-g aliquots of cooked mince were transferred to 50 mL screw-capped jars placed in a bain-marie at 80°C. The samples, coded with random numbers, were served singly to panelist according to a suitable design. Warm distilled water and dry unsalted crackers were presented between samples. Evaluations were made in individual positive air-pressured sensory booths at 22°C and under subdued red lighting.

The 12 panelists were asked to score for 'sweet', 'foreign', 'sweet', 'stale', 'stored', 'sour', 'sheepmeat', 'painty', 'livery', 'meaty', 'grassy', 'cardboardy', 'bitter', 'barnyard' and 'rancid' odor, immediately after the screw cap was removed from the jar. They were then asked to score for same descriptors in terms of flavor. These attributes were deemed the most important descriptors that explained the odor and flavor changes that occurred in previous experiments.

Definition of descriptors were as follows:
Sheepmeat: intensity of sheepmeat
Meaty: intensity of meat 'fullness/full-bodied' associated with cooked meats
Barnyard: e.g. sweaty, musty, animal, manure, dung, stale hay
Grassy: green, fresh, pasture-fed
Sweet: associated with the sweetness of fresh lamb
Foreign sweetness: associated with the sweetness of added sugar
Stale: lacks flavor, old
Stored: old, refrigerated/freezer burn, odors of fridge, musty, sweaty
Cardboardy: wet cardboard/paper (a component of stale)
Rancid: fat breakdown due to oxidation, rancid oils, linseed oil
Painty: rancid oils, linseed oil, old-paint, drying paint, off, rancid component
Livery: associated with offal e.g. liver, kidney
Sour: vinegar-like, lemony
Bitter: sharp, biting, tonic water, caffeine

All attributes were scored on an unstructured line scale with end points of no odor or flavor to extreme. Panelists were also asked to record self-generated descriptors of the odors and flavors. Data was collected using Compusense *five*, version 3.8 (Compusense, Guelph, Canada).

2.3. Microbiology, pH and Chemical Analyses

2.3.1. Microbiology

Meat surface swabs for the determination of aerobic plate count were shaken in culture media and diluted onto agar plates for numeration of colonies.

2.3.2. Homogenate meat pH measurements

Samples (approximately 1-g each) of excised muscle or minced meat were homogenised in 9.0 mL of distilled water at 20°C. The pH of the slurry was measured with a meter fitted with a glass pH probe.

2.3.3. Chemical Analyses

Glucose, lactate, pyruvate, glucose-6-phosphate, and glycogen levels were measured in the supernatant of a 1-g sample of excised muscle precipitated with 9 mL of ice-cold 0.6N perchloric acid. The total free amino acid content in meat drip was measured by the method of Baer *et al.* (1996). Individual free amino acids were determined on samples (2 g) homogenised with cold high purity water to a final 4mL. Norleucine was added as internal standard. The homogenate was stood on ice for 15 minutes before double centrifugation.

An aliquot of supernatant was retained for determination of soluble protein. To a further 1.5 mL of supernatant was added 0.75 mL of 30% (w/v) trichloroacetic acid. The mixture was mixed by vortex, frozen, thawed, re-vortexed, then finally centrifuged and filtered (0.2 μm). Aliquots were stored at –80°C for free amino acid and glucose assays, respectively.

For individual free amino acid analyses, samples were analysed using a Picotag (Waters, USA) precolumn derivitization HPLC method to quantify 28 individual amino acids and the peptide carnosine.

2.3.3a Gel Electrophoresis. The volume of drip recovered from packaging bags was measured, and an aliquot stored at –80°C until needed. Thawed drip was diluted four-fold with normal saline, then were centrifuged to remove any precipitated material generated by the freeze-thaw cycle. The supernatant was analysed for total protein content, and then was diluted and heated in a urea- and thiourea- based SDS buffer before conventional polyacrylamide electrophoresis. A range of molecular weight standards was also run with each gel. After Commassie-blue staining, protein band intensities were determined with a densitometer.

2.3.3b Free fatty acids and fat content. Lipids were extracted from 4 g of mince (Folch *et al.*, 1957), and an aliquot evaporated to dryness with an internal standard. The Folch solvent and an Amberlyst A26 resin were added to bind the free fatty acids. The resin was washed free of other fatty matter, and the free fatty acids methylated in situ with boron trifluoride. The fatty acid methyl esters were extracted with iso-octane and quantified by gas chromatography (Gandemer *et al.*, 1991) with a BPX70 column (SGE, Australia). An aliquot of the total Folch extract was also used for gravimetric measurement of mince fat content.

A series of fatty acid methyl ester external standards was used to determine methyl ester concentrations from each sample.

2.3.3c Malondialdehyde assay. Malondialdehyde, an indicator of lipid peroxidation, was measured after precipitation of meat proteins with trichloroacetic acid (Bergamo *et al.*, 1998).

2.3.3d Gas chromatography/mass spectrometry of fat volatiles. For this work we examined the extremes of treatment on limited numbers of animals. Analyses were in duplicate. After storage at –80°C, fat samples were melted by placing vials in 60°C water before analysis. One gram of melted fat were placed in the bottom of a 50-mL glass-purge tube fitted with a ground-glass stopper joint. An internal standard (2.05 μg of 2-octanone in pentane) was injected deep into the fat, and the tube was stoppered, gently agitated, and left to equilibrate at room temperature for 5 minutes. A glass nitrogen-gas purge tube was positioned 5 mm above the surface of the fat, and a Tenax TA collection trap was attached to the outlet of the purge tube. (The Tenax trap had been previously purged at 260°C in a helium flow.) The purge vessel was immersed in an oil bath maintained at 100 °C. Instrument grade nitrogen was passed through molecular sieve and activated charcoal filters, and a tube filled with purged Tenax before finally passing over the fat surface. Volatile compounds from the heated fat were collected on the Tenax TA over 30 min. Controls were exercised to make sure there was neither contamination nor loss of volatiles.

The volatile compounds were thermally desorbed at 250°C with a flow of redirected gas chromatograph helium carrier gas onto the head (–10°C) of a DB5-MS capillary column housed in a Fisons 8000 gas chromatograph. The chromatography conditions, which ranged between −10 and 280°C, were designed to resolve an alkane series to octadecane with roughly equal time intervals between adjacent alkanes.

A Fisons MD 800 mass spectrometer recorded from 40 to 350 mass range in the total ion monitoring mode. Volatiles were assigned Kovats' indices based on the alkane series. Spectra were compared with a number of mass spectral libraries. An aldehyde series (C4 to C12) was also used to verify aldehyde identification.

2.4. Data Analysis

Sensory panel data, pH and chemical analyses data were analysed for variance by the Residual Maximum Likelihood (REML) routine (GENSTAT Ltd). For some data a separate Principal Components Analysis (PCA) was done for sensory panel scores, free amino acids data, and for free fatty acid data. In addition, a PCA was performed on the scaled correlation matrix of all data (Unistat version 4.5, Unistat, UK).

3.   RESULTS AND DISCUSSION

3.1. Flavor, pH, Amino Acid and Fatty Acid Changes in Sheepmeat During Extended Chilled Storage in a CO_2 Atmosphere

3.1.1.   Changes in Aerobic Plate Count During Storage
There was no significant difference in aerobic plate counts between storage treatments at each of storage time. The aerobic plate count varied between 1.5 and 2.5 ($\log_{10}$) colony forming units (CFU). cm^{-2}. These low values indicate good processing and storage practices throughout and were not high enough to have any effect on meat pH or cooked meat flavor and odor (Gill *et al.,* 1979).

3.1.2.   Meat pH Changes During Storage
The mean meat pH values of the legs assigned to the CO_2-chilled and vacuum-frozen treatments were not significantly different at the start of the storage trial.

Semimembranosus muscle pH, measured by direct probe insertion, decreased significantly ($P< 0.001$) between 0 and 4 weeks in CO_2 packs compared to frozen controls. The mean maximum decrease was 0.15 pH unit at 4 weeks of storage. This effect progressively diminished, and disappeared by 16 weeks (Fig. 1). This decrease in pH could be the result of dissolved carbon dioxide lowering the pH of the meat surface at least to the depth (~1.5 cm) the probe was inserted. After 4 weeks storage, metabolic changes that cause an increase in meat pH could have begun to reduce the effect of the dissolved carbon dioxide. This is also shown in Figure 1, where the slope of the probe curve increased in parallel to the curves for meat pH measured on homogenates immediately and 2 hours after mincing.

The pH of minced tissue (measured on homogenates) from lamb legs stored in CO_2 increased almost linearly during the storage period to a maximum of 0.3 unit at 16 weeks compared to minced meat from legs stored frozen for the same time.

The present results are consistent with earlier work. An increase in meat pH during aging was observed for chilled (2°C) vacuum-packed beef after 28 days storage (Parrish *et al.*, 1969), and for unprotected and vacuum-packed beef after 16 days at 2°C (Boakye and Mittal, 1993). In the latter experiment, there was a greater change in meat pH with the vacuum-packed samples. A study of lamb shoulders, vacuum-packed or packed in a range of modified atmospheres, showed a significant increase in meat pH over 28 days of storage where oxygen was excluded (Doherty *et al.*, 1996). Moore and Gill (1987) also reported an increase in pH for CO_2 and vacuum-packed lamb loins stored chilled for up to 16 weeks.

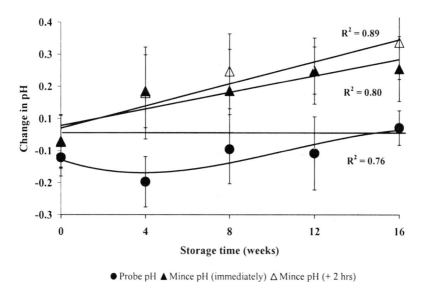

Figure 1. Difference in meat pH between chilled and frozen *semimembranosus* muscles assigned to five storage times. pH was measured using a pH probe inserted directly into the muscle (•), or on a homogenate: ▲ = pH immediately after mincing; △ = pH 2 hours after mincing.

3.1.3. Changes in Sensory Attributes During Storage

There was significant increase ($P< 0.05$) in livery odor for cooked meat from the CO_2-packed legs compared with the frozen legs. This confirms results of a preliminary study (Braggins, 1996). However, there was no significant change in other odor attributes measured. This is in contrast to results of the preliminary investigation (Braggins, 1996), which showed a significant decrease ($P< 0.05$) in overall, sheepmeat and sweet odor attributes over time.

Sheepmeat flavor decreased in both packaging treatments. However, the decrease was statistically significant ($P< 0.05$) only for the CO_2-packed meat. Panelists detected a significant increase ($P< 0.05$) in storage flavor (staleness/rancidity) with storage time, with the CO_2-packed meat showing the greater increase.

Figure 2 shows the most significant flavor changes that occurred during chilled storage. First, there was a decrease ($P< 0.001$) in sweetness, with chilled CO_2-packed meat being half as sweet as frozen meat at 16 weeks. Second, there was a marked increase ($P< 0.001$) in livery flavor intensity, with livery flavor intensity increasing almost 8-fold during 16 weeks chilled storage in the CO_2 packs. Even after 4 weeks chilled storage, livery flavor had increased almost 4-fold. Livery flavor did not change significantly during frozen storage (Fig. 2).

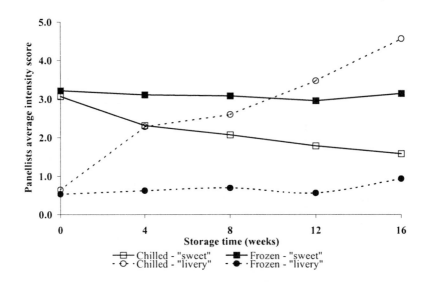

Figure 2. Panelists' average intensity scores for 'sweetness' flavor and 'livery' flavor of chilled CO_2-packed and frozen vacuum-packed lamb legs stored for various times.

3.1.4. Changes in Major Metabolites During Storage

Total protein content was similar for the two storage regimes and did not change significantly during storage. For non-protein nitrogen (NPN), the difference between treatments increased linearly with storage time, reaching a maximum difference of 0.05% at 16 weeks storage (Table 1).

There also was a significant ($P < 0.001$) increase in total soluble protein and myofibrillar protein (MP) content in CO_2-packed meat compared with frozen controls over the storage period (Table 1). There was also a slight but significant ($P < 0.05$) increase in sarcoplasmic protein (SM) for both treatments over time.

The volume of drip (mL 100 g^{-1}) increased for CO_2-packed meat compared with frozen meat over the storage period (Table 1, Fig. 3).

Table 1. Mean amounts of total protein, nonprotein nitrogen (NPN), total soluble protein (TSP), sarcoplasmic protein (SP), myofibrillar protein (MP) and drip, expressed as a percentage for chilled CO_2-packed and frozen vacuum-packed lamb legs stored for 0, 4, 8, 12 and 16 weeks.

	Storage treatment Chilled (C) Frozen (F)	Storage time in (weeks)					Storage method (mean)	Significance (P value)		
		0	4	8	12	16		Storage time	Storage method	Storage time x storage method
% Total prot.	C	18.9	18.0	18.7	18.4	18.7	18.6	0.129	0.807	0.721
	F	18.9	18.3	18.7	18.3	18.7	18.6			
% NPN	C	0.16	0.22	0.19	0.24	0.19	0.19	0.001	0.001	0.001
	F	0.15	0.20	0.16	0.20	0.14	0.17			
% TSP	C	13.8	15.6	16.0	18.3	19.2	16.4	0.001	0.001	0.001
	F	13.8	15.4	15.4	15.5	16.2	15.2			
% SP	C	5.2	5.0	5.6	5.3	6.1	5.4	0.001	0.016	0.023
	F	5.0	5.3	5.8	5.7	6.3	5.6			
% MP	C	8.5	10.7	10.4	13.0	12.7	10.9	0.001	0.001	0.001
	F	8.8	10.2	9.6	9.7	9.9	9.6			
% Drip	C	0.0	0.4	0.5	1.3	2.2	1.11	0.002	0.001	0.001
	F	0.0	0.1	0.1	0.2	0.2	0.13			

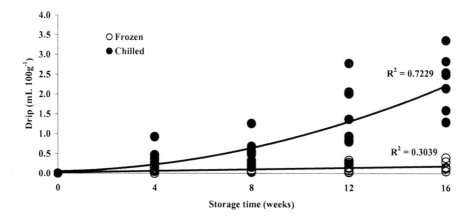

Figure 3. Volume of drip from individual legs of lamb stored chilled in a CO_2 pack or thawed from frozen in a vacuum pack after storage for 0, 4, 8, 12 and 16 weeks. The regression lines represent the means of samples at each of the storage times.

3.1.5. Changes in Free Amino Acids During Storage

Total free amino acid levels increased about two-fold in the chilled legs after 16 weeks (Table 2). Of the 29 free amino acids measured, only eight (hydroxyproline, β-alanine, citrulline, 3-methyl histidine, α-amino butyric acid, cystine, taurine and ornithine) either remained the same or decreased in concentration with storage time. The amino acid hydroxyproline is found only in connective tissue (collagen). Therefore, it appears that collagen was not broken down under chilled or frozen storage. The dipeptide carnosine did not change significantly during either chilled or frozen storage. The concentration of β-alanine, a degradation product of carnosine, also remained unchanged. Carnosine and anserine (not measured) are reported to be main contributors to maintaining the buffering capacity of meat.

The remaining 21 amino acids increased in concentration ($P< 0.001$) with storage time in the CO_2 stored legs (Table 2). Toldrá *et al.* (1995) attribute such an increase to proteolytic degradation of sarcoplasmic and myofibrillar muscle proteins by an aminopeptidase enzyme complex. The pool of amino acids produced by the aminopeptidase complex contributes to the formation of flavor compounds during cooking. The concentration or ratio of particular free amino acids or peptides formed by excessive proteolysis over prolonged periods may contribute to the development of 'livery' flavor notes and decrease the intensity of 'sweet' and 'sheepmeat' flavor.

3.1.6. Changes in Free Fatty Acids During Storage

The majority of individual free fatty acids from chilled lamb packed in CO_2 increased in concentration over the 16 weeks of storage. This increase contributed to a large overall increase in total free fatty acids at 16 weeks (Table 3). The molar increases in these acids were much lower than those of the free amino acids. Therefore, the free fatty acids would contribute far less to the final pH of the stored chilled legs than the free amino acids.

Free fatty acids, especially the C18 unsaturated acids, released from glycerides by lipolytic enzymes during aging, are the main source of carbonyl compounds that contribute to cooked meat flavor.

Table 2. Mean amounts of total and individual free amino acids, expressed as a μmol 100 g^{-1} of meat, for chilled CO_2 -packed and frozen (–35°C) vacuum-packed lamb legs stored for 0, 4, 8, and 16 weeks. Note that week 12 samples were not tested for free amino acid content.

Amino acid	Storage treatment	Storage time in (weeks)				Ratio of C:F at 16 weeks	Storage method (mean)	Significance (P)		
		0	4	8	16			Storage time	Storage method	Storage time x storage method
Total	C	3330	4190	4970	6750	1.78	4930	0.001	0.001	0.001
	F	3280	3280	3670	3800		3520			
Pser	C	8.4	12.6	14.1	18.4	2.17	13.4	0.001	0.001	0.001
	F	7.4	7.6	8.6	8.5		8.0			
Asp	C	8.4	17	36.4	72.2	7.43	36.9	0.001	0.001	0.001
	F	8.2	8.2	8.8	9.7		8.9			
Glu	C	195	365	516	417	1.92	340	0.001	0.001	0.001
	F	169	203	220	217		198			
OHPro	C	4.5	7	4.6	4.2	0.9	4.7	0.026	0.234	0.113
	F	4.4	5.9	9.9	4.7		5.4			
Ser	C	31.1	94.3	173	300	6.14	157	0.001	0.001	0.001
	F	32.7	32.6	42.5	48.8		39.9			
Aspn	C	25.7	61.2	98.9	186	5.34	99.5	0.001	0.001	0.001
	F	25.2	28.6	31.4	34.9		30.0			
Gly	C	126	195	274	375	2.45	247	0.001	0.001	0.001
	F	129	144	178	153		146			
Gln	C	295	370	451	338	1.01	340	0.001	0.181	0.842
	F	280	362	448	335		332			
β-Ala	C	10.9	8	11	12	0.85	10.9	0.129	0.13	0.465
	F	11.1	7.4	13	13.9		11.9			
Tau	C	640	544	584	660	0.89	628	0.381	0.02	0.097
	F	641	538	607	744		662			
His	C	14.7	30.3	47.6	94.8	5.01	50.8	0.001	0.001	0.001
	F	13.9	15.7	17.8	18.9		16.5			
Cit	C	73.9	55.9	52.6	68.5	1.34	66.9	0.142	0.173	0.074
	F	72.5	44.0	70.8	51		60.7			
Thr	C	22.9	60	110	207	5.91	108	0.001	0.001	0.001
	F	25.8	27.2	34.3	35.1		30.5			
Ala	C	445	519	590	785	1.68	600	0.001	0.001	0.001
	F	424	413	442	466		441			
Carn	C	514	477	415	530	0.87	503	0.485	0.022	0.054
	F	511	482	537	610		548			

Table 2. (continued)

Arg	C	34	81	124	219	4.89	120	0.001	0.001	0.001
	F	33	37	45	45		39			
Pro	C	30	51	86	158	4.84	88	0.001	0.001	0.001
	F	27	28	34	33		30			
3MH	C	537	593	489	534	1	537	0.458	0.01	0.092
	F	543	609	604	616		586			
α-ABA	C	12	12	9	16	1.09	13	0.018	0.303	0.272
	F	10	12	12	15		12			
Tyro	C	11	43	72	136	8.42	70	0.001	0.001	0.001
	F	11	10	15	16		13			
Val	C	33	94	153	271	6.31	145	0.001	0.001	0.001
	F	33	32	38	43		37			
Met	C	7	15	38	112	8.87	53	0.001	0.001	0.001
	F	6	5	11	13		9			
Cysteine	C	9	8	16	22	1.29	15	0.004	0.884	0.42
	F	11	8	16	17		14			
Cystine	C	38	25	22	41	1.81	36	0.007	0.002	0.004
	F	42	12	12	22		28			
iso-Leu	C	20	53	91	172	8.45	90	0.001	0.001	0.001
	F	18	21	20	20		20			
Leu	C	28	103	179	318	8.23	165	0.001	0.001	0.001
	F	28	30	35	39		33			
Phe	C	10	44	79	143	8.48	73	0.001	0.001	0.001
	F	10	11	17	17		13			
Tryp	C	8	14	27	45	3.93	25	0.001	0.001	0.001
	F	9	7	11	11		10			
Orn	C	12	12	19	15	1.19	14	0.143	0.001	0.034
	F	12	11	13	13		12			
Lys	C	28	88	154	291	8.32	150	0.001	0.001	0.001
	F	29	34	35	35		33			

Table 3. Mean levels of percent lipid, and total and individual free fatty acids (expressed as a $\mu mol\ 100\ g^{-1}$ of meat), for chilled CO_2-packed and frozen vacuum-packed lamb legs stored for 0, 4, 8, 12 and 16 weeks.

Fatty acid name	Carbon no.	Storage (chilled or frozen)	Storage time (weeks)					Storage method mean	Significance (P values)		
			0	4	8	12	16		Storage time	Storage method	Storage time x storage method
% Lipid		C	14.0	15.9	13.7	12.3	13.0	13.8	0.073	0.962	0.47
		F	13.3	16.1	13.3	13.3	13.1	13.8			
Total free fatty acids		C	51	86	102	164	195	115	0.001	0.001	0.001
		F	50	102	56	29	80	62.4			
Caprylic	8:0	C	0.1	0.1	0.1	0.1	0.1	0.1	0.327	0.009	0.158
		F	0.1	0.1	0.1	0.1	0.0	0.1			
Capric	10:0	C	0.2	0.4	0.5	0.7	0.7	0.5	0.001	0.001	0.001
		F	0.2	0.3	0.2	0.2	0.3	0.2			
Lauric	12:0	C	0.6	0.8	1.4	2.2	0.8	1.1	0.001	0.001	0.035
		F	0.6	0.6	1.2	1.6	0.5	0.9			
Myristic	14:0	C	1.9	3.3	3.4	5.3	5.5	3.8	0.001	0.001	0.001
		F	1.8	4.8	1.8	1.5	2.4	2.4			
Myristoleic	14:1	C	0.1	0.1	0.2	0.3	0.3	0.2	0.001	0.001	0.001
		F	0.1	0.1	0.1	0.1	0.1	0.1			
Pentadecanoic	15:1	C	0.4	0.5	0.6	0.9	1.0	0.7	0.001	0.001	0.001
		F	0.3	0.8	0.4	0.3	0.5	0.4			
Palmitic	16:0	C	11.6	18.0	19.7	31.3	35.9	22.6	0.001	0.001	0.001
		F	11.2	24.9	11.8	10.0	17.1	14.8			
Palmitoleic	16:1	C	0.8	1.6	2.1	3.3	3.7	2.2	0.001	0.001	0.001
		F	0.8	1.4	0.8	0.6	1.2	0.9			
Heptadecanoic	17:1	C	0.7	1.1	1.1	1.8	2.0	1.3	0.001	0.001	0.001
		F	0.7	1.6	0.8	0.6	1.1	1.0			
Stearic	18:0	C	11.2	14.5	14.4	21.6	24.8	16.9	0.001	0.002	0.001
		F	10.9	24.1	11.9	9.9	16.0	14.3			
Oleic	18:1	C	17.2	32.2	40.0	65.7	80.7	45.3	0.001	0.001	0.001
		F	17.1	31.6	19.0	15.6	29.1	22.1			
Elaidic	18:1t	C	0.3	0.5	0.7	1.1	1.2	0.8	0.001	0.001	0.001
		F	0.3	0.6	0.4	0.3	0.5	0.4			
Linoleic	18:2	C	2.0	4.5	6.4	10.9	13.1	7.0	0.001	0.001	0.001
		F	2.0	3.9	2.6	2.3	3.7	2.8			
Linolenic	18:3	C	1.3	3.1	4.3	7.1	8.8	4.7	0.001	0.001	0.001
		F	1.3	2.4	1.6	1.3	2.6	1.8			

Table 3. (continued)

γ- Linolenic	18:3	C	0.1	0.1	0.1	0.2	0.2	0.1	0.364	0.007	0.116
		F	0.0	0.1	0.0	0.0	0.1	0.0			
Arachidic	20:0	C	0.2	0.2	0.2	0.2	0.2	0.2	0.001	0.524	0.001
		F	0.2	0.3	0.2	0.2	0.2	0.2			
cis 11, Eicosenoic	20:1	C	0.1	0.1	0.1	0.4	0.3	0.2	0.002	0.001	0.001
		F	0.0	0.1	0.0	0.0	0.0	0.0			
cis,8,11,14, Elicosatrienoic	20:2	C	0.6	1.4	2.0	3.5	5.1	2.4	0.001	0.001	0.001
		F	0.6	1.0	0.7	0.6	1.3	0.8			
cis,11,14, Eicosadienoic	20:2	C	0.1	0.1	0.3	0.6	0.9	0.4	0.001	0.001	0.001
		F	0.1	0.1	0.2	0.1	0.3	0.2			
cis 11,14,17, Eicosatrienoic	20:3	C	0.0	0.0	0.0	0.1	0.1	0.0	0.001	0.001	0.001
		F	0.0	0.0	0.0	0.0	0.0	0.0			
Arachidonic	20:4	C	0.0	0.2	0.3	0.5	0.8	0.3	0.001	0.001	0.001
		F	0.0	0.2	0.1	0.0	0.2	0.1			
Behenic	22:0	C	0.4	0.5	0.4	0.5	0.7	0.5	0.001	0.006	0.001
		F	0.4	0.8	0.3	0.3	0.4	0.5			
16,1-9- Docosahexanoic	22:06	C	0.4	0.9	1.0	1.7	2.6	1.3	0.001	0.001	0.001
		F	0.5	0.7	0.5	0.3	1.0	0.6			
Erucic	22:1	C	0.6	1.5	2.1	3.9	5.5	2.6	0.001	0.001	0.001
		F	0.6	1.0	0.8	0.6	1.3	0.8			
cis 13,16, Docosadienoic	22:2	C	0.1	0.1	0.1	0.1	0.1	0.1	0.036	0.738	0.027
		F	0.1	0.2	0.1	0.1	0.1	0.1			
Lignoceric	24:0	C	0.2	0.2	0.1	0.1	0.2	0.2	0.001	0.06	0.001
		F	0.2	0.3	0.1	0.1	0.1	0.2			

3.1.7. *Principal Components Analysis of sensory, free fatty acids and free amino acids*

For investigations involving a large number of variables, it is often useful to simplify the analysis by considering a smaller number of linear combinations of the original variables. Principal Components Analysis (PCA) finds a set of linear combinations, called principal components, that, taken together, explain all the variance of the original data. The first principal component (x-axis) has the largest variance among all the linear combinations of data. The second principal component (y-axis) has the largest variance among all the linear combinations of data completely uncorrelated with the first principal component, and so on. There are as many principal components as there are variables. However, it is usual to consider only the first few principal components, which together explain most of the original variation.

In the case where observations are on a common scale, PCA is done on the covariate matrix of the data. Therefore, in this study, a separate Principal Components Analysis was done for sensory panel scores, as free amino acids data and for free fatty acid data. In addition, a PCA was done on the scaled correlation matrix of all data.

For the sensory data, 40% of the variance was explained by the first principal component and a further 20% of the variance was explained by the second principal component (Fig. 4). Changes in sensory attributes were observed for all the storage times for carbon dioxide atmosphere packed meat stored at $-1.5°C$. Data for frozen samples were also included in the PCA analyses. For all storage times, data points for frozen legs grouped close to the zero storage time of carbon dioxide packed meat (data not shown for clarity). This close association with the carbon dioxide-packed samples indicates that the sensory attribute intensities of frozen samples did not change during storage.

The PCA of sensory panel intensity data of carbon dioxide-packed legs showed clear discrimination between zero and 16 weeks storage (Fig. 4). Although data points for 4, 8 and 12 weeks storage overlapped somewhat, there clearly was a progression from zero weeks storage, on the left of the graph, to 16 weeks storage on the right. Loading (arbitrarily rescaled to make a reasonably overlaid plot and depicted by arrows and squares on the graph) for each of the attributes indicate that for legs stored at $-1.5°C$ in carbon dioxide, attributes such as sweetness odor and flavor, and sheepmeat flavor, were associated with very short (overnight) storage (zero weeks of storage). In contrast, for this packaging and storage treatment attributes, such as storage (rancidity and spoilage) odor and flavor, other flavor, and livery odor and flavor, were associated with prolonged storage. Livery flavor, furthest from the origin (0,0) of the graph, had the highest loading and the most influence on the principal components. Overall and sheepmeat odor and overall flavor attributes remained close to the origin and thus did not change with storage.

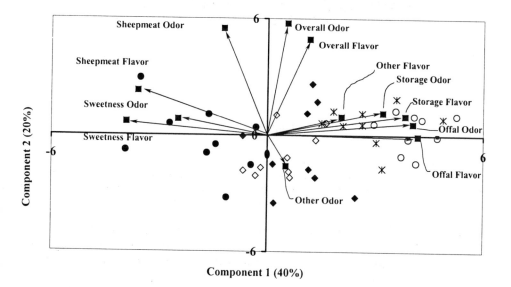

Figure 4. Principal Components Analysis of sensory panel data for all storage times and both storage treatments. For clarity, however, only samples packed in carbon dioxide atmosphere for 0 (•), 4 (◊), 8 (♦), 12 (✳) and 16 (O) weeks at $-1.5°C$ are shown. Attribute loading (arbitrarily rescaled to make a reasonably overlaid plot) for sensory attributes are represented by arrows and a box (■).

Free amino acid concentrations processed by PCA showed the same trend (Fig. 5) as that observed for the PCA analysis of sensory data. For free amino acids measurements, all zero and 16 week storage time samples were tested but only two samples from each of the 4 and 8 weeks storage samples and none of the 12 week storage samples were tested. However, the graph clearly shows good discrimination as storage time increases. Most of the variance between storage times is explained by the first principal component (70%). The amino acids leucine, *iso*-leucine, phenylalanine, tryptophan, lysine, arginine, proline, alanine, threonine, histidine, glycine, aspartate, valine, phosphoserine, methionine, and tyrosine clustered in the direction of the lamb legs packed in carbon dioxide atmosphere and stored at −1.5°C for 16 weeks. To a lesser degree, cystathione, cysteine, alpha amino butyric acid, glutamine, ornothine and glutamine loadings also tended toward the longer storage times. Taurine and hydroxyproline were more closely associated with the zero weeks storage. Beta-alanine, 3-methylhistidine and carnosine were more related to the second principal component and were not associated with any of the storage times.

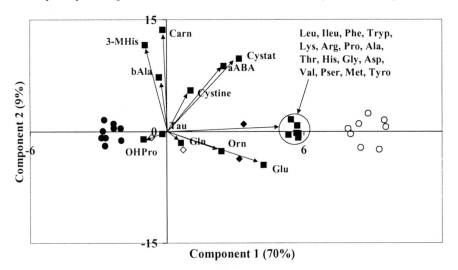

Figure 5. Principal Components Analysis of free amino acid data for all storage times and both storage treatments. For clarity, only samples packed in carbon dioxide atmosphere for 0 (•), 4 (◇), 8 (♦), 12 (✳) and 16 (O) weeks at −1.5°C are shown. Only two samples for weeks 4 and 8 were tested for free amino acid content. Attribute loading (the eigenvector) for each amino acid is represented by (■).

PCA of the free fatty acids data (Fig. 6) showed a similar progression from the left to right of the graph as storage time increased. Most of the variance between storage times was explained by the first principal component (67%). Loadings for all the free fatty acids trended toward the longer storage times for lamb legs packed in carbon dioxide atmosphere and stored at −1.5°C. Unsaturated free fatty acids and the saturated free fatty acids caprylic (F8:0), capric (F10:0) and lauric (F12:0) associated with the longer storage times in the lower right quadrant of the graph, whereas the saturated fatty acids F14:0, F15:0 F16:0 and F17:0 associated with the longer storage time in the upper right quadrant. The reason for this discrimination along the y-axis (second principal component) is unknown.

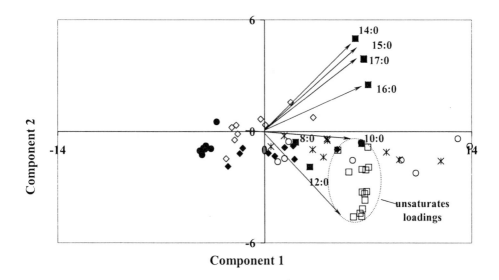

Figure 6. Principal Components Analysis of free fatty acid data for all storage times and both storage treatments. For simplicity, only samples packed in carbon dioxide atmosphere for 0 (●), 4 (◇), 8 (♦), 12 (✳) and 16 (O) weeks at minus 1.5°C are shown. Attribute loading (the eignvector) for each saturated free fatty acid is represented by (■) and for each unsaturated free fatty acid, by (□).

In summary, storage time effects of lamb legs packed in carbon dioxide atmosphere and stored at –1.5°C are explained individually by either sensory scores, free fatty acids content or free amino acids content using Principal Components Analysis.

3.1.8. Principal Components Analysis of Correlations of Sensory, Major Metabolites, Free Amino Acids, and Free Fatty Acids Data

Sensory scores, concentrations of major metabolites, free amino acid concentration and free fatty acid concentration are not expressed on a common scale. Therefore, they cannot be used collectively to derive a combined PCA analysis from their covariance matrices. In cases where observations are of difference types, principal components are calculated on a scaled correlation matrix. This was done in this study to help determine the relative influence that sensory scores, major metabolites, free fatty acids and free amino acids have in explaining the variance in the original data and to indicate which metabolites might be responsible for certain sensory attributes - that increase in intensity with storage time.

Principal Components Analysis results are shown in Figure 7. Certain sensory attributes (sheepmeat and sweetness odor and flavor) were again positioned to the left of the graph (Fig. 7a) and were associated with the amino acids taurine, hydroxyproline, β-alanine, 3-methyl histidine, citrulline, cysteine and carnosine.

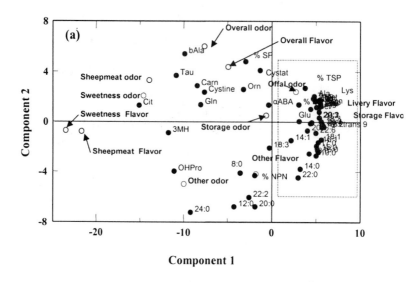

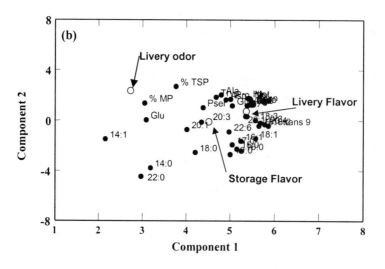

Figure 7. Principal Components Analysis of correlations. (a) complete data set, (b) first expanded view bounded by the dashed area. Metabolites are represented by (•) and sensory attributes by (○).

Closer inspection of an expanded view of the PCA analysis of the correlation matrix reveals that livery flavor associated more with the amino acids lysine, histidine, arginine, methionine, threonine, leucine, iso-leucine, phenylalanine, valine, tyrosine, asparagine and serine; and with unsaturated fatty acids F18:1; F18:1trans; F18:3; F20:2; F20:3; and F22:1.

This data analysis shows that significant increases in certain flavor attributes recorded by sensory panelists were associated with increases in the concentration of certain free amino acids and free fatty acids (particularly the unsaturated fatty acids) when lamb legs were stored in carbon dioxide controlled atmosphere packaging for up to sixteen weeks ($-1.5°C$) compared with vacuum packed frozen controls. These changes in metabolites would have been due to extensive protein and lipid hydrolysis in the stored meat.

3.2. Effect of Rate of Rigor Onset and Different Packing Methods

The previous studies showed that significant sensory and metabolic changes occur in sheepmeat during long term-chilled storage in CO_2 atmosphere. Therefore, we investigated possible ways of manipulating the these flavor and metabolic changes.

This study compared the effect of different rates of rigor onset (fast vs slow) on odor and flavor of chilled ($-1.5°C$) lamb legs stored for 12 weeks in either a CO_2 atmosphere or a vacuum pack. The chilled treatments were also compared with a frozen control. This study also measured glycolytic metabolites in the meat, the amount of drip loss, and drip pH and total free amino acid content, and measured pH at various times and in several ways on meat samples from the different treatments. The aim of these measurements was to investigate pH increases observed during chilled storage.

3.2.1. Sensory Analysis

The trained sensory panel could not detect a significant difference between fast and slow rigor treatments for any of the odor or flavor attributes assessed (data not presented).

The different packaging treatments had a significant effect on 'livery', 'spoilage', and 'sweetness' odor and flavor, and 'rancidity' flavor. The panelists detected significant changes over storage time for most odor (Table 4) and flavor (Table 5) attributes tested.

'Livery' flavor intensity increased significantly with storage time in CO_2- and vacuum-packed lamb legs compared with frozen controls. A 'liver' flavor also developed in CO_2-packed chilled pork (Jeremiah et al., 1992). Ford and Park (1980) found that panelists could detect an off flavor described as bitter - or liverish - after 14-16 weeks storage of vacuum-packed beef.

'Sweetness' and 'sheepmeat' flavors decreased in both CO_2- and vacuum-packed lamb compared to frozen controls. Gill (1988) proposed that a CO_2 atmosphere somehow strips sheepmeat flavor from the meat during chilled storage. The fact that flavor loss also occurs in conventional vacuum packs makes Gill's theory suspect.

Table 4. Mean panel scores for odor attributes of chilled-CO_2 packed, chilled-vacuum packed and frozen ($-35°C$)-vacuum packed lamb legs stored for 3, 6, 9 and 12 weeks. Statistical significance of: Storage time; Storage method; and Storage Time x Storage Method. Values for time (t) are the mean values of all three packaging treatments at each storage time.

	Storage: CO_2 (C) Vacuum (V) Frozen (F) Time (t)	Storage time (weeks)				Significance (P values)		
		3	6	9	12	Storage time	Storage method	Storage time x Storage method
Overall odor	C	5.4	5.3	5.0	5.3			
	V	5.6	5.4	5.1	5.4			
	F	5.6	5.4	5.3	5.5			
	t	5.5	5.4	5.1	5.4	0.05	NS	NS
Sheepmeat odor	C	5.0	4.7	4.6	4.6			
	V	5.1	4.7	4.5	4.4			
	F	5.0	4.7	4.7	4.6			
	t	5.0	4.7	4.6	4.6	0.01	NS	NS
Sweetness odor	C	3.0	2.8	2.7	2.7			
	V	3.0	2.9	2.4	2.4			
	F	3.3	3.1	3.2	3.0			
	t	3.0	2.9	2.7	2.7	0.05	0.001	0.01
Livery odor	C	0.1	0.2	0.4	0.7			
	V	0.2	0.2	0.4	0.9			
	F	0.2	0.1	0.2	0.2			
	t	0.2	0.2	0.3	0.6	0.05	0.01	0.001
Rancidity odor	C	0.3	0.8	1.0	1.7			
	V	0.4	0.9	1.1	1.7			
	F	0.4	0.7	1.0	1.8			
	t	0.3	0.8	1.0	1.7	0.001	NS	NS
Other odor	C	0.3	0.5	0.5	0.8			
	V	0.3	0.4	0.6	0.9			
	F	0.3	0.6	0.7	1.1			
	t	0.3	0.5	0.6	0.9	0.05	NS	NS
Spoilage odor	C	0.2	0.8	0.7	1.5			
	V	0.3	1.0	1.2	2.5			
	F	0.3	0.5	0.4	1.2			
	t	0.3	0.8	0.8	1.7	0.001	0.001	0.001

Rancid flavors also increased in both chilled packaging treatments over the 12 weeks of storage. A similar increase in rancidity was observed in the frozen controls. This result was surprising since the legs for the latter treatment were vacuum-packed in oxygen-impermeable foil-laminated bags and stored at $-35°C$. These conditions would not be expected to encourage rancidity onset over such a short storage period. Panelists detected spoilage odor and flavor increases in the week 12 frozen samples. This effect was unexpected but still significantly lower than the intensity levels detected in the chilled treatments.

Table 5. Mean panel scores for flavor attributes of chilled-CO_2 packed, chilled-vacuum packed and frozen ($-35°C$)-vacuum packed lamb legs stored for 3, 6, 9 and 12 weeks. Statistical significance of: Storage time; Storage method; and Storage Time x Storage Method. Values for time (t) are the mean values of all three packaging treatments at each storage time.

	Storage: CO₂ (C) Vacuum (V) Frozen (F) Time (t)	Storage time (weeks)				Significance (P values)		
		3	6	9	12	Storage Time	Storage method	Storage time x Storage method
Overall Flavor	C	5.9	5.7	5.4	5.6			
	V	5.9	5.8	5.4	5.7			
	F	5.8	5.5	5.5	5.7			
	t	5.8	5.7	5.4	5.7	0.05	NS	0.05
Sheepmeat flavor	C	5.5	5.0	4.9	4.8			
	V	5.5	5.1	4.7	4.4			
	F	5.3	5.1	4.9	5.0			
	t	5.4	5.0	4.8	4.7	0.001	NS	0.001
Sweetness flavor	C	3.9	3.3	3.0	2.9			
	V	3.7	3.3	2.8	2.6			
	F	4.0	3.8	3.7	3.6			
	t	3.9	3.5	3.1	3.0	0.001	0.001	0.01
Livery flavor	C	0.7	0.9	1.2	2.1			
	V	0.7	0.9	1.0	2.6			
	F	0.4	0.3	0.5	0.6			
	t	0.6	0.7	0.9	1.8	0.001	0.001	0.001
Rancidity flavor	C	0.5	1.2	1.4	2.3			
	V	0.5	1.1	1.6	2.3			
	F	0.2	0.7	1.1	2.1			
	t	0.4	1.0	1.3	2.3	0.001	0.01	NS
Other flavor	C	0.7	0.8	0.8	1.0			
	V	0.7	0.6	0.8	1.0			
	F	0.5	0.9	0.8	1.1			
	t	0.6	0.8	0.8	1.1	NS	NS	NS
Spoilage flavor	C	0.4	1.4	1.5	2.4			
	V	0.4	1.7	1.9	3.4			
	F	0.4	0.9	0.8	1.7			
	t	0.4	1.3	1.4	2.5	0.001	0.001	0.001

3.2.2. Storage- and Rigor-Related Changes in Meat pH

3.2.2a   Rigor-Related Changes. Semimembranosus pH measurement taken at 24 hours post-slaughter with a pH probe inserted directly into the muscle showed no significant difference between groups of legs assigned to different storage treatments and storage times. However, the ultimate pH of the semimembranosus, measured before storage, was significantly different ($P< 0.001$) for the two rigor treatments (5.65 and 5.74 for fast and slow treatments, respectively). It is possible that the legs subjected to the slow rigor treatment had not quite reached their ultimate pH after 24 hours because after long-term storage, pH values were unaffected by rigor treatment. Thus, given enough time, the values became the same. However, this does not explain why both the SM muscle and the whole mince pH measurements on homogenized samples (rather than with a pH probe directly into the muscle) were significantly ($P< 0.001$) lower in the fast rigor treatment group after long-term storage, a pattern similar to that seen for ultimate pH measured before storage.

3.2.3. Storage-Related Changes

When the pH of the *semimembranosus* was measured by a pH probe inserted directly into the muscle, meat pH did not change in a linear way over the 12 week storage period (Fig. 8). In contrast, when excised samples were taken from the interior of the SM muscle and homogenised in distilled water, the meat pH increased with storage time, and this increase was significantly ($P< 0.001$) greater for chilled legs in either packaging than frozen controls (Fig. 8). The difference between probe and homogenised-meat pH may be due to the fact that in whole muscle measurements the probe is measuring the extracellular pH only while in homogenate measurements the probe is measuring the combined intracellular and extracellular pH levels. The maximum increase, 0.32 pH unit, was recorded for meat in chilled CO_2 packs at 12 weeks. These results confirm previous reports of elevated pH after meat is placed in CO_2 atmosphere or vacuum packed and stored at $-1.5°C$ for extended periods (Moore and Gill, 1987; Boakye and Mittal, 1993; Braggins, 1996; Doherty *et al.*, 1996). Similar increases, but to a lesser extent, were observed in the present study in homogenates of samples of whole leg mince (Fig. 8).

At each storage time, vacuum-packed samples had a higher muscle pH by between 0.07 and 0.08 pH units than the CO_2-packed or frozen samples (Fig. 8) measured on the muscle rather than homogenate. One possible cause of an elevated meat pH in vacuum-packed meat is anaerobic microbial growth on the meat surface. Microbial growth was significantly higher on the chilled vacuum-packed legs than the CO_2-packed chilled and frozen control legs from 6 to 12 weeks (Fig. 10). However, differences in microbial growth cannot explain the difference observed at 3 weeks storage. In any event, aerobic plate counts of 10^5 $CFU.cm^{-2}$ are probably not high enough to increase meat pH, particularly at a depth of 1 to 1.5 cm.

When the SM pH was measured with the pH probe at the freshly exposed inner surface of the muscle ($SM_{interior}$ pH) immediately after the SM muscle was cut, the frozen samples had elevated pH levels at each of the storage times compared to the CO_2- and vacuum-packed legs (Fig. 8).

The pH of drip from chilled CO_2- and vacuum-packed legs was higher ($P< 0.001$) than that of drips from frozen controls after 9 and 12 weeks storage (Table 3). Data were not recorded for 3 and 6 weeks storage. The pH values were similar to those recorded for the homogenised $SM_{interior}$ meat pH.

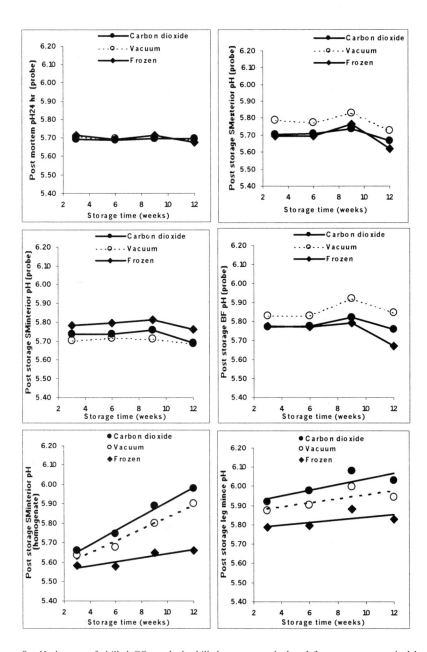

Figure 8. pH changes of chilled CO_2-packed, chilled vacuum-packed and frozen vacuum-packed lamb legs stored for 3, 6, 9, and 12 weeks.

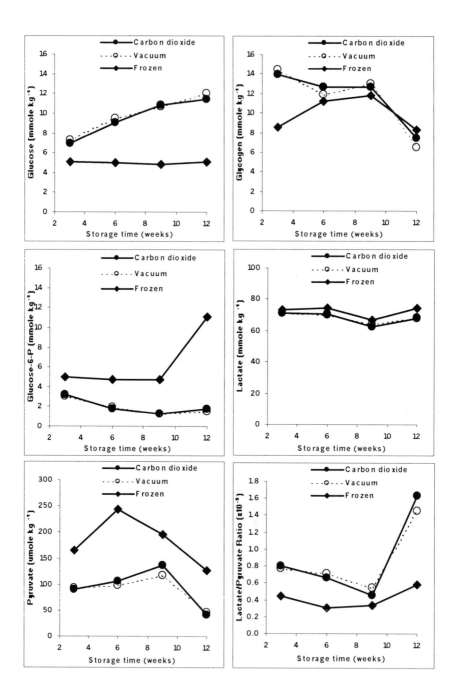

Figure 9. Metabolite changes of chilled CO_2-packed, chilled vacuum-packed and frozen vacuum-packed lamb legs stored for 3, 6, 9 and 12 weeks.

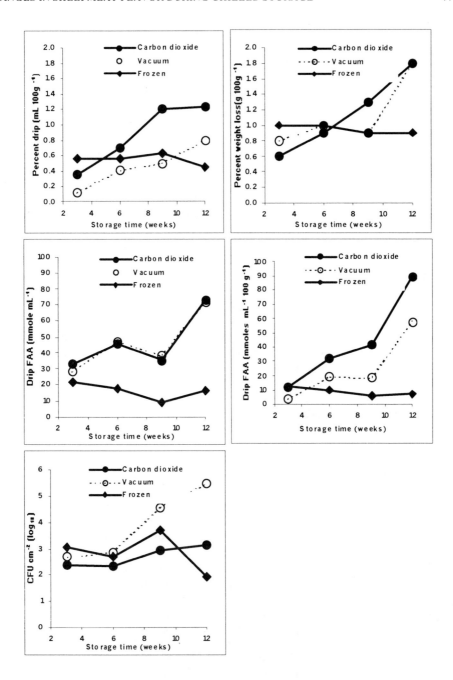

Figure 10. Percent drip (v/wt), weight loss (wt/wt), drip free amino acid content (μmol) and aerobic plate count (10^2 cfu) changes of chilled CO_2-packed, chilled vacuum-packed and frozen vacuum-packed lamb legs stored for 3, 6, 9 and 12 weeks.

3.2.4. Storage- and Rigor-Related Changes Meat Metabolites

Concentrations of the metabolites glycogen, glucose, glucose-6-phosphate, lactate and pyruvate were all influenced by storage time, storage method, and their interactions (Fig. 9). Glucose concentrations were significantly higher in CO_2- and vacuum-packed legs compared with frozen controls after 3 weeks storage, and glucose continued to increase in concentration for the remainder of the storage period (Fig. 9).

Similarly, glycogen levels (expressed as glucose units) were higher in the chilled treatments than the frozen controls at 3 weeks. Glycogen levels decreased significantly by about 6 mmole kg^{-1} between 3 and 12 weeks storage for both chilled treatments (Fig. 9).

Legs from both chilled storage treatments had similar concentrations of glucose-6-phosphate, an intermediary metabolite in glycolysis, and the concentrations of this metabolite were significantly lower than those in the frozen controls at each storage time (Fig. 9). Glucose-6-phosphate concentrations declined slightly in CO_2- and vacuum-packed meat over the remaining 9-week storage period. The glucose-6-phosphate concentrations in the frozen controls remained constant from 3 to 9 weeks of storage but doubled at 12 weeks. This rapid change cannot be explained at this stage.

This study found an overall decrease in meat glycogen concentrations, and a concomitant increase in glucose concentration, with storage time for both chilled treatments. The increase in glucose could be due to its liberation during the debranching of glycogen. Glycogen phosphorylase catalyses the removal of terminal glucose residues from the nonreducing end of the glycogen chain in the form of glucose-1-phosphate, until four glucose residues away from a glycogen chain branch point.

Further breakdown of glycogen requires action of a debranching enzyme, that in the process releases free glucose. This free glucose must undergo phosphorylation to form glucose-6-phosphate before entering the glycolytic pathway. This reaction requires ATP that would normally not be present so long after rigor. Therefore, as seen in this study, glucose accumulates and does not enter the glycolytic pathway.

Alternatively, some type of hydrolase activity might produce glucose. While a specific hydrolytic activity has not been reported, Sharpe (1958) found that glucose accumulates in meat of all the species tested (horse, rabbits, rats, pigs and chickens). Sheepmeat showed the greatest accumulation per glycogen loss. This finding also warrants further investigation, as high free glucose concentrations might play a significant role in flavor development in sheepmeat.

Lactate concentrations were slightly, but significantly higher in the frozen controls compared with both chilled treatments at each storage time. In each treatment lactate concentration changed slightly with time (Fig. 9). Rigor treatment and the interaction of rigor treatment by storage method had significant effects on the concentrations of lactate. Meat that entered rigor slowly had consistently lower concentrations of lactate than meat which entered the fast rigor quickly. This is consistent with the finding that the slow rigor treatment carcasses had a higher ultimate pH than the fast rigor treatment (i.e. higher lactate concentrations are associated with lower meat pH).

Moore (unpublished) observed decreases in lactate concentrations that were inversely related to increases in meat pH in chilled lamb loins stored in CO_2 atmosphere or vacuum packed and held for 12 weeks. Similar changes were not observed in meat from other species. In this present study lactate concentrations were significantly higher in frozen controls compared with both chilled treatments. No correlation was recorded between lactate and pH concentration for frozen controls, but there was a negative correlation between lactate and pH for both chilled treatments. This suggests a possible mechanism that reduces the concentration of lactate and - by some mechanism - the concentration of H+, causing a rise in meat pH during long term storage of sheepmeat at - 1.5°C. The increase in glucose concentration in the chilled treatments might also be involved, by some unknown mechanism, in the changes in lactate concentrations observed.

Lamb legs stored frozen (controls) had consistently higher levels of pyruvate than the chilled legs over the 12 weeks of storage. Pyruvate levels in frozen meat rose to a maximum at 6 weeks storage then decreased to below the initial 3 weeks value by the twelfth week (Fig. 9).

The lactate/pyruvate ratio, an indicator of the reductive/oxidative capacity of the meat, showed that meat stored chilled in CO_2 atmosphere or vacuum packed was in a more reductive state than the frozen controls, suggesting continued metabolic processes were occurring at the higher storage temperatures (Fig. 9).

Neither rigor treatment nor storage time influenced the buffering capacity of either the meat or the drip. However, meat and particularly drip buffering capacity was affected by storage method. This increase in buffering capacity correlated with the rise in pH of $SM_{homogenate}$ meat ($R^2 = 0.83$) and pH of the drip ($R^2 = 0.79$) during storage.

The amount of freeze/thaw drip was constant for frozen legs when measured either as the volume of drip remaining in the packaging after storage or as percent leg weight loss after thawing. Drip from CO_2- and vacuum-packed chilled legs increased significantly with storage time. Drip was significantly higher in CO_2 packed meat only when measured by volume. Drip volume and leg weight losses were greater for the fast rigor treatment regardless storage method (Fig. 10).

Rigor treatment did not affect the concentration of total free amino acids measured in the drip. However, the concentration of total free amino acids increased significantly ($P< 0.001$) with storage time in both chilled treatments, compared with the frozen controls (Fig. 10). This increase was greater for the chilled CO_2 packed legs when the total free amino acids were expressed as concentration per mL of drip produced per 100 g of meat (Fig. 10).

3.2.5. Storage- and Rigor-Related Changes in Microbial Status

Aerobic plate count ($\log_{10}$ CFU cm^{-2}) was not affected by rigor treatment. Chilled vacuum-packed lamb legs had a significantly higher aerobic palate count than CO_2 packed or frozen legs after 6 weeks of storage. The maximum plate count of just over 10^5 CFU.cm^{-2} after 12 weeks of storage, recorded for chilled vacuum-packed legs, was not considered high enough to affect the meat or drip pH (Fig. 10).

3.2.6. Storage-Related Changes in Protein Profiles

Drip protein concentration, expressed as mg.mL^{-1} of drip, declined slightly during chilled storage in both packagings compared to the frozen control (Fig. 11). When drip protein content is expressed as mg per mL of drip and normalised to meat weight (mg mL^{-1} 100 g^{-1}), the chilled CO_2-packed legs had significantly higher total protein loss in the drip than from vacuum and frozen stored legs at 9 and 12 weeks of storage. Drip protein was not measured at 6 weeks.

No obvious difference in electrophoretic profiles was observed for any of the treatments after 3 weeks storage. However, after 9 weeks storage there was a significant difference in the electrophoretic profiles between chilled and frozen meat. This difference was even more marked after 12 weeks (Fig. 12). The significant relative differences between the chilled treatments and the frozen controls were as follows:

Peaks 1 to 5 present in chilled but absent in the frozen sample

Peak 6 much lower in the frozen sample

Peak 7 and 8 absent in the chilled samples

Peak 9, 10 and 11 absent in the frozen sample

A large peak (12) in the frozen sample completely absent in the chilled treatments

Peak 13 elevated in chilled treatments (particularly in CO_2 treatment)

Peak 14 absent in the chilled treatments

A relative reduction in peaks 21 and 23 and increase in peaks 18 and 22 for the chilled treatments

Peak 24 absent in the chilled treatments

These changes are considered to be due to enzymic breakdown of sarcoplasmic and possibly myofibrillar proteins during storage at $-1.5°C$ and not because of freeze-thaw damage of the samples. This is because no difference in electrophoretic profiles was observed between the three treatments after 3 weeks storage. For example, the large peak 12 was present in samples from all storage treatments at 3 weeks but was greatly reduced by 9 and 12 weeks in samples from both chilled-storage treatments. The appearance of higher molecular weight proteins with storage time in the chilled treatments suggests breakdown of high molecular weight structural proteins. A more in-depth study of drip protein changes is required to better elucidate these differences between storage treatments.

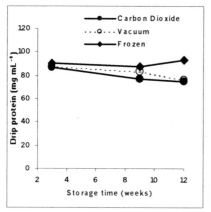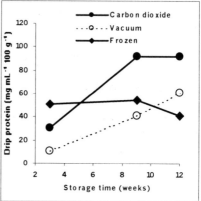

Figure 11 Drip protein content, expressed as mg mL^{-1} and mg mL^{-1} 100 g^{-1}, changes of chilled CO_2-packed, chilled vacuum-packed and frozen vacuum-packed lamb legs stored for 3, 6, 9 and 12 weeks.

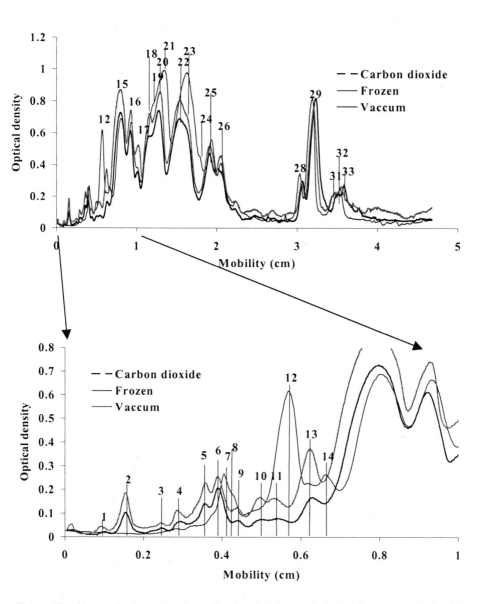

Figure 12. Changes in electrophoretic profile of chilled CO_2-packed, chilled vacuum-packed and frozen vacuum-packed lamb legs stored for 12 weeks. The upper graph shows the whole electrophoretogram and the lower graph the expanded view of the first cm of mobility, showing the changes in proteins between the chilled treatments and the frozen controls.

3.3. Effect of Rate of Rigor Attainment and Early Oxygen Exclusion on Warmed Over Flavor

Principal Components Analysis of results from the previous study showed that significant increases in certain flavor attributes recorded by trained sensory panelists were associated with increased concentrations of certain free amino acids and free fatty acids during prolonged chilled storage. We suggested that these changes in metabolites were due to protein and lipid hydrolysis in the chilled-stored meat and that these changes probably caused the undesirable flavor development observed.

In unrelated work, Ledward (1985) and Young *et al.* (1999) showed that high temperature rigor attainment could cause poor color stability in meat on display after storage. Since color problems are linked to tissue damage it seemed possible that high temperature rigor attainment might also exacerbate flavor deterioration after long term storage.

One hypothetical mechanism is that damage to cellular membranes would – by mixing substrates and enzymes not normally in contact – accelerate a range of hydrolytic reactions leading to off-flavors. This possibility was tested in the present study, where two very different attainment regimes were compared with respect to sensory properties. Hind legs of lambs were used for this work with three storage periods, 0, 5 and 10 weeks. Other measurements made included microbiological status, drip, concentrations of various glycolytic and hydrolytic metabolites, and pH. Drip is a gross indicator of cellular damage while metabolite concentrations give an insight into the causes of damage and flavor changes.

Another area of interest was the effect of long term storage on the development of rancidity in cooked meat. Thus in an extension of the rigor attainment experiment, we explored the development of rancidity in cooked meat over seven days after each of the three storage periods.

With loins from the same carcasses we also performed another experiment: the effect of early oxygen exposure on meat flavor after long term storage. The origins of this experiment stem from work by Reid *et al.* (1993) researching the flavor of sheepmeat. Total exclusion of oxygen from slaughter to presentation to panelists a few days later resulted in enhanced sheepmeat flavor and reduced 'other' flavor. Reid *et al.*, argued that 'other' flavors were simply oxidation products of fats, but not to the point of rancidity. This result raised the possibility that early exclusion of oxygen might have pronounced effects on flavor many weeks later. This possibility was particularly interesting in that new technologies coming to the attention of industry, hot-boning and immersion cooling, will exclude oxygen from meat packs very early in processing. This work was carried out with the striploins of the same lambs over the same 0, 5 and 10 week storage periods.

Thus the main aims of this study were to examine the effects of early product temperature abuse and oxygen exposure on flavor and other qualities after long term storage.

3.3.1. Storage- and Rigor-Related Changes in Meat pH

The amputation procedure for legs results in a significant fraction of the *semimembranosus* (SM) muscle being exposed to air. The possible effects of this were examined in relation to the rate of rigor attainment from fast and slow cooling.

pH measurement taken by probe in exterior muscle surface of the *semimembranosus* increased by an average of 0.145 pH units with storage time ($P< 0.001$) (Fig. 13). Equivalent measurements made in the interior of *semimembranosus* showed a similar significant trend over the 10 weeks.

When excised samples were taken at 0.5 and then 1.0 cm sections from the surface into the centre of the SM muscle and homogenised in distilled water, the meat pH at 0 weeks was about 0.12 pH units lower that the values measured by direct insertion of the pH probe into the meat. This difference has been observed before and is probably because the cell membranes are ruptured during the homogenisation procedure giving an averaged intracellular and extracellular pH measurement. These strata pH values, at each of the sampling depths, also increased with storage time ($P< 0.05$) but the change was less than that measured by direct probe insertion.

When the strata pH data were plotted against the distance from the surface of the SM muscle for each of the storage times, the pH within 0.5 cm of the surface was higher than deeper in the muscle (Fig. 14). Glycolysis is inhibited by oxygen at the meat surface causing reduced production of lactate and associated hydrogen ions resulting in a higher ultimate pH. There would probably be enough available oxygen at the meat surface even when the meat was vacuum-packed within 10 minutes of amputation. Microbial growth on meat surfaces can also raise meat surface pH. However, this would not explain the elevated pH levels at week 0 when the samples were tested only 36 hours after slaughter and the measured CFU was less than 20 per gram.

The pH of slurry samples of whole leg minces was about 0.28 units higher than slurries of the SM muscles. This is because the whole leg mince is a composite of a number of leg muscles of varying pH. Although minced leg meat pH increased with storage time, the increase was barely significant ($P< 0.1$).

For the fast rigor attainment treatment, the ultimate pH was not significantly different from the slow treatment for all pH measurement methods except when the surface 0.5 cm of muscle was tested as a meat slurry and even here the difference was subtle ($P< 0.1$).

3.3.2. Storage- and Rigor- Related Changes of Meat Metabolites

Expressed as glucose equivalents, mean post-rigor muscle glycogen concentration in *semimembranosus* was higher in the fast rigor treatment (9.2 mmole.kg^{-1}) than in the slow (6.1) ($P< 0.001$). Although there was no significant change in glycogen levels with weeks of storage, there was a significant interaction with rigor treatment ($P< 0.05$) and weeks of storage (Fig. 15); glycogen dropped in the fast rigor treatment.

Glucose-6-phosphate concentration decreased with storage time ($P< 0.001$) and was higher ($P< 0.001$) in the slow rigor treatment at each storage time. Conversely, glucose concentration increased ($P< 0.001$) about 2.5 fold with storage time. Values were higher ($P< 0.001$) in the slow rigor treatment. These results suggest that the main source of glucose accumulation in post-rigor muscle during long term-chilled storage is derived from glucose-6-phosphate. However, the hexokinase-mediated conversion of glucose to glucose-6-phosphate is essentially irreversible. One likely explanation is that hydrolysis of glucose-6-phosphate to glucose is catalysed by a (lysosomal) phosphatase enzyme during long-term storage at $-1.5°C$. A small amount of glycogen hydrolysis might also contribute to muscle glucose.

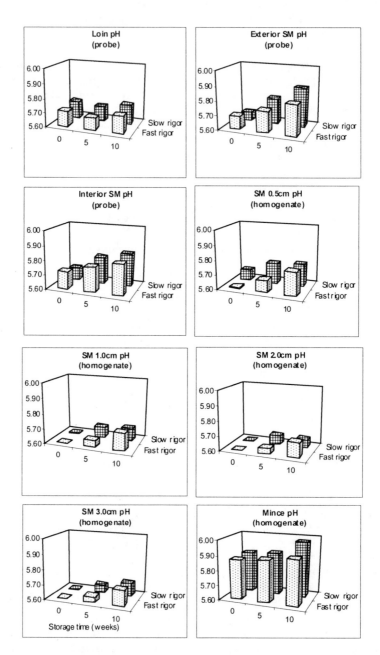

Figure 13. pH changes of vacuum packed lamb legs from fast and slow rigor attainment treatments stored chilled for 0, 5, and 10 weeks.

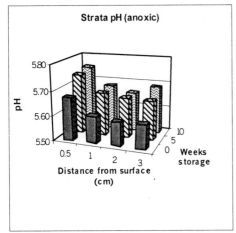
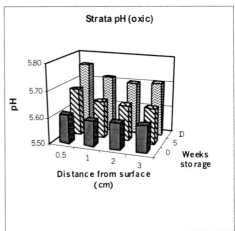

Figure 14. pH changes of *semimembranosus* sections of vacuum packed lamb legs from fast and slow rigor attainment treatments stored chilled for 0, 5, and 10 weeks.

Legs from the fast rigor attainment treatment (slow chill) had twice the glycogen concentration of the slow rigor treatment at week 0. Conversely, glucose-6-phosphate levels were much lower in the fast rigor treatment than the slow. This effect might be explained by the relative rates of glycolysis of the two treatments during and shortly after rigor onset. Low pH conditions will be encountered at different temperatures and the regulatory enzymes of glycolysis phsophorylase and phosphofructokinase are likely to be affected in different ways. Moreover, the concentration kinetics of metabolites that have a regulatory role through allosterism are likely to be different for the two rigor attainment treatments. In the case of fast rigor attainment, glucose-6-phosphate did not accumulate. The fact that glycogen concentration was higher at the same time suggests that the upstream enzyme, phosphorylase, was inhibited – allosterically, for example. By contrast, with slow rigor attainment, glucose-6-phosphate accumulates, suggesting inhibition at the level of phosphofructokinase. At the same time glycogen concentration was lower suggesting low inhibition of phosphorylase in this treatment.

There was no significant difference in ultimate pH between fast and slow rigor treatments.

A small decrease in lactate concentration (about 4 mmole.kg^{-1}, $P< 0.001$) occurred during the 10 weeks storage but there was no difference between rigor treatments. Pyruvate levels were quite variable at such low concentrations and showed no significant difference between storage times or rigor treatments.

Malondialdehyde (MDA), a measure of lipid peroxidation, was determined in minces before and after cooking at each of the storage times and after subsequent storage of the cooked minces at 4°C. Raw mince MDA levels increased with storage time ($P< 0.001$). After 10 weeks, this increase was nearly five-fold in the rapid rigor (slow-chill air-exposed) treatment, and three-fold in the other treatment. (The interaction between weeks of storage and rigor treatment was $P< 0.01$) (Fig. 16).

For analysis of holding effects, data were pooled across rigor attainment treatments. MDA levels of cooked minces from lamb legs stored chilled for up to 10 weeks also increased ($P< 0.001$) and when subsequently stored for 7 days in the refrigerator ($P<$

0.001). There was a highly significant ($P<$ 0.001) interaction between weeks of storage and days of holding after cooking. This confirms panelists' observations that extended chilled storage of lamb legs produces greater rancidity in cooked stored minces than fresh cooked lamb.

Total free amino acids measured in raw minces after chilled storage showed no significant change in concentration over 10 weeks (Fig. 16). This was surprising since results in Section 3.1.5 showed significant increases in free amino acids during chilled storage, presumably due to proteolysis.

In cooked minces (again pooled across rigor attainment treatment) total free amino acids increased ($P<$ 0.001) when measured from legs stored chilled for 10 weeks (Fig. 16). However, no further change was recorded when the cooked minces were stored at 4°C for up to 7 days. This is understandable as bacteria and proteolytic enzymes are killed/denatured by heat so would not be active after cooking.

Rendered fat from the cooked minces served to panelists was also tested for volatile chemicals. These were liberated from the reheated fat by purge-and-trap gas chromatography-mass spectrometry. Aldehydes - dominant products of lipid oxidation and indicative of rancidity - were monitored in 6 of the 9 samples at week 0 and week 10, at 0 days and 7 days after cooking. *iso*-Pentanal and hexanal increased significantly ($P<$ 0.001, $<$ 0.05, respectively) on storage to 10 weeks. The remaining six aldehydes monitored followed the same trend but the difference was not significant (Table 5).

All aldehydes, with the exception of *iso*-butanal, showed highly significant ($P<$ 0.001) increases in concentration after 7 days storage post cooking. Hexanal showed the greatest (460-fold) increase (Table 6). *iso*-Pentanal and hexanal showed an interaction between weeks of storage and days of holding after cooking. *iso*-Pentanal levels were significantly higher after 7 days holding at week 10 compared with the same period at week 0. However, the opposite was true for hexanal. In any event, the aldehyde results confirm that significant lipid oxidation occurs in cooked sheepmeat stored at refrigerator temperatures after cooking.

3.3.3. *Storage- and Holding- Related Changes to Meat Odor and Flavor*

For analysis of holding effects after cooking, data were pooled across rigor attainment treatments.

Livery odor in lamb legs markedly increased with storage time ($P<$ 0.001). There was an interaction ($P<$ 0.01) between weeks of chilled storage and days of refrigerated holding after cooking (Fig. 17) but the interaction was not simple. 'Barnyard' odor, an attribute associated with pastoral-fed animals, did not change with storage time before and after cooking, suggesting the chemical(s) responsible for this odor is not influenced by lipid oxidation or proteolysis.

'Meaty', 'sheepmeat' and 'sweet' odors, decreased in intensity after storage and after holding ($P<$ 0.001 for each) (Fig. 17). Panelists found that these odor attributes were at least half as intense after 10 weeks storage and 7 day-holding after cooking than the freshly cooked products at 0 weeks. There was also an interaction between weeks of storage and holding after cooking for meaty ($P<$ 0.001), sheepmeat ($P<0.001$) and sweet ($P<$ 0.01) odor. Generally, the loss in odor intensity was greater at 0 weeks.

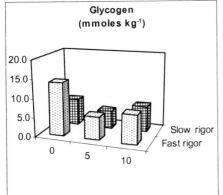

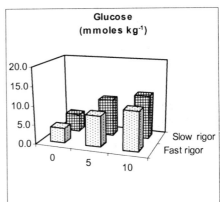

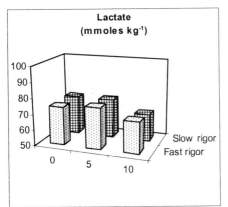

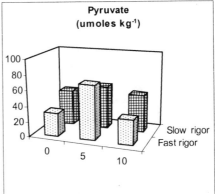

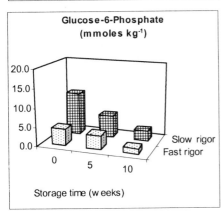

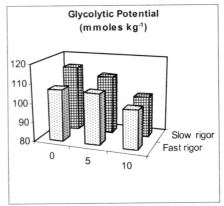

Figure 15. Metabolite changes of vacuum packed lamb legs from fast and slow rigor attainment treatments stored chilled for 0, 5 and 10 weeks.

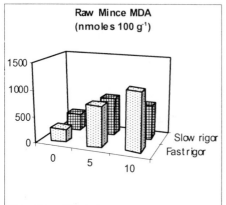

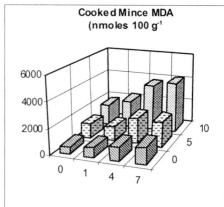

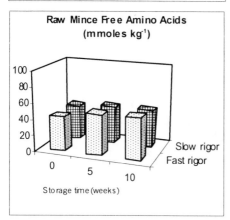

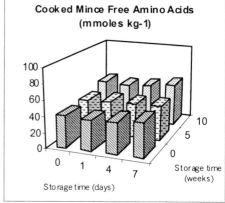

Figure 16. Metabolite changes of vacuum-packed lamb legs from fast and slow rigor attainment treatments stored chilled for 0, 5, and 10 weeks, then cooked and held refrigerated for seven days

Table 5. Effect of rigor attainment treatment, long term storage and holding after cooking on some volatile aldehydes characteristic important in oxidative rancidity. Data are raw mass spectral counts normalised to an internal standard, 2-octanone.

Storage	Vacuum and fast chilling (slow rigor) Days of holding after cooking		Exposure to air and slow chilling (fast rigor) Days of holding after cooking		Statistical effect Primary			Interactions		
	0	7	0	7	Wks of storage (W)	Days of holding (D)	Early treatment (E)	W x D	W x E	W x D x E
iso-Butanal										
0 weeks	7,850	15,600	9,010	15,000						
10 weeks	20,600	100,000	20,700	54,200						
Butanal										
0 weeks	12,600	27,100	20,300	23,500		**				
10 weeks	21,900	41,700	18,400	25,000						
iso-Pentanal										
0 weeks	3,890	12,300	4,160	13,900	***	***		**		
10 weeks	24,700	144,000	26,600	86,600						
Pentanal										
0 weeks	126,000	815,000	122,000	758,000		***				
10 weeks	223,000	941,000	267,000	605,000						
Hexanal										
0 weeks	7,000	30.1 M	9,100	3.50 M	*	***		**		
10 weeks	15,300	1.94 M	24,400	1.30 M						
Heptanal										
0 weeks	4,700	164,000	4,800	126,000		***				
10 weeks	10,000	159,000	18,100	133,000						
Octanal										
0 weeks	6,900	290,000	7,400	248,000		***				
10 weeks	11,100	198,000	18,400	187,000						
Nonanal										
0 weeks	10,500	484,000	10,600	352,000		***				
10 weeks	10,500	175,000	24,400	450,000						

$P<0.001 = ***$, $P<0.01 = **$, $P<0.05 = *$

Table 6. Effect of rigor attainment treatment, long term storage and holding after cooking on some volatile aldehydes characteristically important in oxidative rancidity. Data from Table 5 have been normalised to the area counts for the vacuum treatment at 0 weeks and 0 days of holding.

Storage	Vacuum and fast chilling (slow rigor)		Exposure to air and slow chilling (fast rigor)	
	Days of holding after cooking		Days of holding after cooking	
	0	7	0	7
iso-Butanal				
0 weeks	1	2	1	2
10 weeks	3	13	3	7
Butanal				
0 weeks	1	2	2	2
10 weeks	2	3	1	2
iso-Pentanal				
0 weeks	1	3	1	4
10 weeks	6	37	7	22
Pentanal				
0 weeks	1	6	1	6
10 weeks	2	7	2	5
Hexanal				
0 weeks	1	393	1	457
10 weeks	2	254	3	170
Heptanal				
0 weeks	1	35	1	27
10 weeks	2	34	4	28
Octanal				
0 weeks	1	42	1	36
10 weeks	2	28	3	27
Nonanal				
0 weeks	1	46	1	34
10 weeks	1	17	2	43

'Grassy' odor was one of only two attributes affected by rigor attainment treatment ($P< 0.01$). Although significant, the intensity of grassy odor in the slow rigor attainment was only slightly higher (2.0) than the other treatment (1.9). More importantly, this attribute, which is also commonly associated with pastoral diets, decreased in intensity by more than half over the 10 weeks of chilled storage ($P< 0.001$). The intensity also decreased ($P< 0.05$) after the cooked minces were stored, but most of the decrease occurred within one day of storage.

The intensities of all storage odors – 'painty', 'rancid', 'stale', 'stored' and 'sour' – except 'cardboardy', increased in lamb legs stored chilled for 10 weeks ($P< 0.001$) (Fig. 18). These storage odors also increased in intensity when the cooked minces were stored in the refrigerator ($P< 0.001$). Painty odor was the other attribute affected by rigor treatment, where the intensity was higher in the fast rigor attainment treatment (2.8) than in the slow attainment equivalent (2.5) ($P< 0.05$).

The intensity of rancid, cardboardy, stale, and stored odors were higher after seven days storage of cooked minces the longer the raw lamb legs were stored before cooking ($P< 0.05, 0.01, 0.01$ and 0.1, respectively).

'Livery' flavor of cooked minces from lamb legs increased in intensity ($P< 0.001$) progressively during the 10 weeks storage at $-1.5°C$. After each of the storage times, further storage after cooking did not change the intensity of livery flavor (Fig. 19). This indicates that the chemicals and reaction pathways responsible for the generation of the livery note are not related to lipid oxidation. The most probable cause is proteolysis of the lean during storage at $-1.5°C$. The proteolytic enzymes would be denatured during the cooking process.

Meaty, grassy and sweet flavor intensities decreased ($P<$ 0.001, 0.01, and 0.05, respectively) with storage at $-1.5°C$, reflecting the changes in odor attributes. Sheepmeat flavor intensity followed the same trend but was not significant. Meaty, grassy and sweet flavor intensities decreased further with subsequent storage after cooking ($P<$ 0.001, 0.001 and 0.05, respectively). On average, grassy and sweet flavor intensity were slightly higher (2.2 and 2.3, respectively) in the slow rigor attainment treatment than the fast (2.1 and 2.1 respectively) ($P<$ 0.01). This small difference is not commercially important.

There was an interaction between weeks of storage at $-1.5°C$ and subsequent holding after cooking ($P<$ 0.01) for sheepmeat and sweet flavor intensity. The decrease in flavor on refrigerated holding was more marked at week 0 than at weeks 5 and 10. For these attributes, holding after cooking was the bigger effect in deterioration.

Chilled storage or holding after cooking did not influence barnyard flavor.

All storage flavor intensities, except sour and bitter, increased (P at least $<$ 0.01 and mostly 0.001) when lamb legs were stored chilled for up to 10 weeks. Subsequent holding of cooked minces under refrigeration for 7 days further increased ($P<$ 0.001) storage flavor intensities (Fig. 20) for all attributes except bitter flavor (data not shown).

The results are expressed as three-dimensional graphs. 'Liver' odor increased with storage time ($P<$ 0.01). 'Other' odor showed a parallel but non-significant increase. Sheepmeat odor declined ($P<$ 0.05) (Fig. 21). Parallel effects were obtained for the equivalent flavors, for which the increase in liver flavor was highly significant ($P<$ 0.001) (Fig. 21).

Early exposure to oxygen had no effect on odor and flavor at any time. Reid *et al.* (1993) showed that total exclusion of oxygen to the point of presentation to panelists markedly affected flavor. We conclude that exposure to oxygen during early processing is unimportant for flavor development, while exposure to oxygen during cooking is very important.

Flavor effects on storage were entirely consistent with the results of the rigor attainment and rancidity experiment performed on leg muscles. Storage-induced flavor changes will almost certainly occur in all muscles.

The changes were not attributable to microbiological deterioration as microbial numbers were well below spoilage levels (data not shown).

The average pH across all loin muscles was 5.72 and was not affected by early exposure to oxygen or time of storage (data not shown).

Bloomed colour was unaffected by early exposure to oxygen. Storage had no effect on L* values, whereas a* and b* values significantly increased with storage time ($P =$ 0.06 for a*, $P<$ 0.001 for b*) (Fig. 22). These values translated to an increase in hue – particularly between 0 and 5 weeks – and a steady increase in chroma with weeks of storage. Both these effects are consistent with the known behaviour of lamb on display after storage.

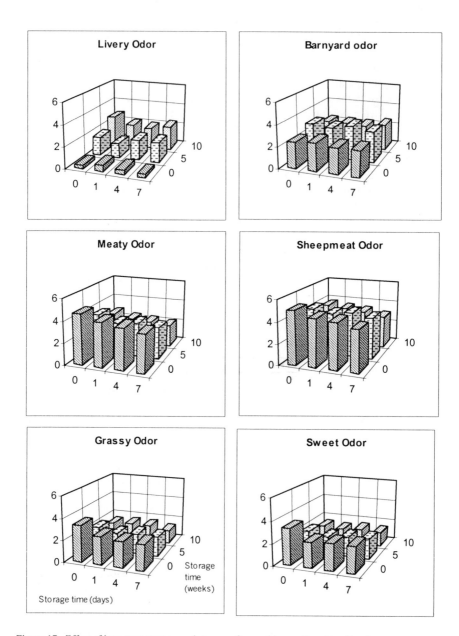

Figure 17. Effect of long term storage, and storage after cooking on the odor of lamb.

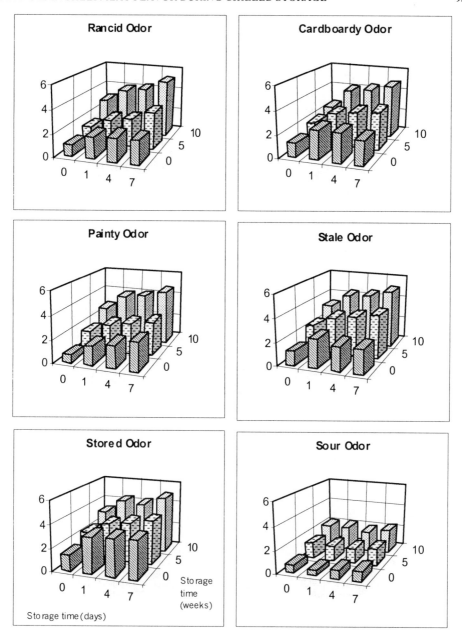

Figure 18. Effect of long term storage, and storage after cooking on the odor of lamb.

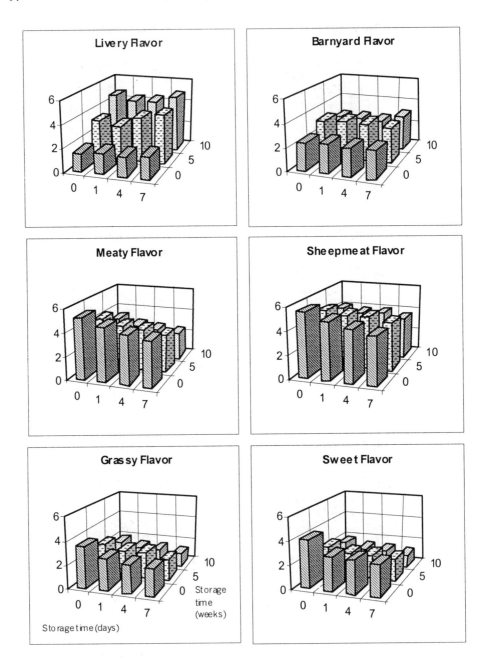

Figure 19. Effect of long term storage, and holding after cooking on the flavor of lamb.

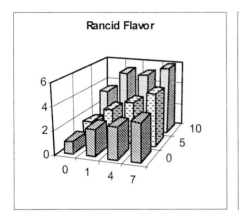

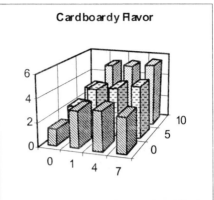

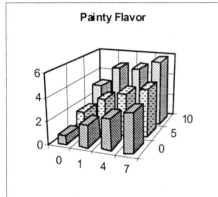

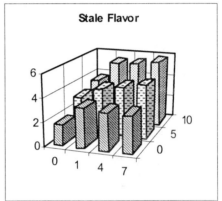

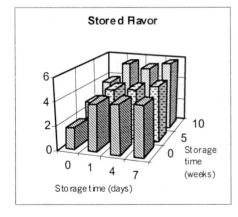

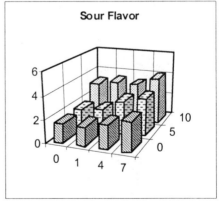

Figure 20. Effect of long term storage, and holding after cooking on the flavor of lamb.

3.3.4. Early Exposure to Oxygen Experiment

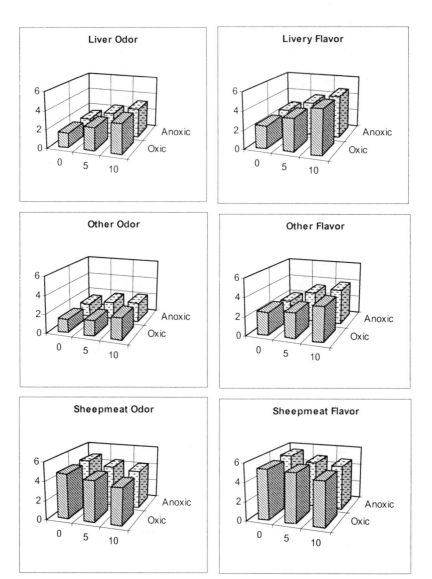

Figure 21. Effect of early exposure to air and long term storage on the odor and flavor of cooked lamb

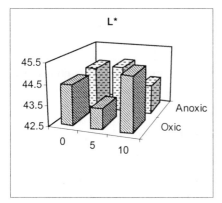

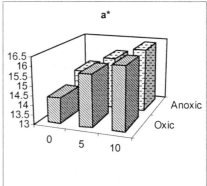

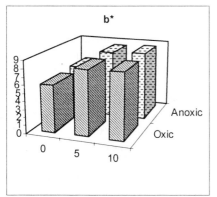

Figure 22. Effect of early exposure to air and long term storage on the colour of raw lamb.

4. CONCLUSIONS

Meat pH increased by over 0.3 units in lamb legs stored for up to 16 weeks chilled in aCO$_2$ or a vacuum pack when tested on meat sample homogenates but not when tested using a pH probe inserted directly into the muscle. The difference between probe and homogenised-meat pH may be due to the fact that in whole muscle measurements the probe is measuring the extracellular pH only while in homogenate measurements the probe is measuring the combined intracellular and extracellular pH levels. The results of measuring meat pH at various depths from the surface also confirm that the increase is not caused by microbial growth. An increase in pH was not observed in the longissimus dorsi muscle. The increase in meat and concomitant drip pH could be caused by increases in basic free amino acids resulting from enzymic breakdown of sarcoplasmic and some myofibrillar proteins. Alternatively, this increase in pH might be related to changes in lactate concentration observed in sheepmeat stored at −1.5°C. The fast and slow rigor treatments had little effect on the rate of increase of meat pH during extended storage. In any event, the increase in meat pH during long-term storage, also observed in beef

(Parrish *et al.*, 1969, Boakye and Mittal, 1993), did not promote microbial growth and therefore would not pose a risk during consumption.

Intriguing increases in glucose concentration in both CO_2- and vacuum- chilled treatments might be caused by hydrolytic degradation and/or be related to changes in lactate concentration observed in this study. Glycogen, glucose and glucose-6-phosphate concentrations were significantly influenced by the rate of rigor attainment. The results suggest that the rate of rigor attainment differentially affects the regulatory enzymes of glycolysis such that glucose-6-phosphate accumulates more with slow rigor attainment whereas more residual glycogen remains with fast rigor attainment. Glucose accumulation during chilled storage probably arises from phosphatase-mediated hydrolysis of glucose-6-phosphate.

Significant increases in 'livery', and decreases in 'sweet' and 'sheepmeat' odor and flavor intensities occurred in lamb legs stored chilled ($-1.5°C$) in a carbon dioxide atmosphere or a vacuum pack for up to 16 weeks. No significant difference in flavor intensities was observed for any of the storage treatments between carcasses subjected to fast or slow rates of rigor attainment. The 'livery' flavor that increases in intensity with extended chilled storage was not influenced by rigor attainment rates nor by early exclusion of oxygen. Holding at $4°C$ for up to 7 days did not influence livery flavor after cooking. These results suggest that livery flavor is derived from protein breakdown and not from lipid oxidation.

The Principal Components Analysis showed that significant increases in certain flavor attributes recorded by trained sensory panelists were associated with increases in the concentration of certain free amino acids and free fatty acids (particularly the unsaturated fatty acids) during prolonged chilled storage. These changes did not occur in vacuum packed frozen controls ($-35°C$). Changes in metabolites would have been due to extensive protein breakdown and hydrolysis of lipid components in the chill-stored meat. These chemical reactions probably cause the flavor developments noted in these studies.

Sensory panelists and data from instrumental analysis showed that the longer meat is stored chilled prior to cooking, the greater the rancidity development in the resultant cooked meat. Rancidity development is responsible for warmed-over-flavors (WOF) detected by sensory panelists. Rigor attainment treatments did not influence the subsequent generation of WOF in cooked meat stored at refrigerator temperatures for 7 days. Similar changes have been recorded for meat from other species (Allen and Foegeding, 1981).

Although these flavor and odor changes were found to be significant by a trained sensory panel in these studies, market information on consumers of chilled CO_2 –packed sheepmeat has not revealed any negative impact on cooked sheepmeat acceptability. Indeed, reduced sheepmeat flavor is more acceptable than strong mutton-flavored meat in many markets.

5. REFERENCES

Baer, A., Ruba, J., Meyer, J., and Bütikofer, U., 1996, Microplate assay of free amino acids in swiss cheese, *Lebensm.-Wiss. U.-Technol.*, **29**:58-61.

Bergmeyer, H. U., 1983, Methods of Enzymatic Analysis: Third Edition (Ed. H.U. Bergmeyer), Verlay Chemie, Florida, Vol. II, 86.

Bergamo, G. A., Fedele, E., Balestrieri., Abrescia, P., and Ferrara, L., 1998, Measurement of malondialdehyde levels, in: Food By High-Performance Liquid Chromatography With Fluorometric Detection, *J. Agric. . Food Chem.*, **46**:2171-2178.

Boakye, K., Mittal, G. S., 1993, Changes in pH and water holding properties of longissimus dorsi muscle during beef ageing, *Meat Sci.* **34**:335-349.

Braggins, T. J., 1996, The effect of extended chilled storage on the odor and flavor of sheepmeat. Meat New Zealand report. 93MZ 8/2.3 (POF) Milestone 8.

Doherty, A. M., Sheridan, J. J., Allen, P., McDowell, D.A., and Blair, I. S., 1996, Physical characteristics of lamb primals packaged under vacuum or modified atmospheres, *Meat Sci.*, **42**:315-324.

Folch, J., Less, M. and Stanley, G. H. S., 1957, A simple method for the isolation and purification of total lipids from animal tissue, *J. Biol. Chem.*, **226**:497-508.

Ford, A. L. and Park, R. J., 1980, Odor and Flavors in Meat, in: *Developments in Meat Science-1*, R. Lawrie, ed., Applied Scientific Publishers, London, pp. 219-248.

Gandemer, G., Morgan-Magi, B., Meyner, A. and Lepercq, M., 1991, Quantitative and qualitative analysis of free fatty acids in meat and meat products using ion exchange resin, 37th Int. Cong. Meat Sci. Technol., Kulmbach, pp. 1139-1142.

Gill, C. O., Newton, K. G. and Nottingham, P. M., 1979, Microbiology in the meat industry, Meat Ind. Res. Inst. N.Z., Hamilton, New Zealand.

Gill, C. O., 1988, CO_2 packaging - the technical background. Proc. 25th Meat Ind. Res. Conf., Meat Ind. Res. Inst. N.Z., Hamilton, N.Z., 181-185.

Jeremiah, L. E., Gill, C. O., and Penny, N., 1992, The effects on pork storage life of oxygen contamination in normally anoxic packagings, *J. Muscle Foods,* **3**:263-272.

Ledward, D. A., 1985, Post-slaughter influences on the formation of metmyoglobin in beef muscles, *Meat Sci.* **15**:149-158.

Moore, V. J., Gill, C. O., 1987, The pH and display life of chilled lamb after prolonged storage under vacuum or CO_2, N.Z., *J. Agric. Res.,* **30**:449.

Parrish, F. C., Goll, D. E., Newcomb, W. J., de Lumen, B. O., Chaudhry, H. M. and Kline, E.A., 1969, Molecular properties of post-mortem muscle. 7. Changes in nonprotein nitrogen and free amino cids of bovine muscle, *J. Food Sci.*, **34**:196-202.

Reid, D. H., Young, O. A., and Braggins, T. J., 1993, The effects of antioxidative treatments on mutton flavor/odor intensity and species flavor differentiation, *Meat Sci.*, **35**:171-178.

Sharpe, J. G., 1958, Ann. Rept. Fd. Invset. Bd., Lond., No. 7, 6.

Toldrá, F., Flores, M. and Aristoy, M-C., 1995, Enzyme generation of free amino acids and its nutritional significance in processed pork meats, in: *Food Flavors: Generation, Analysis and Process Influences,* pp 1303-1322.

Young, O. A., Priolo, A. Simmons, N. J. and West, J., 1999, Effects of rigor attainment temperature on meat blooming and colour on display, *Meat Sci.* **52**:47-56.

EFFECT OF ELECTRON BEAM IRRADIATION ON MICROBIAL GROWTH, LIPID OXIDATION AND COLOR OF GROUND BEEF PATTIES UPON REFRIGERATED STORAGE

Patty W. B.Poon[1], Paula Dubeski[2] and David D. Kitts[1]*

1. INTRODUCTION

Ground beef is a perishable product with noted food safety concerns because it provides a favorable medium for the growth of both spoilage and food-borne microorganisms. It is frequently contaminated by microorganisms due to excessive handling, (ie. slaughtering, processing and transporting). Recently, the U.S. Department of Agriculture's Food Safety and Inspection Service announced voluntarily recall of approximately 46,000 pounds of ground beef for possible *E.coli* O157:H7 contamination. Detection of pathogenic bacteria, such as *E.coli* O157:H7 in ground beef resulted in a decreased beef consumption with tremendous economic losses. Therefore, the demand for wholesome and safe, as well as nutritious, ground beef is continuously increasing.

An alternative method to the standard practices to ensure food safety of processed beef products is the use of irradiation (Lefebvre *et al.,* 1994). Irradiation has been shown to initiate a chain of events leading to the impairment of structural or metabolic functions, such as fragmentation of DNA, and the eventual death of microbial cells (Diehl, 1990). Processing of fresh or frozen raw ground beef with low γ-irradiation (^{60}Co) was approved recently by the U.S. Department of Agriculture (USDA) on February 22, 2000, to extend product shelf-life and to protect the consumer against illnesses from food-borne

[1]* Food, Nutrition and Health, Faculty of Agricultural Sciences, University of British Columbia, 6650 NW Marine Drive, Canada, V6T 1Z4,
[2] LaCombe Research Station, Agricultural Canada, LaCombe, Alberta.

Quality of Fresh and Processed Foods, edited by Shahidi et al.
Kluwer Academic/Plenum Publishers, 2004.

pathogens. Electron beam irradiation, an alternative to γ-irradiation using ^{60}Co, is much less costly than the traditional ^{60}Co irradiation. Researchers have reported the effectiveness of EBI at a dose of 3 kGy to reduce bacteria as much as 2-3 log cycles (Chung *et al.,* 2000), without affecting the flavour of foods. This method may also have greater consumer acceptance since it does not require a radioactive source. Thus, EBI technology has the potential to add value to traditional meat products by reducing the pathogenic microbial load and enhancing consumer confidence in beef safety.

In addition to food safety, other attributes of food quality including color and flavor must also be considered in irradiated ground beef. Numerous studies have indicated that radiolytic products resembling products of lipid oxidation reactions are formed in irradiated beef, and are detrimental to flavour and texture (Chen *et al.,* 1999; Lee *et al.,* 1999; Kitts, 1997). The potential for generation of unsafe radiolytic products in beef during irradiation processing is relatively high, due to the high iron-pigment content (which enhances the oxidation reactions) and a relatively low catalase level (an enzyme required to decompose peroxides), compared to other muscle foods such as chicken or pork. Consequently, γ-irradiation accelerates lipid and cholesterol oxidation (Lefebvre *et al.,* 1994; Lea *et al.,* 1960), and peroxide values increase in meats as a function of the irradiation dose (Lefebvre *et al.,* 1994). The purpose of the present study was to determine the efficacy of electron beam irradiation processing on lethality of food-borne bacterial organisms in ground beef patties, lipid oxidation and color of ground beef patties during 21 days storage at 4°C.

2. MATERIALS & METHODS

2.1. Preparation of ground beef patties

Thirty-six steers (Hereford x Angus, Charolais x MainAnjou), were slaughtered at the Lacombe Research Station Meat Lab. Meats were sliced into 2.5 cm steaks, identified by treatment, and then grounded into hamburger using a ratio of 72-73% lean trimmed neck muscle to 27-28% subcutaneous fat. A final total crude lipid content of 30 % was achieved. The subcutaneous fat used for the hamburger was collected from fat trimmed off the neck, ribs, butt, or rounds, and was identified by treatment. Beef patties were made at the Food Processing Development Centre (Alberta Agriculture, Food and Rural Development, Leduc, Alberta, Canada), and vacuum-packaged. Frozen hamburger patties were transported from the Leduc processing plant to the University of British Columbia laboratory (Vancouver, B.C., Canada) in a freezer truck.

2.2. Irradiation Treatment

Frozen ground beef patties were individually packed into Ziploc bags (S.C. Johnson and Son Ltd., Canada) and irradiated frozen using an electron beam linear accelerator (Iotron Technologies Inc., Port Coquitlam, Canada). Irradiation was performed at 10 MeV energy level and 60 kW of power on single layer of beef patties at ambient temperature (22°C). Beef patties were irradiated to 5 irradiation doses of 0, 2, 5, 10 and 20 kGy, respectively. To minimize the effect of transport or handling conditions, non-irradiated patties were subjected to the same transport and handling conditions as the irradiated patties.   Following irradiation, patties were stored at 4 °C for microbial analysis.

2.3. Microbial Analysis

TAPC and Total Coliform Counts were determined by aseptically weighing 10 g of individual pattie, and placing each sample inside a Stomacher bag containing 90 ml of sterile 0.1% peptone water (Difco).   Each sample was homogenized in a Stomacher lab blender for 1 min, and serially diluted in 0.1% peptone.   Diluted samples (1ml) were pour-plated onto 3M Aerobic Plate Count and 3M Total Coliform Counts Petrifilms (3M Canada Inc., Ontario, Canada).   The plates were incubated at $37^{\circ}C$ for 48 h to determine the TAPC and Total Coliform Counts, and at $20^{\circ}C$ for 7-10 days for numerating the psychrotrophic bacteria.   For all microbial analyses, three beef patties from each treatment were used and each pattie was plated in triplicate. Values were expressed as colony forming units (CFU)/g of beef pattie.   Samples were considered to be too numerous to count (TNTC) when the total count was <250 CFU/petrifilm at a dilution factor of 10^{-7}.

2.4. pH Measurement

pH of beef patties was measured using a glass pH electrode (Accumet pH meter 25, Fisher Scientific, Nepean, ON) after homogenizing 2g of meat in 8ml of double deionized water.

2.5. Water Activity (a_w)

Water activity was determined using a water activity meter (AquaLab Series CX2, Decagon, WA).

2.6. Thiobarbituric acid reactive substances (TBARS)

TBARS value, an index of lipid oxidation, was determined using the procedure outlined by Wijewickreme and Kitts (1997).   A standard curve was prepared using 1,1,3,3-tetraethoxypropane (TEP) [malonaldehyde bis (diethyl acetal)].   A 2.5 g beef sample was homogenized in 25 ml extracting solution made up of 20% TCA in 1.6% o-phosphoric acid. 2ml of the homogenate was mixed with 1ml of TBA reagent (0.5% TBA in 0.025M NaOH with 0.02% BHT/BHA) and heated in boiling water bath for 15 min. The solution was cooled overnight at room temp.   After centrifugation for 10 min at 2400 rpm, absorbance readings were taken at 532 nm using a spectrophotometer (UV 1600 PC, Shimadsu, Tokyo, Japan) and compared to the standard curve to determine mg of malonaldehyde per kg of meat.

2.7. Color Measurement

Instrumental color evaluation were determined using a Hunterlab "Labscan 6000' 0°/45°" Spectro Colorimeter (Hunter Associated Laboratories, Inc., Reston, VA) that had been calibrated against white and black reference tiles. Meat color was read directly on beef sample in a petri plate (8.5 cm diameter and 1.4 cm depth). Hunter L (lightness), a

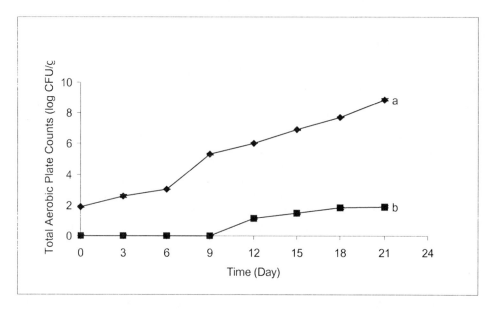

Figure 1. Effect of irradiation dose on Total Aerobic Plate Counts (log CFU/g) of beef patties stored at 4°C for 21 days. Complete inhibition of bacterial growth occurred at 5kGy or higher dosage of EBI. Values represent means ± SEM (<1%). [a-b]Means on the same day with different letter are significantly different (P ≤ 0.05) ♦ 0 kGy ■ 2 kGy

(redness) and b (yellowness) values were obtained at three random locations on each ground beef sample surface.

2.8. Statistical Analysis

All data were analyzed by ANOVA using the General Linear Model Procedure of the Statistical Analysis System (SAS Institute, Inc., 1998), with a significance level of α set at 0.05. A multiple range test (Tukey test, SAS, 1998) was used to identify significant treatment means (P ≤ 0.05).

3.   RESULTS AND DISCUSSION

3.1. Total Aerobic Plate Counts

Effect of irradiation processing on Total Aerobic Plate Counts (TAPC) of beef patties stored for 21 days is presented in Figure 1. The initial TAPC of non-irradiated patties was averaged to be 1.88 ± 0.66 log CFU/g. In every 3-day storage interval, there was an approximately 1-log increase in aerobic microbial growth in non-irradiated patties. Samples without irradiation became unacceptable for consumption after 21 days of storage at 4°C (TAPC reached 8.87 log CFU/g at 10^{-7} dilution), according to the Canadian microbial level limit of 10^{-7} CFU/g of raw ground beef.   With 2.0 kGy irradiation treatment, however, the TAPC was uncountable at 10^{-1} dilution until after 9 days of 4°C storage.   During storage at 4°C, TAPC slowly increased. This result is supported by the findings of Lee *et al.* (1999) which showed that storage at 5°C, TAPC increased in patties irradiated at doses of 0 and 1.5 kGy. However, Chung *et al.* (2000) reported that the TAPC on patties irradiated at both 1.5 and 3.0 kGy recovered after 4 days at 5°C storage, and the counts reached 4.5 and 3.8 log CFU/g by 8 days of storage. The fact that we did not observe any growth until after 12 days of storage was because beef patties used in this study had a very low initial bacterial count (e.g. 1.88 CFU/g vs. 4.3 CFU/g in Chung *et al.* (2000)). Freshly prepared commercial ground beef has been reported to contain aerobic mesophiles between 4.51-6.48 log CFU/g (Jay, 1996). Using a low dose of irradiation (2 kGy) used in our experiment resulted in a significant reduction in microbial growth by approximately 4 log CFU/g. Aerobic bacterial growth in beef patties was completely inhibited at 5 kGy, or higher upon storage for 21 days at 4°C.

3.2. Psychrotrophic Counts

Psychrotrophs   include   *Pseudomonas,   Achromobacter,   Flavobacterium   and Alcaligenes* species.   These organisms are cold-tolerant micro-organisms which are capable at growing in foods at refrigeration temperatures.   When present in large numbers, these organisms contribute to spoilage during extended storage and may also cause a variety of off-flavors and physical defects in refrigerated foods.   The effect of irradiation processing on Psychrotrophic Counts of beef patties stored for 21 days at 4°C is presented in Figure 2. For every 3-day storage interval, there was approximately a 3 log CFU/g increase in the Psychrotrophic bacterial growth in the non-irradiated samples. Samples without irradiation became too numerous to count (TNTC) at a 10^{-7} dilution after 15 days of storage at 4°C. Applying 2.0 kGy of irradiation significantly reduced microbial growth in beef patties by approximately 4 log CFU/g after 6 days of storage. However, 2.0 kGy was not sufficient to totally eliminate Psychrotrophic bacterial growth immediately following irradiation, and the count was only reduced by 0.5 log CFU/g initially.   This result is supported by the findings of Chung *et al.* (2000), where Psychrotrophs had a better ability to recover from a low irradiation stress compared to total aerobic organisms. These workers demonstrated that Psychrotrophs in beef samples irradiated by electron-beam at 1.5 and 3.0 kGy recovered to 4.0 and 3.5 log CFU/g, respectively, after 2 day storage at 5°C, as compared to 8 days of recovery period in the case of total aerobic organisms. This trend was also observed in our study, but rather than recovering on Day 2, our results demonstrated that Psychrotrophic bacteria growth was partially suppressed immediately following irradiation, and exhibited a slower (P≤0.05) growth compared to the non-irradiated samples during 4°C storage. Psychrotrophic bacterial growth in beef patties was completely inhibited at 5 kGy, or higher, upon storage 4°C storage for 21 days. When using bacterial counts of 7 log CFU/g as an index of spoilage, irradiation treatments may extend shelf life of the beef patties at least 14 days, based on the initial bacterial counts of the control samples measured at Day 0.

In this study, psychrotrophic bacteria appeared to recover faster from irradiation stress, and were able to grow more rapidly upon storage than aerobic bacteria.   The results showed that psychrotrophs were more resistant to electron-beam irradiation, or had a more efficient repair system as demonstrated by the fast recovery after irradiation

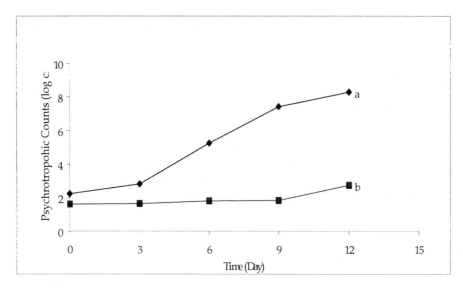

Figure 2. Effect of irradiation dose on Psychrotrophic Counts (log CFU/g) of beef patties stored at 4°C for 21 days. Psychrotrophic Counts of non-irradiated patties became too numerous to count (TNTC) after 12 days of storage. Complete inhibition of bacterial growth occurred at 5kGy or higher dosage of EBI. Values represent means ± SEM (<1%). [a-b]Means on the same day with different letter are significantly different (P ≤ 0.05). ◆ 0 kGy ■ 2 kGy

treatment.   The aerobic bacterial cells could have undergone sublethal injury, or more severe damage, which probably requires a more complex repair mechanism and thus, cannot be repaired until environmental conditions become more favorable (Tarte *et al.*, 1996).

3.3. Total Coliforms and E. coli Counts

Total coliforms and E.coli counts can be used to indicate potential safety and quality of food.   The presence of coliforms in processed foods is a useful indicator of post-sanitization and post-processing (pasteurization) contamination.   A limitation of our study todate, however, has been the presence of relatively low initial coliform and E. coli counts in beef samples, which we attempted to extrapolate to a serious food safety issue. It will be important to extend our study to further examine electron-beam irradiation in beef processing by inoculating a higher load of coliforms and E.coli in beef patties before exposing samples to irradiation treatment.

3.4. Water activity and pH

Water activity (a_w) and pH are both reliable indicators of food stability associated with microbial growth and chemical reactions to cause decomposition.   The optimal a_w and pH for bacterial growth are about 1-0.75 and 6.0-8.0 respectively.   In this study, changes in pH and water activity measurements in beef patties were not significantly altered by both irradiation and storage treatments. This enables us to conclude that the influence of both parameters on irradiation inactivation of microorganisms was neglectable.

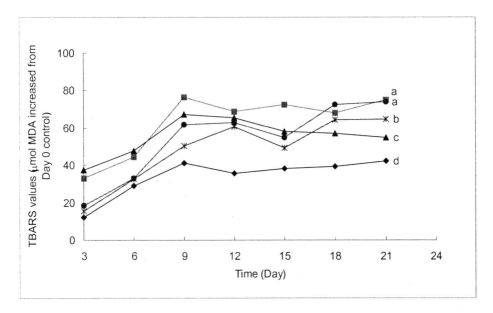

Figure 3. Effect of irradiation dose on TBARS values (μmol MDA increased from Day 0 control) of beef patties stored at 4°C for 21 days. Values represent means of nine determinations. [a-d]Means on the same day with different letter are significantly different (P ≤ 0.05). ♦ 0 kGy ■ 2 kGy ▲ 5 kGy ● 10 kGy ✳ 20 kGy

3.5. Thiobarbituric acid reactive substances (TBARS)

Malonaldehydes are secondary products of lipid oxidation, derived upon ionizing radiation treatment in presence of oxygen. Malonaldehydes are undesirable because of the oxidative changes that occur with various food components leading to poor organoleptic qualities. The effect of irradiation processing on TBARS values of beef patties stored for 21 days is presented in Figure 3. Following Day 3, TBARS values of non-irradiated beef patties increased (P ≤ 0.05) progressively, and reached a plateau (ranged 54.87 to 62.69) on Day 9. After 9 days of storage, the TBARS values of beef patties decreased (P ≤ 0.05). A similar trend in irradiation effect was observed in beef patties irradiated at 2, 5 and 10 kGy. With irradiation, a higher (P ≤ 0.05) level of TBARS was detected in beef patties, compared to the non-irradiated control, and this effect was observed immediately after irradiation (Day 0). This observation was supported by Lefebvre *et al.* (1994), who also reported an increase in hydroperoxides in beef patties immediately after irradiation treatment on Day 0. Among the four different doses of irradiation studied, 2 and 5 kGy treatments led to a consistently higher level of TBARS production up to Day 9 before gradually decreasing. The observed decrease in TBARS values was probably due to the instability of malonaldehyde in the beef samples. An alternative explanation suggested by Murano *et al.* (1998) has malonaldehyde being metabolized by a large population of spoilage bacteria in non-irradiated controls, which led to a decrease in TBARS towards the end of storage period. The increase in TBARS value was significantly reduced (P ≤ 0.05) by higher irradiation doses (10, 20 kGy) up to day 15, and the level continued to increased gradually up to 21 days storage. Hampson *et al.* (1996) reported that irradiation-induced oxidative chemical changes in turkey meat were dose dependent over a range of 1.5 to 10 kGy. However, little research has been conducted on the oxidative chemical changes in beef patties irradiated at > 10 kGy.

3.6. Correlation of TBARS and Total Aerobic Plate Counts

Linear regression coefficient for TBARS and Total Aerobic Plate Counts was determined to be 0.8486 (Figure 4), with n=8 and p=0.05, which suggested a positive correlation between TBARS and Total Aerobic Plate Counts in nonirradiated beef patties. This correlation might be due to the oxygen permeability of the Zip-lock package, which allowed a slight increase in oxygen content inside the package overtime.

In this study, we measured the level of lipid oxidation in raw beef patties rather than in cooked samples. Our reasoning was that the level of lipid oxidation in raw pattie before cooking would determine the initial oxidation status of cooked meat. The progress of lipid oxidation in cooked meat during storage may also be highly dependent on the amount of primary lipid oxidation products in raw meat (Ahn *et al.,* 1998). Therefore, controlling the degree of lipid oxidation in raw pattie would help to improve the oxidative stability and storability of both raw and cooked beef patties.

3.7. Color

Color appearance is one of the most important factors involved in consumer judgement of meat quality. In this study, electron-beam irradiation processing was shown to impose no significant effect on beef pattie lightness (L values). This finding was in agreement with other studies on color of irradiated meats that irradiation treatment did not affect L values of meat during refrigerated storage (Chen *et al.*, 1999; Nanke *et al.*, 1998). The effect of irradiation on HunterLab a values of beef patties, stored at 4°C for 21 days is presented in Fig. 5. HunterLab a values of non-irradiated beef patties decreased (P ≤ 0.05) overtime until Day 12 and then gradually increased (P ≤ 0.05). During storage, the desirable cherry red colour of fresh beef was lost and beef became brown due to an accumulation of metmyoglobin. With all four levels of irradiation, HunterLab a values of beef patties decreased (P ≤ 0.05) significantly, and this effect could be observed immediately after irradiation. However, irradiated beef patties did not show an increase in HunterLab a values towards the end of the 21 days storage period as observed in the non-irradiated samples. It is proposed that myoglobin facilitates O_2 diffusion from the cell wall to the mitochondria, and that myoglobin oxidation occurs at this point (Baseman, 1982). Metmyoglobin, the oxidized myoglobin, could be reduced by the enzymic system attached to the mitochondria after O_2 release (Baseman, 1982). This enzymatic reaction has been shown to involve metmyoglobin reductase (NADH-cytochrome b5 reductase), which catalyses the reduction of metmyoglobin by using mitochondrial cytochrome b5 (Arihara *et al.*, 1995). Once metmyoglobin was reduced to myoglobin, it could be further oxygenated to the cherry red oxymyoglobin. That was a proposed reason why we observed a decrease, followed by an increase in redness, in the non-irradiated beef patties. This enzymatic system might be inactivated by irradiation, therefore, the cherry red color would not be re-generated in the irradiated beef patties. However, this explanation could only partly explain our observations since mincing could also disrupt the functional properties of mitochondria, which in turn would affect the

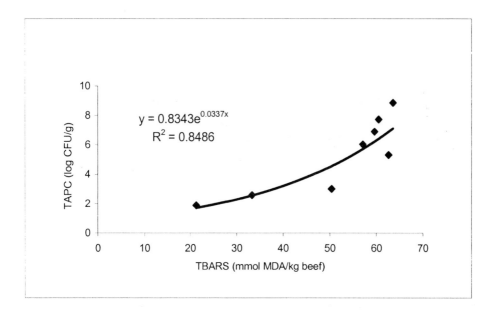

Figure 4. Correlation of TBARS (μmol MDA/kg beef) and Total Aerobic Plate Counts (log CFU/g)

reducing power in beef patties. Further research on the activity of metmyoglobin reductase (NADH-cytochrome b5 reductase) and metmyoglobin concentration in electron-beam irradiated beef is needed to support this explanation.

Irradiation at a high dose (20 kGy) had a small effect on HunterLab a values of beef patties, compared to low doses (2, 5, 10 kGy) on the first 9 days of storage. Nanke *et al.* (1998) also found a decrease in redness values for beef, as irradiation dose levels increased from 0 kGy to 4.5 kGy. As dose levels increased from 4.5 kGy to 10.5 kGy, a values decreased to a lesser extent (Nanke *et al.*, 1998), which also indicated a smaller effect with higher doses. However, Chen *et al.* (1999) reported a greater a value with irradiated pork patties compared with non-irradiated samples. The associated decrease in beef color may therefore be also due to the interaction between irradiation-induced free radicals and heme pigments in beef patties. Effect of irradiation on HunterLab b values of beef patties irradiated at all 4 doses was not significant. Studies reporting changes in yellowness due to irradiation have been inconsistent. Chen *et al.* (1999) reported a lower yellowness value was associated with irradiated pork patties, whereas Nanke *et al.* (1998) found that the b values remained relatively unchanged until a dose of 10.5 kGy was reached.

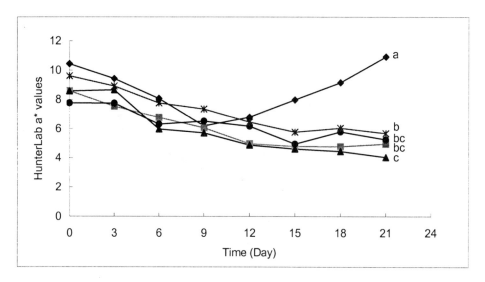

Figure 5. Effect of irradiation dose on HunterLab a values of beef patties stored at 4°C for 21 days. Values represent means of nine determinations. [a-c]Means on the same day with different letter are significantly different (P ≤ 0.05).
♦ 0 kGy   ■ 2 kGy   ▲ 5 kGy   ● 10 kGy   ✳ 20 kGy

4. CONCLUSION

This study demonstrated that Electron-beam irradiation treatment offers a good alternative to conventional processing methods for eliminating aerobic and psychrotrophic bacterial growth in beef patties during extended storage at 4°C. Irradiation to 2 kGy eliminated microbial growth approximately 3-4 log CFU/g without adversely affecting beef color attributes. Complete inhibition of bacteria occurred at 5 kGy, or higher dosage of electron-beam irradiation. However, electron-beam irradiation treatment led to higher level of TBARS production in beef patties upon storage. TBARS values were also found to be positively correlated with TAPC.

5. REFERENCES

Arihara, K., Cassens, R. G., Greaser, M. L., Luchansky, J. B., Mozdziak, P. E., 1995, Localization of metmyoglobin-reducing enzyme (NADH-cytochrome b5 reductase) system components in bovine skeletal muscle, *Meat Sci.,* **39**:205-213.
Ahn, D. U., Sell, J. L., Jo. C., Chen, X., Wu, C., Lee, J. I., 1998, Effects of dietary vitamin E supplementation on lipid oxidation and volatiles content of irradiated, cooked turkey meat patties with different packaging, *Poultry Sci.,* **77**:912-920.
Baseman, K.J., 1982, Metmyoglobin reducing systems in red meat. *Dissertation Abstracts Int.,* **42**:4730-4731.
Chen, X., Jo, C., Lee, J. I., Ahn, D. U., 1999, Lipid oxidation, volatiles and color changes of irradiated pork patties as affected by antioxidants, *J. Food Sci.,* **64**:16-19.
Chung, M. S., Ko, Y. T., Kim, W. S., 2000, Survival of Pseudomonasfluorescens and Salmonella typhimurium after electron beam and gamma irradiation of refrigerated beef., *J. Food Prot.,* **63**:162-166.

Diehl, J. F., 1990, Biological effects of ionizing radiation, in: *Safety of Irradiated Foods,* Marcel Dekker, Inc., New York, pp. 95-136.

Hampson, J. W., Fox, J. B., Lakritz, L., Thayer, D. W., 1996, Effects of low dose gamma radiation on lipids in five different meats, *Meat Sci.,* **42**:271-276.

Kitts, D. D., Giroux, M., LaCroix, M., 1997, Chemical and physical changes from irradiation. Ground beef irradiation petition to Health Canada. Canadian Cattlemen's Association, Chapter 7, pp. 24-29.

Jay, J. M., 1996, Microorganisms in fresh ground meats: the relative safety of products with low versus high numbers, *Meat Sci.,* **43**:S59-S66.

Lea, C. H. J., Macfarlane, J., Parr, L. J., 1960, Treatment of meats with ionizing radiation. II Radiation pasteurization of beef for chilled storage, *J. Sci. Food Agric.,* **11**:690.

Lee, J. W., Yook, H. S., Kim, S. A., Lee, K. H., Byun, M. W., 1999, Effects of antioxidants and gamma irradiation on the shelf life of beef patties, *J. Food Prot.,* **62**:619-624.

Lefebrve, N., Thibault, C., Charbouneau, R., Piette, I. P. G., 1994, Improvement on shelf-life and wholesomeness of ground beef by irradiation – Chemical analysis and sensory evaluation, *Meat Sci.,* **36**:371-380.

Moseley, B. E. B., 1989, Ionizing radiation: action and repair, in: *Mechanism of action of food preservation procedures,* Elsevier Applied Science Publishers Ltd., London.

Murano, P. S., Murano, E. A., Olson, D. G., 1988, Irradiated ground beef: sensory and quality changes during storage under various packaging conditions, *J. Food Sci.,* **63**:548-551.

Nanke, K. E., Sebranek, J. G., Olson, D. G., 1998, Color characteristics of irradiated vacuum-packaged pork, beef, and turkey, *J. Food Sci.,* **3**:1001-1006.

Rocelle, M., Clavero, S., Monk, J. D., Beuchat, L. R., Doyle, M. P., Brackett, R. E., 1994, Inactivation of Escherichia coli O157:H7, Salmonellae, and Campylobacter jejuni in raw ground beef by gamma irradiation, *Applied Environ. Microbiol.,* **60**:2069-2075.

Tarte, R., Murano, E. A., Olson, D. G., 1996, Survival and injury of Listeria monocytogenes, Listeria innocua and Listeria ivanovii in ground pork following electron bean irradiation, *J. Food Prot.,* **59**:596-600.

Wijewickreme, A. N., Kitts, D.D., 1997, Influence of reaction conditions on the oxidative behavior of model Maillard reaction products, *J. Agric. Food Chem.,* **45**:4571-4576.

ENZYMATIC HYDROLYSIS OF LIPIDS IN MUSCLE OF FISH AND SHELLFISH DURING COLD STORAGE

Masaki Kaneniwa, Masahito Yokoyama, Yuko Murata, and Ryuji Kuwahara[*]

1. INTRODUCTION

In previous research (Olley and Lovern, 1960; Olley et al., 1962; Wu et al., 1974; Toyomizu et al., 1977; Hanaoka and Toyomizu, 1979; Ohshima and Koizumi, 1983; Ohshima et al., 1983a, 1983b, 1984a; Hwang and Regenstein, 1993; Aubourg et al., 1998; Ingemansson et al., 1995; Ben-Gigirey et al., 1999), enzymatic hydrolysis of lipids in fish muscle during cold storage has been reported in some lean and fatty fish such as cod, skipjack, carp, sardine, and rainbow trout. It is known that free fatty acids (FFA) accumulate in muscle lipids due to enzymatic hydrolysis of lipids, and the increase of FFA's reduce the quality of the fish muscle (Dyer, 1951; Dyer and Fraser, 1959; Ohshima et al., 1984b). The functional fatty acid components such as EPA (20:5n-3) and DHA (22:6n-3) are also released from polar lipids or triacylglycerols to the free fatty acid fraction by hydrolysis, and it is presumed that the lipid hydrolysis affects their trophic value. The enzymatic hydrolysis was mainly caused by phospholipase in muscle, and Hirano et al. (1997) partially purified phospholipase A_1 from skipjack muscle. It is necessary to elucidate properties of enzymes for the prevention of enzymatic hydrolysis occurring in fish muscle. In this study, muscle of nine species of fish and shellfish (yellowtail, rainbow trout, bigeye tuna, chum salmon, skipjack, butterfly bream, Pacific cod, kuruma prawn, and scallop) were stored at -10°C for 30 days, and their lipid and fatty acid compositions analyzed before and after the storage for determination of the properties of enzymes concerned with lipid hydrolysis in muscles.

2. MATERIALS AND METHODS

Nine species of fish and shellfishes [yellow tail (*Seriola quinqueradiata*), rainbow trout (*Oncorhynchus mykiss*), bigeye tuna (*Thunnus obesus*), chum salmon (*Oncorhynchus keta*), skipjack (*Euthynnus pelamis*), butterfly bream (*Nemipterus virgatus*), Pacific cod (*Gadus*

---

[*] Masaki Kaneniwa, Masahito Yokoyama, Yuko Murata, and Ryuji Kuwahara, Marine Biochemistry Division, National Research Institute of Fisheries Science, 2-12-4, Fukuura, Kanazawa, Yokohama, Kanagawa, 236-8648, Japan. Masahito Yokoyama, Present address: Fisheries Division, Japan International Research Center for Agricultural Sciences, Ohwashi, Tsukuba, Ibaragi, 305-8686, Japan.

Quality of Fresh and Processed Foods, edited by Shahidi et al.
Kluwer Academic/Plenum Publishers, 2004.

macrocephalus), kuruma prawn (*Penaeus japonicus*), and scallop (*Patinopecten yessoensis*)] purchased at a market in Yokohama City were used in this study. The muscles of fish, including white and dark tissues, and kuruma prawn, and adductor muscle of scallop were minced, packed, and stored at -10°C for 30 days. To deactivate the enzymes in the muscle, a portion of minced muscle were heated in boiling water for 10 min, and after cooling, the heated muscles were packed and stored as the unheated samples. Total lipids (TL) were extracted from the minced muscle according to the procedure of Bligh and Dyer (1959).

The FFA fractionation was separated from TL by TLC (thin-layer chromatography) on silica gel 60 plates (Merck, Darmstadt, Germany) with n-hexane/diethylether/acetic acid (70:30:1, v/v/v) as the developing solvent.

The lipid compositions were analyzed by the TLC-FID (flame ionizative detection) method (Ohshima *et al.*, 1987) using a Chromarod SIII and an Iatroscan MK-5 (Iatron Laboratories, Tokyo, Japan) with the same solvent system of TLC. Peak area percentages were obtained with an Iatrocorder TC-11 (Iatron Laboratories, Tokyo, Japan).

The lipids were converted into fatty acid methyl esters using 5% HCl-methanol. The gas chromatographic analysis of the methyl esters was conducted using a Shimadzu GC14A instrument (Shimadzu Seisakusho Co., Kyoto, Japan), with a flame ionization detector on a fused-silica capillary column coated with Omegawax 320 (30m x 0.32mm i.d.). The carrier gas was helium. The column temperature was 210°C, and the injector and detector temperature was 230°C. Peak area percentages were obtained with a Shimadzu integrator C-R6A (Shimadzu Seisakusho Co., Kyoto, Japan). The fatty acid component of each peak of the gas chromatogram was identified on the basis of the agreement of retention time data with those of the reference specimens. The tricosanoic acid (23:0) was used as internal standard.

3. RESULTS AND DISCUSSION

The content and composition of lipids from muscle of fish samples examined in this study are shown in Table 1. The TL contents of Pacific cod, kuruma prawn, and scallop muscle were less than 1%, and the major component of their TL was polar lipids (PL, over 80%).

Table 1.   Lipid contents in the minced muscle of the samples.

Scientific name	English name	TL* contents in the minced muscle (%)	Contents in TL (%)			
			TG	FFA	ST	PL
Seriola quinqueradiata	Yellowtail	13.7	87.1	0.4	0.4	12.0
Oncorhynchus mykiss	Rainbow trout	4.0	76.6	0.2	1.3	21.9
Thunnus obesus	Bigeye tuna	2.7	83.3	-**	0.8	15.8
Oncorhynchus keta	Chum salmon	1.8	47.3	1.9	2.0	48.8
Euthynnus pelamis	Skipjack	1.9	71.9	1.7	2.6	23.8
Nemipterus virgatus	Butterfly bream	1.5	58.9	-	3.2	37.8
Gadus macrocephalus	Pacific cod	0.7	1.9	1.7	5.5	90.9
Penaeus japonicus	Kuruma prawn	0.9	-	0.1	12.0	87.9
Patinopecten yessoensis	Scallop	0.6	0.1	-	18.5	81.4

*TL: Total lipids; TG: Triacylglycerols; FFA: Free fatty acids; ST: Sterols; PL: Polar lipids.
**-: Not detected.

Contrary to this, the TL contents of skipjack, bigeye tuna, rainbow trout, and yellowtail were 1.9-13.7%, and the major component of their TL was triacylglycerols (TG, over 70%). The TL contents of butterfly bream and chum salmon were 1.5 and 1.8 %, respectively, and the major components of the TL were TG (58.9% and 47.3%) and PL (37.8% and 48.8%). Free fatty acids (FFA) contents of all the sample fish were under 2%.

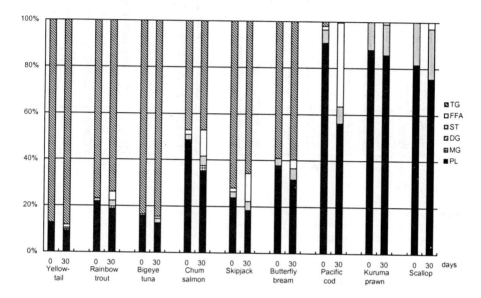

Figure 1. Changes in the lipid class of unheated minced muscle of samples during storage at -10°C.
TG: Triacylglycerols; FFA: Free fatty acids; ST: Sterols; DG: Diacylglycerols; MG: Monoacylglycerols; PL: Polar lipids.

Changes of the lipid classes of minced muscle during storage at -10°C for 30 days are shown in Figure 1. During storage, the content of FFA increased, while that of PL decreased in all species examined. The changes of FFA contents were from 0.4 to 1.5% (yellowtail), from 0.2 to 3.7% (rainbow trout), from 0.0 to 1.0% (bigeye tuna), from 1.9 to 11.1% (chum salmon), from 1.7 to 12.0% (skipjack), from 0.0 to 3.7% (butterfly bream), from 1.7 to 36.3% (Pacific cod), from 0.1 to 0.9% (kuruma prawn), and from 0.0 to 3.0% (scallop). The changes of PL contents were from 12.0 to 9.1% (yellowtail), from 21.9 to 18.7% (rainbow trout), from 15.9 to 12.6% (bigeye tuna), from 48.8 to 35.5% (chum salmon), from 23.8 to 18.3% (skipjack), from 37.8 to 31.9% (butterfly bream), from 90.9 to 55.9% (Pacific cod), from 87.9 to 85.8% (kuruma prawn), and from 81.2 to 75.5% (scallop). Furthermore, the content of TG decreased in rainbow trout (from 76.6 to 73.9%), skipjack (from 71.9 to 65.7%) and Pacific cod (from 1.9 to 0.2%). These phenomena were inhibited by heating of the muscle before storage as shown in Figure 2. These results suggested that the muscle lipids were hydrolyzed by lipase and phospholipase in rainbow trout, skipjack and Pacific cod and mainly by phospholipase in the other six species.

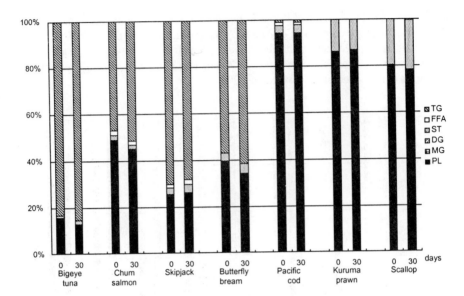

Figure 2. Changes in the lipid class of heated minced muscle of samples during storage at -10°C. heated: heated in boiling water for 10 min.; TG: Triacylglycerols; FFA: Free fatty acids; ST: Sterols; DG: Diacylglycerols; MG: Monoacylglycerols; PL: Polar lipids.

For comparison of the lipid hydrolase activities between these nine species, increments of FFA per 100g of muscle were calculated from data of lipid contents and compositions, and shown in Figure 3. The maximum increment of FFA was 206.3mg/100g of muscle observed in skipjack and the minimum increment was 7.9mg/100g of muscle observed in kuruma prawn. In this study, the relationship between increments of FFA and lipid contents or composition was not observed. It is presumed that differences of lipid hydrolase activities are due to species of fish and shellfish.

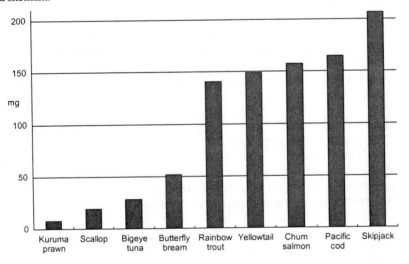

Figure 3.   Increments of free fatty acids in 100g of muscle of the samples during storage at -10°C for 30 days.

Table 2 . Fatty acid compositions of the total lipids from the samples.

	Yellow-tail	Rainbow trout	Bigeye tuna	Chum salmon	Skipjack	Butterfly bream	Pacific cod	Kuruma prawn	Scallop
14:0	3.8	2.0	1.9	4.4	3.8	2.7	0.8	0.5	2.2
16:0	17.4	21.0	17.6	12.1	19.1	22.6	15.9	16.3	20.6
16:1n-7	5.3	6.5	4.7	4.9	4.6	6.7	2.1	2.9	2.6
18:0	4.7	5.6	4.1	2.8	6.0	6.0	3.8	8.4	4.4
18:1	19.7	30.9	24.7	17.9	12.8	17.4	13.6	14.3	4.3
18:2n-6	1.6	10.3	0.8	0.8	1.2	0.6	0.8	11.2	0.4
18:4n-3	1.2	0.2	0.2	0.9	0.8	0.1	0.5	0.1	1.0
20:1	3.6	2.2	2.1	9.4	3.8	2.1	3.0	1.5	5.0
20:4n-6	1.3	0.9	2.5	0.4	1.9	3.3	1.7	2.7	2.0
20:5n-3	5.9	1.7	4.9	7.7	8.4	5.1	17.3	13.5	22.9
22:1n-11	2.6	0.5	0.5	5.0	3.3	0.5	0.7	0.2	0.1
22:1n-9	0.5	0.3	0.4	3.8	0.3	0.2	0.2	0.1	0.1
22:4n-6	0.3	-*	0.4	-	0.1	1.4	-	0.1	0.2
22:5n-6	0.8	0.3	1.5	0.2	1.0	1.0	0.3	0.3	0.4
22:5n-3	2.5	0.8	1.6	2.9	1.5	4.3	1.5	0.9	0.8
22:6n-3	20.4	12.0	24.1	20.4	22.7	18.1	32.0	16.9	21.0
Others	8.5	5.0	8.0	6.3	8.5	7.8	6.0	10.1	12.1

*-: Not detected.

The fatty acid compositions of TL from the samples are shown in Table 2. The content of 20:5n-3 and 22:6n-3 in the fatty acid of TL from the muscle of samples examined in this study were 1.7-22.9% and 12.0-32.0%, respectively. The ratio of hydrolyzed 20:5n-3 and 22:6n-3 to those fatty acids in the TL is calculated from fatty acid compositions of TL and FFA fractions contained internal standard as shown in Figure 4. In all the samples, the ratios of hydrolyzed 20:5n-3 were higher than those of 22:6n-3. These results show that 20:5n-3 is much more prone to enzymatic hydrolysis than 22:6n-3. In the Pacific cod muscle, over 40% of 20:5n-3 and over 30% of 22:6n-3 were hydrolyzed during storage for 30 days.

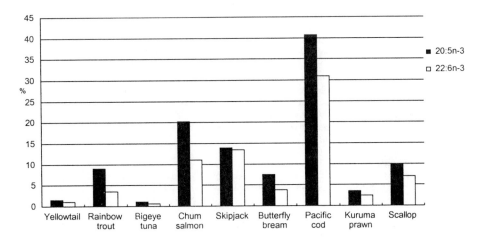

Figure 4. The ratio of the hydrolyzed EPA and DHA to those fatty acids in the TL of the muscle of the samples during storage at -10°C for 30 days.

In this study, we determined that the enzymatic hydrolysis of lipids in fish and shellfish muscle occurred during storage at -10°C. For the prevention of enzymatic hydrolysis of fish and shellfish muscle lipids, it is necessary to purify lipid hydrolases in fish and shellfish muscle and elucidate their properties. In most previous studies, examinations were carried out at low temperatures (<5°C) and over long-term storage periods (>one week). But these conditions are unsuitable for purification of the enzymes. Wu *et al.* (1974) elucidated the formation of glycerylphosphorylcholine (GPC) by enzymatic decomposition of phosphatidylcholine in carp ordinary muscle during incubation at 37°C for 2 h using ^{14}C labeled phosphatidylcholine. In a previous study (Kaneniwa *et al.* 2000), we determined that lipid hydrolysis of silver carp muscle was detectable by incubation at 20°C within one week. It is necessary to evaluate the high temperature and short-term storage or incubation to determine the activity of enzymes concerned with lipid hydrolysis in fish and shellfish muscle. Furthermore, Ohshima *et al.* (1983b) determined that over 200mg of FFA were accumulated in 100g of cod flesh by enzymatic hydrolysis of the phospholipids during storage at -16°C for 24 days. They also determined that about 180mg of FFA were accumulated in cod flesh during storage with ice at 4°C for 30 days (Ohshima *et al.*, 1984a). Lovern and Olley (1962) showed that the highest accumulation of FFA in cod flesh occurred at -4°C. The difference of storage temperature can be affecting lipid hydrolysis. It is therefore important to determine the differences of lipid hydrolase activities at various temperatures.

4. REFERENCES

Aubourg, S. P., Sotelo, C. G., and Pérez-Martin, R., 1998, Assessment of quality changes in frozen sardine (*Sardina pilchardus*) by fluorescence detection, *J. Am. Oil Chem. Soc.*, **75**:575-580.

Ben-Gigirey, B., Vieites Baptista De Sousa, J. M., Villa, T. G., and Barros-Velazquez, J., 1999, Chemical changes and visual appearance of albacore tuna as related to frozen storage, *J. Food Sci.*, **64**:20-24.

Bligh, E. G., and Dyer, W. J., 1959, A rapid method of total lipids extraction and purification, *Can. J Biochem. Physiol.*, **37**:911-917.

Dyer, W. J., 1951, Protein denaturation in frozen and stored fish, *Food Res.*, **16**:522-527.

Dyer, W. J., and Fraser, D. I., 1959, Proteins in fish muscle. 13. Lipid hydrolysis, *J. Fish. Res. Bd. Canada*, **16**:43-52.

Hanaoka, K., and Toyomizu, M., 1979, Acceleration of phospholipid decomposition in fish muscle by freezing, *Bull. Japan. Soc. Sci. Fish.*, **45**:465-468.

Hirano, K., Tanaka, A., Yoshizumi, K., Tanaka, T., and Satouchi, K., 1997, Properties of phospholipase A$_1$/Transacylase in the white muscle of bonito *Euthynnus pelamis* (Linnaeus), *J. Biochem.*, **122**:1160-1166.

Hwang, K. T., and Regenstein, J. M., 1993, Characteristics of mackerel mince lipid hydrolysis, *J. Food Sci.*, **58**:79-83.

Ingemansson, T., Kaufmann, P., and Esktrand, B., 1995, Multivariate evaluation of lipid hydrolysis and oxidation data from light and dark muscle of frozen rainbow trout (*Oncorhynchus mykiss*), *J. Agric. Food Chem.*, **43**:2046-2052.

Kaneniwa, M., Miao, S., Yuan, C., Iida, H., and Fukuda, Y., 2000, Lipid components and enzymatic hydrolysis of lipids in muscle of Chinese freshwater fish., *J. Am. Oil Chem. Soc.*, **77**:825-830.

Lovern, J. A., and Olley, J., 1962, Inhibition and promotion of post-mortem lipid hydrolysis in the flesh of fish, *J. Food Sci.*, **27**:551-559.

Ohshima, T., and Koizumi, C., 1983, Accumulation of lysophosphatidyl-ethanolamine in muscle of fresh skipjack, *Bull. Japan. Soc. Sci. Fish.*, **49**:1205-1212.

Ohshima, T., Wada, S., and Koizumi, C., 1983a, Deterioration of phospholipids of skipjack muscle during ice storage: mainly concerning to enzymatic hydrolysis of phosphatidylcholine, *Bull. Japan. Soc. Sci. Fish.*, **49**:1213-1219.

Ohshima, T., Wada, S., and Koizumi, C., 1983b, Enzymatic hydrolysis of phospholipids in cod flesh during cold storage, *Bull. Japan. Soc. Sci. Fish.*, **49**:1397-1404.

Ohshima, T., Wada, S., and Koizumi, C., 1984a, Enzymatic hydrolysis of phospholipids in cod flesh during storage in ice, *Bull. Japan. Soc. Sci. Fish,*. **50**:107-114.

Ohshima, T., Wada, S., and Koizumi, C., 1984b, Effect of accumulated free fatty acid on reduction of salt soluble protein of cod flesh during frozen storage, *Bull. Japan. Soc. Sci. Fish.*, **50**:1567-1572.

Ohshima, T., Ratnayake, W. M. N., and Ackman, R. G., 1987, Cod lipids, solvent systems and the effect of fatty acid chain length and unsaturation on lipid class analysis by Iatroscan TLC-FID, *J. Am. Oil Chem. Soc.*, **64**:219-223.

Olley, J., and Lovern, J. A., 1960, Phospholipid hydrolysis in cod flesh stored at various temperatures, *J. Sci. Food Agric.*, **11**:644-652.

Olley, J., Pirie, R., and Watson, H., 1962, Lipase and phospholipase activity in fish skeletal muscle and its relationship to protein denaturation, *J. Sci. Food Agric.*, **13**:501-516.

Toyomizu, M., Hanaoka, K., Satake, K., and Nakagawa, H., 1977, Effect of storage temperatures on accumulation of glycerylphosphorylcholine and decomposition of phosphatidylcholine in fish muscle during cold storage, *Bull. Japan. Soc. Sci. Fish.*, **43**:1181-1187.

Wu, C., Nakagawa, H., Satake, K., and Toyomizu, M., 1974, Formation of glycerylphosphorylcholine by enzymatic decomposition of phosphatidylcholine in carp ordinary muscle, *Bull. Japan. Soc. Sci. Fish.*, **40**:835-840.

TEMPERATURE, COLOR, AND TEXTURE PREDICTION MODELS FOR SURIMI SEAFOOD PASTEURIZATION

Jacek Jaczynski and Jae W. Park[*]

1. INTRODUCTION

Surimi is a frozen concentrate of fish myofibrillar proteins stabilized with cryoprotectants. It is a major ingredient, along with other compounds (flavors, starches, protein additives, etc.), for surimi seafood products such as crabmeat analog. Surimi seafood is a ready-to-eat product, which requires precise pasteurization practice for product quality and safety. Due to good sensory characteristics and relatively low price, compared to natural counterparts, surimi seafood has its own, well-established position in the market and is well recognized by consumers (Shie and Park, 1999).

FDA regulations pertaining to the microbial load limits in processed seafood require all manufacturers to comply with microbial safety (Ward and Price, 1992). At the same time, on a consumer driven market, a product with very high sensory quality is required to maintain the manufacturer's market position (Shie and Park, 1999). High quality surimi seafood is distinguished by its firm texture and white meat color. It is generally accepted that the whiter surimi seafood, the higher is its quality (Park, 1995). Current trends in the market are fresh-like products (Shie and Park, 1999). However, excessive heat applied to surimi seafood causes deterioration of texture and color (Bertak and Karahadian, 1995).

Color in food products determines the appearance and therefore consumer acceptance. It is often said that color is the most important aspect of food acceptance since if a food product does not look "right" the consumer will not evaluate it any further for flavor and subsequent texture (Francis, 1994). Color is always judged as a first of all the organoleptic attributes. This particularly applies to surimi seafood that is packaged in a plastic film. Therefore, texture and flavor can only be judged after the package is opened. In this regard, color is a primary attribute, whereas texture and flavor are secondary attributes, upon which the decision whether to buy or not is based.

---

[*] Jae W. Park, Oregon State University Seafood Laboratory, 2001 Marine Dr., Astoria, OR 97103

Quality of Fresh and Processed Foods, edited by Shahidi et al.
Kluwer Academic/Plenum Publishers, 2004.

The surimi seafood market in the U.S. was developed in '70s and '80s. It exhibited steady growth, 10-100% annually, reaching 96,000 MT in 1996 (Park, 2000). In 1991, the price of the raw material, surimi, increased dramatically. Since then manufacturers were forced to replace surimi with cheaper starches and proteins. The new additives, proteins of different origins than fish (egg white, beef plasma) and starches exhibited different properties with respect to thermal processing (Kim and Lee, 1987). This resulted in lower quality of color and texture (Park, 1994). The problem intensified even further when the industry continued to utilize the same pasteurization regimes used for surimi seafood with a higher surimi content (Shie and Park, 1999).

Abusive pasteurization and reduction of fish myofibrillar protein by surimi substitutes seem to be contradictory to the market demand, minimally processed, fresh-like seafood products. This might have been one of the main causes of market decline observed in the '90s. Therefore, it is necessary to develop an optimized pasteurization technique that would provide microbial safety along with maximum retention of color and texture, which are the primary sensory attributes affected by thermal processing.

Thermal treatment is necessary for the microbial safety of surimi seafood products. However, excessive thermal treatment can cause impairment, from a sensory point of view (Rippen and Hackney, 1992). Therefore, there is a vital need to determine parameters that simultaneously guarantee the maximum sensory/physical quality and sufficient microbial safety in surimi seafood. The physical quality and microbial flora changes are governed by the temperature alterations that occur during thermal treatment (Shie and Park, 1999). Furthermore, the temperature changes are affected by the amount of heat to which sample is exposed during the pasteurization period. However, time-temperature exposure calculations present thermodynamic challenges and are not an easy task. Numerous time-temperature exposure models are presented in the literature (McAdams, 1954; Hsu, 1963; Su et al., 1999). To our knowledge, literature information on the pasteurization of surimi seafood is extremely limited. It is necessary to realize that heat penetration data are fundamental whenever it comes to thermal processing of a food commodity (Sadeghi and Swartzel, 1990; Kebede et al., 1996). It allows temperature prediction at different locations across the sample at various time intervals.

From a microbial inactivation standpoint, an idea of "cold spot" requires special consideration. "cold spot" is defined as a region within a food sample that reaches the temperature of the surroundings (water bath) last (Rippen et al., 1993). If there is any bacterium left alive in the cold spot it will proliferate and contaminate the remaining parts of a food sample. This is an example of post-pasteurization contamination.

In pasteurization, where hot water is used to inactivate microorganisms, the "killing factor" is a thermal energy, heat (Kebede et al., 1996). If the cold spot receives enough energy to inactivate bacteria, then all other regions have received at least the same amount of energy. Then, such pasteurization treatment is recognized as sufficient to maintain microbial safety (Rippen et al., 1993).

To maintain the highest quality of color and texture, we are more concerned about a gradient of time-temperature exposure across the sample. This implies a different, more complex approach than for microbial consideration. To predict color and texture changes that happen during pasteurization, it is necessary to monitor time-temperature exposure across the sample. This imposes some difficulties since thermal changes are dynamic. Heat penetrates in a certain distance, but simultaneously it changes over time (Carlsaw and Jaeger, 1959). A dynamic system where time-temperature exposure changes with time and distance appears to be functional (Hsu, 1963; Toledo, 1999). The result is a 3-D temperature profile in the form of surface response that is a function of pasteurization temperature, time, and distance from sample's cold spot and sample's initial temperature. The amount of heat to which sample has been exposed during the pasteurization period is equal to the integrated volume under the three-dimensional surface.

Our overall objective was to develop models to predict various physical properties of surimi seafood under thermal treatment. Details were:

1. Construction of a model predicting temperature at any location of a surimi seafood sample at any pasteurization time.
2. Calculation of time-temperature exposure of surimi seafood slab during pasteurization based on the temperature model.
3. Development of interactive models for color and texture changes during thermal pasteurization.

2. MATERIAS AND METHODS

Unpasteurized commercial surimi seafood sticks, which were cooked as a thin sheet for less than a minute, rolled, and wrapped, were obtained from a local manufacturer and transported in ice slush to the laboratory. They were kept in ice until used. The approximate dimension of an individual stick was 12-13 cm long and 1.4 cm in diameter. Before thermal treatment, the sticks were tightly placed on plastic trays and vacuum packed in a plastic film commonly used for surimi seafood packaging. For model construction, two types of plastic trays were used. A small tray, 15 cm (L) x 11.5 cm (W) x 2.5 cm (D), was made of 0.05 mm polypropylene (5 PP). A single layer of eight sticks was placed onto each small tray. A large tray, 13.5 cm (L) x 10.5 cm (W) x 5 cm (D), was manufactured of 0.02 mm high-density polyethylene (2 HDPE). Four layers of 24 sticks were placed onto each large tray.

An eighteen-liter water bath with water circulation was employed to mimic a typical industrial pasteurization process. Prior to the experiment, it was verified that the water bath was maintaining a uniform temperature in all locations and the temperature reading on the water bath's panel was correct.

2.1. Temperature Prediction Model

A datalogger (CR10X, Campbell Scientific, Logan, UT) was used to monitor the temperature changes in the samples. The datalogger was programmed in Edlog version 6.3 (Campbell Scientific, Inc.) so that five thermocouples (type T) were taking temperature readings every 5 s. The thermocouple wires were connected to temperature reading probes (59 mm long) by a threaded plug-socket system. Before the experiment, the probes were placed into the samples by puncturing the package and inserting the probe inside. To maintain vacuum inside the package a soft rubber washer was applied around the punctured holes and kept in place by the bolts screwed on the probe's thread. Careful attention was made to put the probes in the "cold spot" area of the samples. The geometrical center of the package was assumed as the "cold spot". After the experiment, the package was opened and the position of the probe's tip was verified. The datalogger was connected with a laptop computer to display the temperature profiles of the five thermocouples and the data were recorded by Tcom version 1.2 (Campbell Scientific, Inc.).

For model construction, two different packages (single- and four-layer trays) were heated at 75, 85 and 95°C. For model verification, two different packages (double- and – eight-layer trays) were heated at 65, 80 and 90°C. The temperature probes were inserted in samples. Their positions were verified afterwards.

Gurney-Lurie heat conduction solutions for a block in an open, unsteady-state system (Toledo, 1999; Hsu, 1963; McAdams, 1954) were applied to construct a temperature prediction model. It was assumed that the heat transfer coefficient (h) was approaching infinity (Su et al., 1999; Toledo, 1999; Carslaw and Jaeger, 1959).

2.2. Color and Texture Prediction Models

To construct the model, single-layer trays were heated at 85°C for 6, 12, 18, 24, 30, 36, 42, 48, 54, 60, 66, 72, 78, 84, and 90 min. To verify the model, 2- and 4-layer trays were heated at 65, 75, and 95°C for various time intervals. Texture and color changes during pasteurization were compared with heat to which the sample had been exposed. A temperature prediction model for surimi seafood slab previously constructed was used for this purpose.

2.2.1. Color Prediction Model

A Minolta Chroma Meter CR 300 (Minolta Camera Co. Ltd., Osaka, Japan) was used to determine tristimulus color values L^*, a^*, and b^*(CIE color system). After heat treatment, samples were removed from the water bath and cooled in ice water. After the external (red colored) layer of the stick was peeled off, the remaining stick was ground. The paste was transferred onto a Petri dish for color measurement.

Two formulas were used to calculate whiteness. One method was recommended by NFI (1991):

$$Whiteness(I) = 100 - \sqrt{(100 - L^*)^2 + a^{*2} + b^{*2}} \qquad \text{(Eq. 1)}$$

The second method was a Japanese traditional formula used in the surimi industry to determine whiteness (Park, 1994):

$$Whiteness(II) = L^* - 3b^* \qquad \text{(Eq. 2)}$$

PowerPoint'97 was employed to display possible color changes. The changes were predicted using the color models and then the values were converted to the color and displayed. A side-by-side display between the initial color values (control, no heat treatment) and the heat-treated samples was prepared. The Power Point display is not shown here.

2.2.2. Texture Prediction Model

For texture determination, a Sintech 1/G universal testing instrument (MTS, Cary, NC) was used. A Warner-Bratzler shear device attached to the Sintech 1/G was set up to measure peak load (g) for the breaking strength. Peak load was measured during the test by the load cell. Whole sticks were chilled after heat treatment. The samples were equilibrated to room temperature before the measurement was taken.

3. RESULTS AND DISCUSSION

Gurney-Lurie charts provide a relationship between Fourier modulus (X) and temperature ratio (Y). This relationship allows determination of temperature (T) at any location (x) and at any time (t).

$$X = \frac{\alpha t}{x_1^2} \qquad \text{(Eq. 3)}$$

X - Fourier modulus, dimensionless,
α - thermal diffusivity, m^2/s,
t - time, s,
x_1- distance from the center to the surface of a slab (half the smallest dimension), m,

$$Y = \frac{T_1 - T}{T_1 - T_0} \qquad \text{(Eq. 4)}$$

Y – temperature ratio, dimensionless,
T_1- water bath temperature, °C,
T - temperature at certain location (x) at certain time (t), °C,
T_0 - initial temperature of slab, °C.

The relationship between "Y" and "X" follows the exponential function.

$$Y = 1.32e^{-2.44X} \qquad \text{(Eq. 5)}$$

Y - temperature ratio, dimensionless,
X - Fourier modulus, dimensionless,

$$\alpha = \frac{k}{\rho C_p} \qquad \text{(Eq. 6)}$$

α - thermal diffusivity, m^2/s,
k - thermal conductivity, J/smK,
ρ - density, kg/m^3,
C_p- specific heat, J/kgK.

Gurney and Lurie charts use dimensionless parameters, "m" (1/Biot modulus) and "n".

$$m = \frac{k}{hx_1}$$ (Eq. 7)

$$n = \frac{x}{x_1}$$ (Eq. 8)

m - 1/Biot modulus, dimensionless,
h - heat transfer coefficient, W/m^{2}K,
n - dimensionless, $n = 0$ at slab's center, $n = 1$ at slab's surface,
x - distance from the sample's center to the position where temperature (T) was reached at time (t),
      The following values (constants) were used in model calculations: $k = 0.5$ W/mK (Toledo, 1999), $C_p = 3.18$ kJ/kgK (ASHRAE, 1972, 1967), $h = 1200$ W/m^2 K (Su et al.,1999), $\rho = 1000$ kg/m^3.

      However, the Gurney-Lurie charts did not predict temperature profiles satisfactorily (Figure 1). Thermal diffusivity (α) was calculated based on the temperature profiles obtained in the experiments. Thermal diffusivity was determined to be temperature dependent since it increased as sample temperature increased.

$$\alpha = 0.0604T + 13.811$$ (Eq. 9)

      Once α was incorporated into the model, the accuracy of the temperature prediction improved accordingly. The inaccuracy of the original Gurney-Lurie model stemmed from the assumption of constant "α". However, model predictions below 20°C did not fit the experimental data (Figure 1). Since the α (Eq. 9) was determined for temperatures above 20°C, it resulted in inaccuracy for the model below 20°C. Thermal diffusivity showed inconsistent patterns below 20°C.
      Model verification experiments showed satisfactory agreement between the model with α correction and temperatures recorded by the datalogger (Figure 1). It confirmed that for better accuracy the thermal diffusivity correction (α) in the original Gurney-Lurie model was necessary.
      Relationship between temperature ratio ("Y") and Fourier modulus ("X") allows determination of temperature (T) at any location (x) and at any time (t). Thermal diffusivity (α) correction improved the accuracy. The spreadsheet was designed to divide a surimi seafood package into small parts. Then, for every part, temperatures at different pasteurization times, from the beginning to the end, were calculated. This resulted in a numerical matrix that showed all the temperatures for all the parts (from the surface to the "cold spot") during the entire heating treatment. The matrix was displayed as a

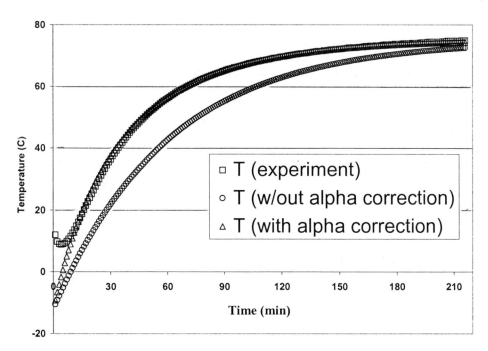

Figure 1. Comparison of temperature predictions at the cold spot by model with and without alpha correction with experimental temperature values

surface response chart (Figure 2) and may be changed by various inputs (i.e., package size, water temperature, pasteurization time and initial temperature of a sample). The program also calculates an absolute time-temperature exposure. The absolute time-temperature exposure was obtained by integration of the volume under the 3-D surface. Division of the absolute time-temperature exposure by the package size removed the dimension factor, resulting in relative time-temperature exposure. Therefore, the relative time-temperature exposure characterizes the thermal exposure of a sample (regardless of its dimension) over the pasteurization period. It may be used as a cumulative value that incorporates all pasteurization settings (sample size, water temperature, pasteurization time, and initial temperature) to compare various surimi seafood samples that were heat-processed under various thermal conditions. The relative time-temperature exposure was used in the color and texture models.

Figure 2 is an example of the temperature predictions for a sample (initial temperature 8°C, heated for 30 min at 90°C). The sample's smallest dimension was 16 mm. The change of internal temperature across the whole sample during the entire heating can be monitored. Additionally, it is possible to determine the holding time of the cold spot once it reaches the desired temperature. If the sample size was changed, for example, for a bigger package, it might be expected that the cold spot of the bigger package may not be sufficiently heated to kill bacteria. In other words, if the size of a sample was changed for a bigger package without extending heating time, then the sample may possess microbial danger. To compensate for the size difference, it would be required to increase the heating time. This kind of pasteurization optimization may be performed using the temperature prediction model.

Pasteurization is a very dynamic system because the sample temperature changes continuously from the surface towards the center during the heating period. To determine the changes of color and texture resulting from thermal processing, the temperature prediction model was used. The model calculates relative time-temperature exposure of a whole sample at given process settings (variables). It also calculates the relative time-temperature exposure for every part in the package, into which, the sample is divided by the model. The relative time-temperature exposure was related to the change of color and texture caused by thermal processing. It resulted in a relationship between relative time-temperature exposure and the change of color and texture. Based on that relationship, a model for color and texture changes was constructed. The relationships for L*, a* and b* values are shown as Eq. 10, 11, and 12, respectively. The relationship for peak load, elongation to break, and energy to break are shown as Eq. 13, 14, and 15, respectively.

$$L* = 7e^{-6}TTE^2 + 0.0041TTE - 0.0051 \qquad \text{(Eq. 10)}$$

$$a* = 4e^{-4}TTE^2 - 0.1395TTE + 0.9617 \qquad \text{(Eq. 11)}$$

$$b* = 0.2195TTE - 1.3958 \qquad \text{(Eq. 12)}$$

$$PkLd = 8e^{-5}TTE^2 - 0.0581TTE - 0.9029 \qquad \text{(Eq. 13)}$$

L*, a*, b* - tristimulus color values
TTE – relative time temperature exposure
Pk Ld – peak load

By changing the settings, the model first calculates relative time-temperature exposure and then color or texture changes (Eq. 10-15). A numerical matrix representing color and texture is displayed as a 3-D surface response (Figure 3 and 4). Whenever the pasteurization settings are changed, the model repeats the procedure and displays a new surface response accordingly.

Color browning during surimi seafood pasteurization was proven. Whiteness II distinguished browning better than whiteness I. Park (1994) drew the same conclusion. The most pronounced change was increased b* value, resulting in a more yellow hue. Shie and Park (1999) studied color changes during heating and reported similar results for b*. A model for b* prediction is shown in Figure 3. The surface of the product receives the most heat during thermal processing. Therefore, the changes of color are most profound at the surface. The model allows monitoring color changes close to surface of the product. From a marketing standpoint it is important to be able to design pasteurization to minimize detrimental color changes at surface of the product since this is the part of the product seen and judged by consumers.

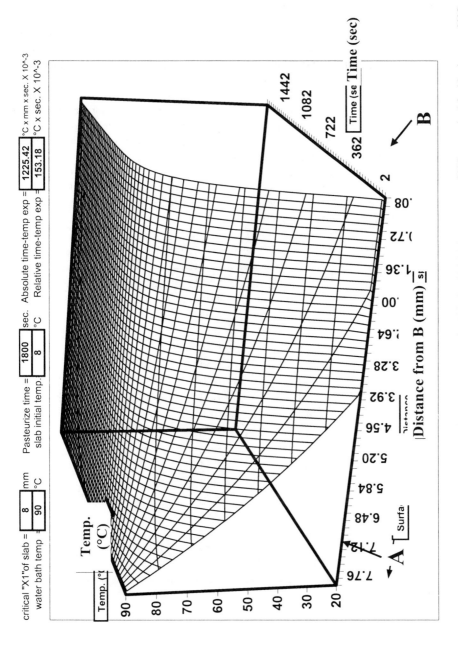

Figure 2. Temperature prediction model for 16 mm surimi seafood block with initial temperature of 8°C pasteurized for 30 minutes at 90°C. A: surface; B: cold spot

The same pasteurization settings (Figure 2) were used to obtain predictions for the b* value. The a* value changed least (data not shown). However, it first decreased, resulting in a green hue, then the a* value shifted towards positive value, which added a slight red hue. The L* value increased slightly, causing the sample to be lighter, which additionally highlighted the yellow hue from the b* change. The color values confirmed the results obtained by Shie and Park (1999). The surimi seafood samples used in our studies were not pasteurized, but they were heated during production for less than 1 min providing enough heat to develop initial texture for sheet formation. Therefore, as a result of starch gelation, the L* value increased upon additional heat applied during pasteurization (Charley, 1982). The L* value increase seems to play an important role since it does not change the color directly, however, it renders the color changes (a* and b* values) more apparent, and therefore more striking to the consumer.

Initial texture deteriorated during pasteurization (Table 1). Texture deterioration was demonstrated by a peak load decrease, which resulted in softer meat. Bertak and Karahadian (1995) concluded that excessive heating reduces firmness. The model for peak load prediction is presented in Figure 4.

Table 1. Texture of samples heated at 85°C for various times.

Time (min)	Peak load (g)
Control*	732
6	654
12	647
18	658
24	658
30	670
36	663
42	656
48	673
54	662
60	654
66	653
72	649
78	645
84	637
90	648

* Control sample, cooked as a thin sheet for less than a minute before formed into sticks, did not receive additional thermal heat.

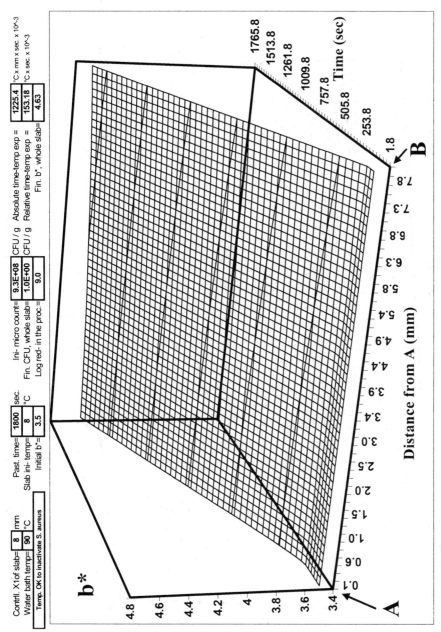

Figure 3. Model for b* prediction for 8 mm surimi seafood block with initial temperature of 8°C pasteurized for 30 minutes at 90°C. A: cold spot; B: surface

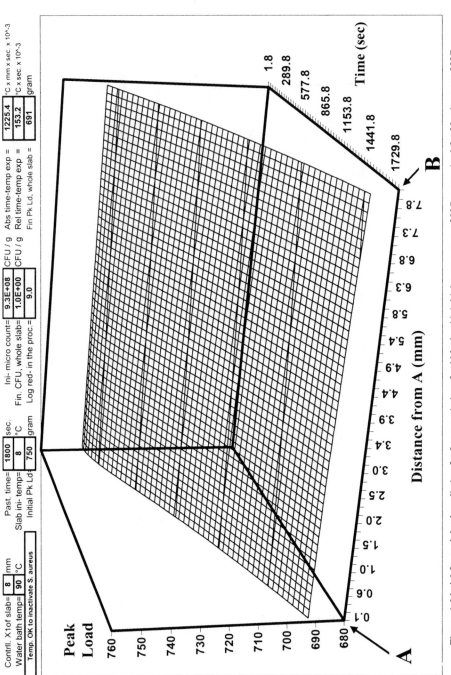

Figure 4. Model for peak load prediction for 8 mm surimi seafood block with initial temperature of 8°C pasteurized for 30 minutes at 90°C. A.: cold spot; B: surface

4.   CONCLUSIONS

The temperature across a surimi seafood package was predicted during thermal treatment. The model predicts the temperature at the cold spot, as well as at the other parts towards the surface. Temperature prediction at the cold spot is particularly important for proper thermal processing, whereas, the temperature gradient across the slab is important for determining changes in the physical properties. The models may be applied for pasteurization optimization or new process design in industry.

5.   REFERENCES

Ashrae, 1972, 1967, *Handbook of Fundamentals*, ASHRAE, New York.

Bertak, J. A., and Karahadian, C., 1995, Surimi–based imitation crab characteristics affected by heating method and end point temperature, *J. Food Sci.,* **60**:292-296.

Charley, H., 1982, *Food Science, 2nd ed.*, John Wiley and Sons, New York.

Carslaw, H. S., and Jaeger, J. C., 1959, *Conduction of Heat in Solids, 2nd ed.*, Oxford University Press, London.

Francis, F. J., 1994, Color analysis, in: *Food Analysis*, S. S. Nielsen, ed., Aspen Publishers, Inc., p. 599.

Hsu, S. T., 1963, *Engineering heat transfer*, Van Nostrand Reinhold, New York.

Kebede, E., Mannheim, C. H., and Miltz, J., 1996, Heat penetration and quality preservation during thermal treatment in plastic trays and metal cans, *J. Food Eng.,* **30**:109-115.

Kim, J. M., and Lee, C. M., 1987, Effect of starch on textural properties of surimi gel, *J. Food Sci.,* **52**:722-725.

Lanier, T.C., 1986, Functional properties of surimi, *Food Technol.,* **40(3)**:107-124.

McAdams, W. H., 1954, *Heat transmission, 3rd ed.*, McGraw-Hill Book Co., New York.

Park, J. W., 2000, *Surimi and Surimi Seafood*, Marcel Dekker, Inc., New York.

Park, J. W., 1995, Surimi gel colors affected by moisture content and physical conditions, *J. Food Sci.,* **60**: 15-18.

Park, J. W., 1994, Functional protein additives in surimi gels, *J. Food Sci.,* **59(3)**:525-527.

Rippen, T. E., and Hackney, C. R., 1992, Pasteurization of seafood: potential for shelf-life extension and pathogen control, *Food Technol.,* **46(12)**:88-94.

Rippen, T. E., Hackney, C. R., Ward, D. R., Martin, R. E., and Croonenberghs, R., 1993, Seafood pasteurization and minimal processing manual, *Virginia cooperative extension publication*, Hampton, Virginia.

Sadeghi, F., and Swartzel, K. R., 1990, Generating kinetic data for use in design and evaluation of high temperature food processing syatems, *J. Food Sci.,* **55**:851-853

Shie, J. S., and Park, J. W., 1999, Physical characteristics of surimi seafood as affected by thermal processing conditions, *J. Food Sci.,* **64**:287-290.

Su, A., Kolbe, E., and Park, J. W., 1999, A model of heat transfer coefficients over steam-cooked surimi paste., *J. Aq. Food Prod. Tech.,* **8(3)**:39-53.

Toledo R. T., 1999, *Fundamentals of food engineering*, Aspen Publishers, Inc., Gaithersburg.

Ward, D. R., and Price, J. R., 1992, Food microbiology: exact use of inexact science, in: *The NFI Green Book*, vol. 3,. National Fisheries Institute, Washington, DC, pp. 34-42.

THE CHEMISTRY OF QUALITY ENHANCEMENT IN LOW-VALUE FISH

Nazlin K. Howell[*]

1. INTRODUCTION

There is major concern, globally, regarding the overfishing of popular lean fish which leads to depletion of fish stocks. In contrast, there are abundant supplies of mainly pelagic fatty fish such as mackerel, herring, sardine and tuna in the Northern Hemisphere. In addition horse mackerel and several other species found in the Southern hemisphere are underutilized as they tend to be small, bony and deteriorate quickly. To utilize low-value species, generally by-catch which are discarded, requires an understanding of their physicochemical properties and additional processing to improve their organoleptic and keeping quality.

1.1. Beneficial Aspects of Fatty Fish

Fish is an excellent source of protein. In many countries, especially developing countries, the average diet of low income groups lacks adequate protein and the increased need for fish as a protein source can be met by efficient utilization of existing catches particularly of small pelagic fish species and demersal trash fish.

In addition, although a high fat intake is generally associated with atherosclerosis, and with colon and breast cancer (Doll and Peto,1981; Kinlen, 1983), there is growing evidence that long chain polyunsaturated fatty acids from fatty fish and fish oils help reduce coronary heart disease and lower blood cholesterol (Harris, 1989).

Fatty fish contain around 3-20% lipids which are composed mainly of triacylglycerols (75%) and phospholipids (25%). About 8-10 major fatty acids are present in fish oil including saturated acids (myristic, palmitic and stearic); monounsaturated acids (palmitoleic, oleic, 11-eicosenoic and 11-docosenoic) and

[*] Nazlin. K. Howell, School of Biomedical & Life Sciences, University of Surrey, Guildford, GU2 7XH, U.K.

Quality of Fresh and Processed Foods, edited by Shahidi et al.
Kluwer Academic/Plenum Publishers, 2004.

polyunsaturated fatty acids (docosahexaenoic acid (DHA) and eicosapentaenoic acid (EPA)). Fatty fish such as mackerel (*Scomber scombrus*) contain about 70% unsaturated fatty acids of which 30% are polyunsaturated (PUFA) with four to six double bonds (Ackman, 1990). In particular, fish oils are rich in the polyunsaturated fatty acids DHA (22:6ω3) and EPA (20:5ω3).

1.2. Problems Associated with Fatty Fish

Because of the large number of double bonds present, the long chain PUFA are susceptible to lipid oxidation. The lipid oxidation products are responsible for rancidity in the fish and give rise to off-flavors and toxic compounds including mutagens (Simic and Karel, 1980; Pryor, 1976); carcinogenic fatty acid hydroperoxides; cholesterol hydroperoxides (Bischoff, 1969), endoperoxides, cholesterol and fatty acid epoxides (Bischoff, 1969; Petrakis *et al.*, 1981; Imai, 1980); enals and other aldehydes; and alkoxy radicals and hydroperoxide (Levin, 1982). Thus it is vitally important to stabilize the PUFA with the aid of antioxidants as well as processing.

In addition, the lipid oxidation products, including free radicals, can be transferred to fish proteins and the resultant protein radicals undergo aggregation leading to protein denaturation and texture changes (Saeed *et al.*, 1999b, Howell and Saeed, 1999a). Lipid oxidation and the subsequent transfer of free radicals from the lipids to proteins are described below in relation to frozen stored mackerel.

2.0. LIPID OXIDATION IN FISH

2.1. Lipid Oxidation Mechanism

Lipid peroxidation occurs in the presence of oxygen and transition metal ions or enzymes, usually in three stages of initiation, propagation and termination as described below:

Initiation stage: When a hydrogen atom is abstracted from a methylene group   (-CH$_2$-) in the polyunsaturated fatty acid by a reactive species such as a hydroxyl radical ($^{.}$OH), it leaves behind an unpaired electron on the carbon (-$^{.}$CH- or lipid radical). The carbon radical is stabilized to form a conjugated diene which can combine with oxygen to form a peroxy radical LOO$^{.}$ or LO$_2^{.}$.

$$LH \longrightarrow L^{.} + H^{.}$$

Propagation stage: The peroxy free radicals abstract and combine with H from another lipid molecule to form lipid hydroperoxides and cyclic peroxides. This is an autocatalytic chain reaction with lipid hydroperoxides being formed at different points on the carbon chain; the peroxidation of arachidonic acid for example is reported to afford six lipid hydroperoxides.

$$L^{.} + O_2 \longrightarrow LO_2^{.}$$
$$LO_2^{.} + LH \longrightarrow LOOH + L^{.}$$

Secondary products: Hydroperoxides are the primary molecular products which are unstable and degrade in the presence of transition metal ions including traces of iron and copper salts. The metal ions can cause fission of an O-O bond to form an alkoxy radical RO$^.$ as well as peroxy radicals RO$_2$$^.$. Thiols or other reducing agents for example ascorbic acid, reduce O$_2$ to superoxide radical anion (O$_2$$^.$ $^-$), which then dismutates to H$_2$O$_2$ or reduces Fe^{3+} to Fe^{2+}. The Fe^{2+} reacts with H$_2$O$_2$ via the Fenton reaction to produce the hydroxyl radical ($^.$OH) and can initiate further chain reactions. Iron, present in the fish myoglobin, reacts with lipid peroxides and can generate a wide range of products which affect flavor, including *pentane*, from linoleic acid and arachidonic acid and *ethane and ethylene* by a similar β-scission reaction from linolenic acid in the presence of Fe^{2+}. Various secondary products are formed including hydroxy-fatty acids, epoxides and scission products such as aldehydes (including malondialdehyde), ketones and lactones, many of which are toxic (Halliwell and Gutteridge, 1995).

Termination stage: The free radicals produced from lipids (LH) can combine with each other and end the chain reaction.

$$L^. + L^. \longrightarrow L - L$$
$$nLO_2^. \longrightarrow (LO_2)n$$
$$LO_2^. + L^. \longrightarrow LO_2L$$

Alternatively, the free radicals can readily combine with proteins (P)

$$HO^. + PH \longrightarrow P^. + H_2O$$
$$P^. + P^. \longrightarrow P - P$$
$$P^. + P - P \longrightarrow P - P - P^.$$

The reaction can also be terminated by antioxidants.

2.2. Effect of Antioxidants

Antioxidants can inhibit lipid oxidation through competitive binding of active forms of oxygen involved in the initiation step of oxidation. Alternatively, antioxidants can react with fatty acid peroxyl free radicals (LOO$^.$) to form stable antioxidant-radicals (TO$^.$) which are either too unreactive for further reactions or form non-radical products. Propagation steps may be retarded by destroying or inhibiting catalysts, or by stabilizing hydroperoxides (Halliwell, 1994).

$$LOO^. + TOH \longrightarrow LOOH + TO^.$$

The most common synthetic food antioxidants are those which contain phenolic groups such as gallic acid esters, butylated hydroxyanisole (BHA), butylated hydroxytoluene (BHT) and tertiary-butylhydroxyquinone (TBHQ). The effectiveness of phenolic antioxidants depends on the resonance stabilization of the phenoxy radicals, this is determined by the substitution at the ortho and para positions on the aromatic ring and by the size of the substituting group (Shahidi *et al.*, 1992). The presence of carbonylic and carboxylic groups in phenolic compounds inhibits oxidative rancidity by metal chelation (Hudson and Lewis, 1983).

Due to increasing concern over the potential toxicity and safety of some phenolic antioxidants (Ito *et al.*, 1986; Thompson and Moldeus, 1988) the replacement of synthetic antioxidants by 'safe, natural' antioxidants such as vitamins E and C, flavonoids and other plant phenolics has received considerable attention. Vitamin E compounds (tocopherols and tocotrienols) are reported to effectively inhibit lipid oxidation in foods and biological systems (Burton and Traber, 1990). In addition, the protective role of ascorbic acid in biological systems via free radical reactions has been studied. In many systems, including lipid oxidation, vitamins E and C act synergistically (Badii and Howell, 2002). Vitamin E is lipophilic and is considered to be the primary antioxidant. Vitamin C reacts with the vitamin E radical to regenerate vitamin E, and the resulting ascorbic acid radical is reduced back to vitamin C by NADH (Packer *et al.*, 1979). The chemistry and antioxidant properties of tocopherol and ascorbic acid have been summarized by Howell and Saeed (1999b).

2.3. Enzymic Peroxidation

Enzymes such as lipoxygenase, cycloxygenase and peroxidase promote th peroxidation of fatty acids to give hydroperoxides and endoperoxides that are stereospecific. For example, lipoxygenase (EC 1.13.11.12) contains one iron atom per molecule and catalyzes peroxidation of polyunsaturated fatty acids to primary and secondary oxidation products (Hildebrand, 1989; German and Crevelling, 1990) by abstracting a hydrogen atom from the unsaturated fatty acid; the bonds rearrange and oxygen is added to form hydroperoxides. Enymic peroxidation is more specific and more complex than the formation of non-enzymic hydroperoxides described above, and the hydroperoxide products may interact further with the enzyme. The action of lipoxygenase may produce co-oxidation of other materials such as carotenoid pigments or proteins. The isolation of the crude lipoxygenase in our laboratory has been studied in relation to its action on the oxidation of fish lipids from Atlantic mackerel *Scomber scombrus* (Saeed and Howell, 2001).

3.0. ANALYSIS OF LIPID OXIDATION PRODUCTS

Lipid oxidation products are so varied and transient that it is necessary to use a number of the most advanced techniques available to monitor the products systematically and to characterize them reliably. Most studies examine final lipid oxidation products such as malondialdehyde and peroxides using traditional methods which lack sensitivity and specificity.

Hydroperoxides from oxidized polyunsaturated fatty acids form conjugated dienes as the double bonds rearrange. Fatty acid hydroperoxide have previously been measured by absorbance at 234 nm (Corongiu and Banni, 1994), iodometric titration (Jessup *et al.*, 1994), xylenol orange reactivity and glutathione oxidation (O'Gara *et al.*, 1989) but these methods are not sensitive.

3.1. High pressure liquid chromatography (HPLC)

We have developed an HPLC method for identifying breakdown products of hydroperoxides (Saeed and Howell, 1999a); this is a simple and very useful method for investigating the initiation, propagation and termination stages of lipid oxidation in detail.

Method: Fish lipids were extracted by the Bligh and Dyer (1959) method. Lipids were transesterified according to Schmarr *et al.,* (1996) and the hydroperoxides were purified by LC fractionation on amino-phase solid phase extractions (SPE) cartridges. The hydroperoxides are very unstable and break down to hydroxides which were separated on a reverse phase HPLC Hichrom Kromasil 100 5C18 column using acetic acid-methanol-water (0.1-65-35) solvents. The development of hydroperoxide alcohol derivatives from methyl linoleate, oxidized under UV radiation for 6 and 24 h are shown in Figure 1. The oxidation products were the 9-, 10-, 12-, and 13-hydroxylinoleate, with the 9-hydroxylinoleate (trans-cis) being the main peak.

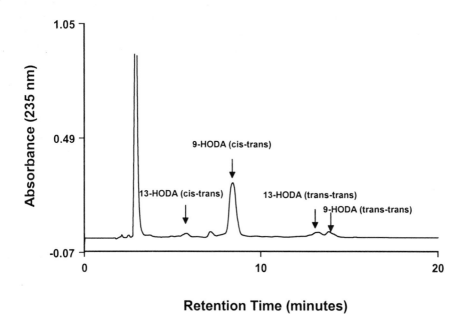

Figure 1. HPLC chromatogram of hydroxides formed in oxidized methyl linoleate HODA refers to hydroxydecadienoic acid

The HPLC method also worked well for complex lipids extracted from mackerel fish. One major peak, assigned as 13-HODA was observed in the chromatograms of oil obtained from fish stored at both -20 and -30°C. However, for mackerel stored at -20°C the peak was bigger and two extra minor peaks were present which were absent in the spectra obtained from control fish fillets stored at -30°C. In the presence of α-tocopherol, vitamin C, BHT and BHA added to the mince fish prior to storage at -20°C, amounts and types of hydroxides were fewer compared with the control fish stored at –20°C without antioxidants.

3.2. Gas chromatography-mass spectroscopy

Gas chromatography-mass spectrometry (GC-MS) chromatograms of oxidized methyl linoleate indicated three GC peaks representing isomers of molecular weight 382 and consistent with the molecular formula $C_{22}H_{42}O_3Si$, which is the alcohol derivative of the hydroperoxide spectra. The major high mass fragment had a mass to charge ratio (m/z) value of 225 (Saeed and Howell, 1999a) as indicated by ion fragmentation.

3.3. Nuclear magnetic resonance spectroscopy

Hydroxides produced from methyl linoleate oxidized by UV radiation were also confirmed by ^{13}C nuclear magnetic resonance (NMR) spectroscopy indicating a peak at 87 ppm but the peak was only observed for samples oxidised for longer periods (24 h and 48 h) (Figure 2). Although ^{13}C NMR appears less sensitive than HPLC for measuring lipid oxidation products generated in the initial stages of oxidation, it is a useful technique for measuring the high concentrations produced in the later stages (Saeed and Howell, 1999a).

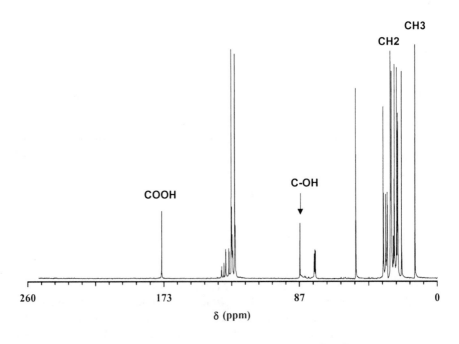

Figure 2. NMR spectra of hydroxides (87 ppm) produced from oxidized fish oil

4.0. PROTEIN-LIPID INTERACTIONS

The production of free radicals and other oxidation products can severely damage proteins and DNA, cytochrome C and hemoglobin (Roubal and Tappel, 1966). The interactions can form covalent bonds which may involve hydroperoxides, saturated and unsaturated aldehydes, ketones, ketols, diketones and epoxides in the oxidized lipid; these can react with amines, thiols, disulfides and phenolic groups in the protein (Gardner, 1979). Non-covalent hydrogen bonds involving hydroxyls and carboxyl groups in proteins and lipids may also be formed (Pokorny, 1987).

Further, protein-lipid interactions result in the unavailability of essential amino acids such as cysteine, lysine, histidine and methionine. These interactions may also produce new toxic products. In addition free radicals can induce cross-linking of proteins and affect the nutritional and functional properties including texture.

An increase in fluorescence has been reported which is attributed to the formation of certain oxidized lipid-protein complexes; the oxidation reaction of linoleate and myosin in frozen salmon yields fluorescent compounds containing phosphorous and C=N functional groups (Braddock and Dugan, 1973). It has also been reported that aldehydes can react with -SH groups and dialdehydes such as malondialdehyde that can attack amino groups to form intramolecular cross-links as well as cross-links between different proteins. The aldehyde-amine Schiff bases can result in non-enzymic browning pigments (Gardner, 1983). In addition, the interaction of free radicals with specific groups on amino acids may be involved; for example alkoxy radicals can attack tryptophan and cysteine residues (Schaich, 1980, Halliwell and Gutteridge, 1995).

4.1. Electron paramagnetic resonance (EPR) spectroscopy

EPR spectroscopy provides direct measurement of free radicals formed and identifying the atoms on which the radical resides. The g value provides enough information to distinguish between the carbon, nitrogen and sulfur centered radical. We have employed both EPR spectroscopy and fluorescence spectroscopies to provide direct evidence of the transfer of free radicals from the oxidized lipid to the amino acid or protein carbon or sulfhydryl groups, followed by protein cross-linking (Saeed *et al.*, 1999b).

Method. An emulsion consisting of oxidized methyl linoleate (ML) or extracted oxidized fish lipid and either amino acids (arginine and lysine); or proteins (lysozyme, ovalbumin and myosin), was freeze-dried. Control systems of either amino acids or proteins in the absence of lipid were similarly prepared. The effect of antioxidants was also tested using BHT (200 ppm), vitamin C (500 ppm), and vitamin E (500 ppm). Samples were analyzed periodically in a Jeol RE IX X-band EPR spectrometer with 100 kHz modulation. First derivative spectra were recorded everyday for the first week and once a week subsequently for five weeks. Manganese oxide was used as a reference marker to calculate the g value (1.981, 2.034).[20]

Results. A tranfer of radicals from the oxidized lipid to the proteins was shown by a strong central singlet signal by EPR spectroscopy and assigned to the carbon radical (g value 2.0021) Downfield shoulders were also observed which were due to the radical on sulfydryl groups (g value 2.014). With the addition of antioxidants a reduction in signal intensity of free radicals was observed (Figure 3).

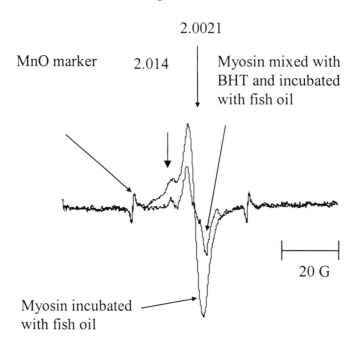

Figure 3. Electron paramagnetic resonance spectra of mackerel myosin incubated with oxidized fish oil and treated with or without vitamin E.

4.2. Fluorescence spectroscopy

The above changes in the proteins were accompanied by an increase in fluorescence indicating the formation of cross-links between the individual amino acids as well as conformational changes in the proteins (Figure 4) (Saeed *et al.*, 1999b). Recent studies in the author's laboratory have confirmed conformational changes induced by the lipids on myosin by differential scanning calorimetry (Badii and Howell, submitted) and Raman spectroscopy. Antioxidants such as BHT, BHA, tocopherol and ascorbic acid inhibited the development of both the free radical peak and fluorescence when added to the proteins prior to incubation with oxidized lipids.

4.3. Effect on texture

In whole mackerel fillets stored at -10°C, the extracted lipid was oxidized, the proteins were more denatured and the muscle was tougher compared with the control stored at -30°C as indicated by rheological analysis (Saeed and Howell, 2002a,b). With the addition of antioxidants to fish fillets the elastic modulus decreased thus indicating the importance of lipid oxidation on protein denaturation and subsequent muscle toughening.

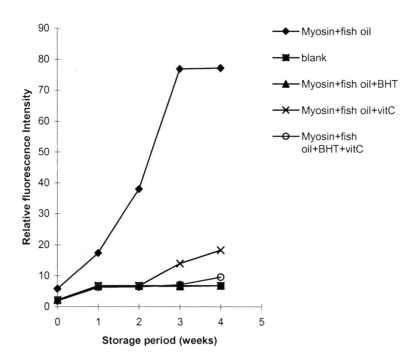

Figure 4. Fluorescence formation in myosin incubated with oxidized fish oil and treated either with butylated hydroxytoluene (BHT) or vitamin C (Vit C) or a combination of BHT and vitamin C

5.0. CONCLUSIONS

Lipid oxidation in fish including pelagic fatty fish can be monitored successfully and specifically using HPLC combined with NMR and GC-MS techniques. In addition, it is very apparent from numerous studies on both fatty and lean fish, that lipid oxidation products can alter proteins through the transfer of free radicals from oxidized lipids which results in cross-linking of protein. These changes can be monitored by EPR and fluorescence spectroscopy respectively, as well as Raman spectroscopy and differntial scanning calorimetry. The damage caused by lipid oxidation products in whole fish muscle can be minimized by the addition of antioxidants particularly a combination of water soluble ascorbic acid with the lipophilic α-tocopherol. These measures may alleviate the problems associated with many low-value fish species.

6.0. REFERENCES

Ackman, R. G., and Eaton, C. A., 1971, Mackerel lipids and fatty acids, *Can. Inst. Food Sci. Technol. J.*, **4**:169-172.

Ackman, R. G., 1990, Seafood lipids and fatty acids, *Food Rev. Int.*, **6**:617-646.

Badii, F and Howell, N. K., 2002, Effect of antioxidants, citrate and cryoprotectants on protein solubility and texture of frozen cod (*Gadus morhua*), *J. Agric. Food Chem.*, **50**:2053-2061.

Bischoff, F., 1969, Carcinogenic effect of steriods, *Adv. Lipid Res.*, **7**:165.

Bligh, E. G. and Dyer, W. J., 1959, Lipid extraction from fish muscle, *Can. J. Biochem. Physiol.*, **37**:911-913.

Braddock, R. J. and Dugan, L. R.,1973, Reaction of autoxidizing linoleate with Coho salmon myosin, *J. Am. Oil Chem. Soc.*, **50**:343-346.

Burton, G. W. and Traber, M. G., 1990, Vitamin E: Antioxidant activity, biokinetics and bioavaiability, *Ann. Rev. Nutr.*, **10**:357-382.

Corongiu, F. P. and Banni, S., 1994, A sensetive electrochemical method for quantitative hydroperoxide determination, *Methods in Enzymology*, **233**:303-310.

Doll, R. and Peto, J., 1981, The causes of cancer of quantitative estimated of avoidable risks of cancer in the United States today, *J. Nat. Cancer Inst.*, **66**:1192.

Gardner, H. W., 1979, Lipid hydroperoxide reaction with proteins and amino acids, A review, *J. Agric. Food Chem.*, **27**:220-229.

Gardner, H. W., 1983, Effects of lipid hydroperoxides on food components, in: Xenobiotics in food and feeds (American Chemical Society Ed.), pp. 63-84.

German, J. B. and Crevelling, R. K., 1990, Identification and characterisation of a 15-lipoxygenase from fish gills, *J. Agric. Food Chem.*, **38**:2144-2147.

Halliwell, B., 1994, Free radicals and antioxidants: A personal view, *Nutrition Reviews*, **52**:253-265.

Halliwell, B and Gutteridge, J. M. C., 1995, Free radicals in Biology and Medicine, Second edition, Clarendon Press, Oxford.

Harris, W. S., 1989, Fish oils and plasma lipid and lipoprotein metabolism in humans, a critical review, *J. Lipid Res.*, **30**:785-807.

Hilderbrand, D. F.,1989, Lipoxygenases, *Physiologia Plantarum*, **76**:249-253.

Howell, N. K. and Saeed, S., 1999a, The application of electron spin resonance spectroscopy to the detection and transfer of free radicals in protein-lipid systems, in: Applications of Magnetic Resonance in Food Science, P. S. Belton, B. P. Hills and G. A. Webb, ed., The Royal Society of Chemistry, Cambridge, UK. pp. 133-143.

Howell, N. K. and Saeed, S., 1999b, The effect of antioxidants on the production of lipid oxidation products and transfer of free radicals in oxidised lipid-protein systems, in: Antioxidants in Human Health and Disease, T. K. Basu, N. J Temple and M. L. Garg, ed., CAB International, Oxford, UK. pp. 43-54.

Hudson, B. J. and Lewis, J. I.,1983, Polyhydroxy flavonoid antioxidants for edible oils: structural criteria for activity, *Food Chem.*, **10**:47-51.

Imai, H., Weithesser, N. T., Subramosyam, V., Lequesne, P., Suoway, W. A. H. and Kanisawa, M., 1980, Angiotoxicity of oxygenated sterols and possible precursors, *Science*, **207**:651.

Jessup, W., Dean, R. T. and Gebicki, J. M., 1994, Iodometric determination of hydroperoxides in lipids and proteins, *Methods in Enzymology*, **233**:289-303.

Kinlen, L. J., 1983, Fat and cancer, *Brit. Med. J.*, **286**:1081.

Levin, D. E., Hollstein, M., Christman, M. F., Schweirs, E. and Ames, B. N., 1982, A new salmonella tester strain (TA102) with A.T.Base-pairs at the site of mutation defects oxidative mutagens, *Proc. Nat. Acad. Sci. U.S.A.*, **79**:7445-7449.

O'Gara, C. Y. Maddipati, K. R. and Marnett, L. J., 1989, Detection of conjugated dienes by 2nd derivative ultraviolet spectrophotometry, *Methods in Enzymology*, **2**:295.

Packer, J. E., Slater, T. F. and Willison, R. L., 1979, Direct observation of a free radical interaction between vitamin E and vitamin C, *Nature*, **278**:737-738.

Petrakis, N. L., Gruevenke, L. D. and Craig, T. C., 1981, Cholesterol and cholesterol epoxides in nipple aspirates of human breast fluid, *Cancer Research*, **41**:2563

Pokorny, J., 1987, Major Factors Affecting the Autoxidation of lipids, in: Autoxidation of unsaturated lipids, H. W. S. Chan, ed., Academic Press, London, pp. 141-206.

Pryor, W.A., 1976, Free Radicals In Biology, Academic Press, New York.

Roubal, W. T. and Tappel, A. L., 1966, Damage to proteins, enzymes and amino acids by peroxidizing lipids. *Arch. Bioch. Biophys*, **5**:113-117.

Saeed, S. and Howell, N.K., 1999a, High performance liquid chromatography (HPLC) and spectroscopic studies on fish oil oxidation products extracted from frozen Atlantic mackerel, *J. Am. Oil Chem. Soc.*, **76**:391-397.

Saeed, S, Fawthrop, S and Howell, N. K., 1999b, Electron spin resonance (ESR) studies on free radical transfer in fish lipid-protein interactions, *J. Sci.Food Agric.*, **79**:1809-1816.

Saeed, S. and Howell, N. K., 2001, 12-Lipoxygenase activity in the muscle tissue of Atlantic mackerel (*Scomber scombrus*) and its prevention by antioxidants, *J. Sci. Food Agric.*, **81**:745-750.

Saeed, S. and Howell, N. K., 2002a, Effect of lipid oxidation and frozen storage on muscle proteins of Atantic mackerel (*Scomber scombrus*), *J. Sci. Food Agric.*, **82**:579-586.

Saeed, S. and Howell, N. K., 2002b, Biochemical and rheological changes in frozen Atlantic mackerel (*Scomber scombrus*), Unpublished results.

Schaich, K. M, 1980, Free radical initiation in proteins and amino acids by ionizing and ultraviolet radiation and lipid oxidation, CRC *Crit. Rev. Food Sci. Nutr.*, **13**:131-159.

Schmarr, H. G., Gross, H. B. and Shabamoto, T., 1996, Analysis of polar cholesterol oxidation products, evaluation of a new method involving transesterification. Solid-phase extraction and gas chromatography, *J. Agric. Food Chem.*, **44**:512-517.

Shahidi, F. and Wanasundara, P. K. J. P. D., 1992, Phenolic antioxidants, *Crit. Rev. Food Sci. Nutr.*, **32**:67-103.

Simic, M. G. and Karel, M., 1980, Autoxidation in Food and Biological Systems, Plenum Press, New York.

Thompson, M. and Moldeus, P., 1988, Cytotoxicity of butylated hydroxyanisole butylated hydroxytoluene in isolated rat hepatocytes, *Biochem. Pharmacol.*, **37**:2201-2207.

7.0. ACKNOWLEDGEMENT

This research project was financed by The Commission of the European Communities within the STD Framework Contract No. TS3*-CT94-0340 awarded to and co-ordinated by Dr. Nazlin K.Howell.

THE INTERACTION OF DISULFIDE FLAVOR COMPOUNDS WITH PROTEINS IN MODEL SYSTEMS

Donald S. Mottram[1], Rachel L. Adams[1], Jane K. Parker[1], and Helen M. Brown[2]

1. INTRODUCTION

Sulfur-containing compounds are extremely important in cooked foods for roast, savory, meaty, and coffee-like aromas and flavors (Gasser and Grosch, 1988; Mottram *et al.*, 1998). Many have low odor threshold values and, therefore, relatively small quantities in a food product can have a major influence on the aroma. Recently it has been shown that irreversible binding of such compounds to food proteins can occur, with a possible change in flavor perception of the food. This has implications for consumer acceptance of certain food products where these compounds are used as flavorings or where they occur naturally.

When the disulfides, bis(2-methyl-3-furyl) disulfide and bis(2-furylmethyl) disulfide, were added to ovalbumin, a significant proportion of the disulfides were broken down to corresponding thiols (2-methyl-3-furanthiol and 2-furanmethanethiol, respectively) and some were lost completely (Mottram *et al.*, 1996). A similar effect was observed in a meat system (minced beef) and, in addition, small amounts of mixed disulfides, 2-methyl-3-furyl methyl disulfide and 2-furylmethyl methyl disulfide, were formed (Mottram *et al.*, 1998). Aqueous blanks, used as controls, and aqueous solutions of maltodextrin showed no breakdown of the disulfides. Further experiments using casein, which is less rich in cysteine residues, showed that there was very little conversion of disulfides to thiols. It was suggested that that covalent interaction between the protein and the disulfides, or the corresponding thiol, had occurred probably through disulfide interchange reactions as described by Whitesides (1983).

This contribution reports on investigations into the effect of heat and pH on the binding of a number of different disulfides to both native and heat denatured protein systems.

[1] The University of Reading, School of Food Biosciences, Whiteknights, Reading RG6 6AP, United Kingdom
[2] Campden and Chorleywood Food Research Association, Chipping Campden, Gloucester GL55 6LD, U.K.

Quality of Fresh and Processed Foods, edited by Shahidi et al.
Kluwer Academic/Plenum Publishers, 2004.

2. EXPERIMENTAL DETAILS

Aliquots (1.00 g ± 0.05 g) of an aqueous solution containing 10 mg/L of each of diethyl, dipropyl, dibutyl, diallyl, 2-furylmethyl methyl disulfides, were added to 1% aqueous solutions of different soluble proteins. A solution containing a similar concentration of a reference standard (propyl propanoate) was added and the mixture was made up to 100 mL to give a concentration of 100 µg/L of each compound. A control system in water was also prepared. The proteins used were ovalbumin, β-lactoglobulin and lysozyme.

Other mixtures were prepared in which the pH was adjusted to 8.0, 5.5 or 2.0 using 0.1M NaOH or 0.1M HCl. The effect of the denaturation was examined by heating solutions of the proteins in closed 100 mL glass bottles in a boiling water bath for 10 min before adding the disulfide mixture.

The headspace volatiles above the mixtures (20 g in 40 ml screw top vials) were collected and analyzed by solid phase microextraction (SPME) using a carboxen-polydimethylsiloxane fiber (phase thickness of 75 µm, Supelco Inc.) as described elsewhere (Adams *et al.*, 2001). The fiber was exposed to the sample headspace for 30 min while the mixtures were held at 37 °C. Analyses were carried out in quadruplicate. After headspace collection the fiber was desorbed in the GC injection port (at 250°C) for 3 min.

GC-MS analyses were also carried out using a Hewlett-Packard 5890/5972 GC-MS using the same column, fiber desorption and operating conditions as those used in the GC analysis. Mass spectra were recorded in the electron impact mode (ionization voltage 70 eV, source temperature of 200 °C, scan range 29-400, scan time 0.69 s). Thiols and disulfides were identified by comparison of retention times and mass spectra with those of authentic compounds.

Sulfhydryl groups in the proteins were determined using Ellman's reagent (3 mM 5,5'-dithiobis(2-nitrobenzoic acid) in 0.1M phosphate buffer at pH 7.3) which was reacted with protein solutions (1%) at room temperature. Some of the protein solutions were subjected to heat (100°C for 10 min), whilst others were used in their native state. The reagent (0.2 mL) was added to 4 mL aliquots of the protein solution, vortexed for 1 min and left for 10 min for the color to develop. The absorbance was then read at 412 nm. The protein solutions were also measured at 280 nm which allowed calculation of the molar concentration of protein (ε ovalbumin = 27306 M^{-1} cm^{-1}; ε b-lactoglobulin = 40682 M^{-1} cm^{-1}; ε lysozyme = 116290 M^{-1} cm^{-1}) (Ellman, 1959).

3. RESULTS AND DISCUSSION

Figure 1 shows relative changes in disulfides in unheated protein systems compared with a water blank. Propyl propanoate was used as an internal standard. This did not bind covalently to the protein, but was used to compensate for changes in the headspace concentration due to other factors, such as hydrophobic bonding or solution viscosity.

The pH of the ovalbumin solution was 6.7, and at this pH there was very little difference in recovery of disulfides between the ovalbumin system and the aqueous blank. However, when the pH was adjusted to 8, using NaOH, greater losses were observed.

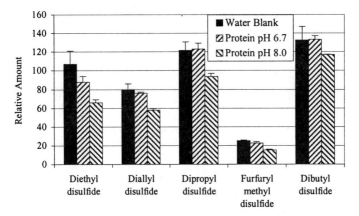

Figure 1. Relative quantities of disulfides in headspace above aqueous solutions of ovalbumin at two pHs compared with water blank (propyl propanoate = 100).

The effect of thermal denaturation of the ovalbumin on the recovery of disulfides from the aqueous systems is shown in Figure 2. The losses of disulfides were greater with the denatured protein than with the native ovalbumin. This indicates that more available sulfhydryl groups were present in the denatured protein than in the native state. The same effect of pH, as seen in the unheated protein, was observed in the heated systems; the systems at pH 6.7 lost fewer disulfides than the similar systems at pH 8.0. This suggests that a pH dependent reaction occurred between the disulfides and the protein systems involving a disulfide interchange reaction, initiated by the sulfhydryl groups in the protein (Figure 3). At pH 8, which is close to the pK_a of cysteine SH in a protein environment (8.80) (Damodaran, 1996), more of the sulfhydryl groups would be present as the more reactive thiolate anion.

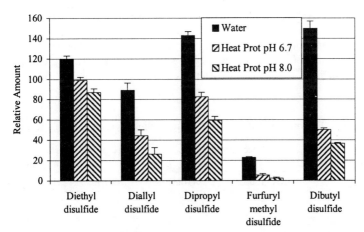

Figure 2. Relative quantities of disulfides in headspace above aqueous solutions of heat denatured ovalbumin compared with water blank (propyl propanoate = 100).

$$PSH \rightleftharpoons PS^- + H^+$$

$$PS^- + RS\text{–}SR \rightleftharpoons PS\text{–}SR + RS^-$$

$$RS^- + PS\text{–}SP \rightleftharpoons RS\text{–}SP + PS^-$$

$$RS^- + H^+ \rightleftharpoons RSH$$

Figure 3. Sulfhydryl – disulfide interchange reaction in aqueous solution. P = protein chain.

The diallyl and the 2-furylmethyl methyl disulfides showed greater reactivity towards the ovalbumin than the alkyl disulfides, presumably because the unsaturation in these molecules made the disulfide group more susceptible to nucleophilic attack.

GC-MS analyses of the headspaces from the two heated protein systems showed the presence of propanethiol, butanethiol, 2-propene-1-thiol, and 2-furanmethanethiol, clearly showing that some of the disulfides in the protein system were converted to their corresponding thiols. However, the amounts were significantly less than the amounts of the corresponding disulfides that were lost from the systems. Ethanethiol may also have been produced, however it eluted too early in the chromatogram for it to be identified under the GC conditions used.

The extent of disulfide loss from the ovalbumin systems appears to depend on the availability of sulfhydryl groups in the protein and on the extent of their ionization to the thiolate anion. Although the ovalbumin molecule has four sulfhydryl groups, in the native state the tertiary protein structure may make some or all unavailable for interchange reaction with disulfides. Available sulfhydryl groups in the native and heat denatured systems were estimated using Ellman's method (Ellman, 1959; Sedlak and Lindsay, 1968). This is based on the reaction of 5,5'-dithiobis(2-nitrobenzoic acid) with free sulfhydryl groups in which the reagent is converted to 2-nitro-5-mercaptobenzoic acid. The nitromercaptobenzoic acid anion has an intense yellow color ($\varepsilon = 13,600$ M^{-1} cm^{-1} at 412 nm) and can be used quantitatively as an indirect measure of SH groups. When reacted with Ellman's reagent, the native ovalbumin was found to have only 0.01 available sulfhydryl groups per molecule (Table 1). This suggests that, in the native state, the sulfhydryl groups of ovalbumin are buried within the tertiary structure of the protein and are unavailable for interchange reactions (Onda *et al.*, 1997). However, in the

Table 1. Determination of sulfhydryl groups in proteins using Ellman's reagent

Protein	Molecular Structure		SH groups determined using Ellman's reagent	
	SH groups	S–S groups	Native	Heat denatured
Ovalbumin	4	1	0.01	1.24
β-Lactoglobulin	1	2	0.3	0.97
Lysozyme	0	4	0.0	0.63

denatured state, Ellman's reagent showed that 1.24 sulfhydryl groups per molecule were exposed which represents a 200-fold excess of protein SH groups compared with the added disulfides. This explains why there was a greater loss of disulfides in the denatured state and supports the hypothesis that the free sulfhydryl groups initiate the disulfide exchange reaction and the binding of the flavor disulfides to the protein.

Two other proteins, β-lactoglobulin and lysozyme, were also examined. β-lactoglobulin has one SH-group and 2 disulfide bonds and Ellman's reagent indicated that 0.3 SH-groups per mole were available in the native protein whilst almost all the SH-groups were available after heat denaturation. The recoveries of the five disulfides from solutions of these proteins, as well as ovalbumin, at three pHs, 8.0, 5.5 and 2.0, were examined to confirm that availability of SH-groups and pH were important factors in the loss of disulfides. Figures 4 – 6 show the losses of the disulfides from these systems. At pH 2.0 no losses of disulfides were observed and, for ovalbumin, greater losses were shown at pH 8.0 than at pH 5.5. Overall, the results support the proposed mechanism of sulfhydryl – disulfide interaction initiated by the formation of the thiolate anion, which is suppressed at lower pH (Figure 3). In the β-lactoglobulin system, there was no difference between pH 8 and pH 5.5, suggesting that the optimum pH for the ionization of the sulfhydryl groups may be different for ovalbumin and β-lactoglobulin. With lysozyme, the only changes in the disulfides observed were small losses at pH 8.0. Lysozyme does not have any free SH-groups, and therefore can only undergo sulfhydryl – disulfide interchange reactions if the disulfide links in the protein are first cleaved.

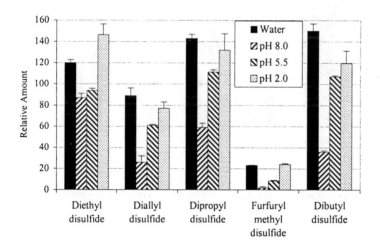

Figure 4. Relative quantities of disulfides in headspace above aqueous solutions of heat-denatured ovalbumin at three pHs compared with water blank (propyl propanoate = 100).

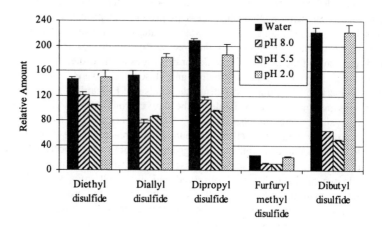

Figure 5. Relative quantities of disulfides in headspace above aqueous solutions of heat-denatured β-lactoglobulin at three pHs compared with water blank (propyl propanoate = 100).

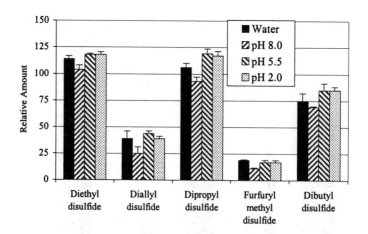

Figure 6. Relative quantities of disulfides in headspace above aqueous solutions of heat-denatured lysozyme at three pHs compared with water blank (propyl propanoate = 100).

4. CONCLUSIONS

Interchange reactions occur between proteins, containing free sulfhydryl groups, and aroma compounds, containing a disulfide link, resulting in a loss of disulfides and the formation of free thiols. The loss of disulfides appeared to be directly related to the availability of the sulfhydryl groups in the denatured protein and on the degree of ionization of the sulfhydryl group. Only small losses of the disulfides were observed in the systems containing native unheated protein, but the losses increased when the protein was heated. The extent of disulfide loss also depended on the substituents attached to the disulfide bond of the aroma compound.

5. REFERENCES

Adams, R. L., Mottram, D. S., and Parker, J. K., 2001, Flavor-protein binding: disulfide interchange reactions between ovalbumin and volatile disulfides, *J. Agric. Food Chem.*, **49**:4333-4336.

Damodaran, S., 1996, Amino acids, peptides and proteins, in *Food Chemistry*, O.R. Fennema ed., Marcel Dekker, New York, pp. 321-429.

Ellman, G. L., 1959, Tissue sulfhydryl groups. *Arch,. Biochem. Biophys.*, **82**:70-77.

Gasser, U., and Grosch, W., 1988, Identification of volatile flavour compounds with high aroma values from cooked beef, *Z. Lebensm. Unters. Forsch.*, **186**:489-494.

Mottram, D. S., Nobrega, I. C. C., and Dodson, A. T., 1998, Extraction of thiol and disulfide aroma compounds from food systems, in: *Flavor Analysis: Developments in Isolation and Determination*, C. J. Mussinan and M. J. Morello, eds., American Chemical Society, Washington, DC, pp. 78-84.

Mottram, D. S., Szauman-Szumski, C., and Dodson, A., 1996, Interaction of thiol and disulfide flavor compounds with food components, *J. Agric. Food Chem.*, **44**:2349-2351.

Onda, M., Tatsumi, E., Takahashi, N., and Hirose, M., 1997, Refolding process of ovalbumin from urea-denatured state - evidence for the involvement of nonproductive side chain interactions in an early intermediate, *J. Biol. Chem.*, **272**:3973-3979.

Sedlak, J., and Lindsay, R.H., 1968, Estimation of total, protein-bound, and nonprotein sulfhydryl groups in tissue with Ellman's reagent, *Anal. Biochem.*, **25**:192-205.

Whitesides, G. M., Houk, J., and Patterson, M. A. K., 1983, Acitvation parameters for thiolate-disulfide interchange reactions in aqueous solution, *J. Org. Chem.*, **48**:112-115.

GAS CHROMATOGRAPHY-OLFACTOMETRY ANALYSIS AND ITS IMPORTANCE IN FOOD QUALITY CONTROL

Influence of assessors' training and sampling methods on gas chromatography-olfactometry data

Saskia M. van Ruth and Jacques P. Roozen[*]

1. INTRODUCTION

Foods are composed of both volatile and non-volatile substances. Some of these compounds contribute to the flavor of foods, in which case they affect its aroma, taste, texture or mouthfeel perception. The flavor stimuli occur when chemicals from the food come into contact with sensory receptor cells in the nose (odor/aroma) and mouth (taste), or when food structures such as emulsions or rigid cell walls affect the chewing process (texture) or interact with mouth mucosa (mouthfeel). Odor/aroma is a broad sensation and encompasses an estimated 10,000 or more different odors (Reineccius, 1993). Therefore, it is not surprising that an important part of flavor research has dealt with the analysis of volatile compounds.

In the early days of flavor research, most emphasis was on development of methods to establish chemical identity of volatile compounds found only in trace quantities. Many of these compounds have extremely low odor thresholds. The analytical task in aroma chemistry is rather complicated, as a relatively simple aroma may have 50-200 constituents, which together contribute to the characteristic flavor of a food (Teranishi, 1998). Currently, the majority of these constituents have been identified, and emphasis is extended to the determination of their biological functionality. In other words, which of these compounds contribute to the sensory properties of a specific food product.

When focusing on the compounds contributing to an aroma, two important steps have to be considered in relation to instrumental characterization. The first step involves

---

[*] Saskia M. van Ruth, University College Cork, Department of Food Science and Technology, Division of Nutritional Sciences, Western Road, Cork, Ireland. Jacques P. Roozen, Wageningen University, Department of Agrotechnology and Food Sciences, Laboratory of Food Chemistry, P.O. Box 8129, 6700 EV Wageningen, the Netherlands.

Quality of Fresh and Processed Foods, edited by Shahidi et al.
Kluwer Academic/Plenum Publishers, 2004.

representative isolation of the volatile compounds. No analytical method is valid unless the isolate represents the material released during eating, and no logical conclusions on sensory perception can be drawn if the isolate does not have the characteristic sensory properties of the food product under study. In the beginning of flavor research, as mentioned above, focus was mainly on identification of compounds in food products. Methods originating from this period are extraction techniques, distillation procedures and combinations thereof. It was attempted to isolate as much of the volatile constituents from the food product as possible to measure the composition of the volatiles in the food product itself.

In the last decade, scientists began to realize that the volatile composition of the food itself was often difficult to relate to the pattern of volatiles released when foods were eaten. The release of volatile compounds from foods is determined by the partitioning of volatile compounds over product and air phase (de Roos and Wolswinkel, 1994). Partitioning is related to both the concentration of the volatile compounds in the food, and the composition and structure of the food itself. However, equilibrium of the volatile compounds in the product and air phase is normally not achieved during eating. Resistance to mass transfer is, therefore, another important release rate determining factor (de Roos and Wolswinkel, 1994). Mass transfer is determined by the food properties, e.g. its viscosity. During eating, factors such as chewing efficiency, saliva flow rate, and airflow rate affect both air-product partitioning and mass transfer factors. Altogether, the amounts of volatile compounds available for perception are determined by several factors, and these amounts can differ considerably from quantities and ratios of volatile compounds in the food product itself. The awareness of these differences resulted in development of methods to measure the composition of the volatile compounds in the air above a food product, i.e. headspace sampling methods. Static headspace techniques, dynamic headspace techniques and more recently model mouth systems have been applied to measure these compounds in the headspace of a food product (van Ruth, 2000). In a model mouth system it is possible to incorporate the effect of chewing and saliva on the release of volatile compounds.

The isolation method of volatile compounds is important because the perception of these compounds depends on the extent of their release from the food matrix. It depends, however, also on the functional properties of the odorous compounds. There are indications that only a small fraction of the large number of volatiles occurring in food actually contributes to the smell and aroma (Guth and Grosch, 1999). Therefore, the distinction between odor active compounds and the whole range of volatiles present in a particular food product is an important second step in aroma research. A usual approach is sniffing the gas chromatographic effluent of a representative isolate of the volatile compounds of a food, in order to associate odor activity with the eluting compounds. It is well-known that many of the chemical/physical detectors are not as sensitive as the human nose for odor active compounds (Acree and Barnard, 1994). The peak pattern obtained by any chemical/physical detector does not necessarily reflect the profile of odor active compounds. Gas chromatography-olfactometry (GC-O) was proposed by Fuller and co-workers as early as 1964 and has shown to be a valuable method for the selection of odor active compounds from complex mixtures since (Grosch, 1993).

Initially, volatiles were sniffed individually when eluting from the analytical column of the GC and a description of the odor was given for each retention time, which corresponded to an odor active compound. In general, it is very difficult to judge the sensory relevance of a volatile profile from a single GC-O run. The technique of GC-O is limited to screening for odor active volatile compounds, unless any quantification of chemical stimuli and of assessors' responses is performed. It should, of course, be kept in mind that in GC-O single compounds are assessed and that this approach does not provide information on the compounds' behavior in a mixture. Recombination of odor active compounds in the food matrix in order to match the original aroma of the food, and subsequent sensory evaluation, can be used to validate the correct selection of odor active compounds as a final step in aroma analysis. Correlations between amounts of odor active compounds determined and the sensory data of a food product also indicate the relevance of the compounds.

Different techniques have been developed to collect and process GC-O data and to estimate the sensory contribution of single odor active compounds. They can be classified as follows:

A. Dilution analysis methods for producing potency values on stepwise dilution to threshold, e.g. CharmAnalysis (Acree *et al.*, 1984) and Aroma Extract Dilution Analysis (Ullrich and Grosch, 1987).
B. Detection frequency methods for recording the number of assessors perceiving an odor, which is considered an estimate for the odor's intensity (Pollien *et al.*, 1997; van Ruth and Roozen, 1994).
C. Posterior intensity methods for producing estimates of perceived intensity, which are recorded after a peak has eluted (Casimir and Whitfield, 1978).
D. Time-intensity methods for producing estimates of perceived intensity recorded simultaneously with the elution of the chromatographic peak, e.g. Osme (Sanchez *et al.*, 1992).

Summarizing, the outcome of a GC-O analysis is affected by factors in the area of sampling and by methodological variables. Experience of assessors is an obvious aspect in this context. The present study deals with the influence of the training of assessors and methods for volatile sampling on GC-O data.

2. MATERIALS AND METHODS

2.1. Influence of Training of Assessors

2.1.1. Materials

For training of the assessors a mixture of 2-butanone, diacetyl, ethyl acetate, 3-methyl-1-butanol, ethyl butyrate, hexanal, 2-heptanone and α-pinene in pentane was used. Each compound was present in the concentration of 1 mg/ml pentane. The solvent and all the compounds but 3-methyl-1-butanol and ethyl butyrate were purchased from Sigma-Aldrich, Steinheim, Germany. 3-Methyl-1-butanol was obtained from Lancaster, Walkerburn, UK, and ethyl butyrate from Merck-Schuchard, Hohenbrunn, Germany.

2.1.2. Instrumental Analysis

An aliquot (0.4 µl) of the mixture was injected onto a glass trap filled with Tenax TA 60/80 (SGE, Kiln Farm Milton Keynes, UK). The volatiles were thermally desorbed from the Tenax at 225 °C for 5 min (SGE concentrator/headspace analysis injector, Kiln Farm Milton Keynes, UK). The compounds were subsequently cryofocused on the analytical column (SGE CTS.LCO2, Kiln Farm Milton Keynes, UK). Gas chromatography was carried out on a Varian Star 3400 CX (JVA Analytical Ltd, Dublin, Ireland) equipped with a BPX5 capillary column (60 m length, 0.32 mm i.d. and 1.0 μm film thickness; SGE Kiln Farm Milton Keynes, UK). The oven temperature was programmed as follows: an initial oven temperature of 40 °C was employed for 4 min, followed by a rate of 2 °C/min to 90 °C, then by 4 °C/min to 130 °C, and finally by 8 °C/min to 250 °C. The GC effluent was split 10:45:45 at the end of the capillary column, for the flame ionization detector (FID, 275 °C), sniff port 1, and sniff port 2, respectively.

Eight assessors (aged 30-50) were selected for GC-O analysis. Assessors used laptop computers with a program in Pascal for data collection (Linssen *et al.*, 1993). They pressed a key on the keyboard when they detected an odor. Tenax tubes without adsorbed volatile compounds were used as dummy samples. The panel assessed the mixture of eight compounds after 1, 2 and 4 months. There were 7 sessions per month between assessments, in which various concentrations of the compounds were analyzed, odor descriptors were generated and intensities were measured. In one of those sessions (after odor detection) assessors chose an odor descriptor from a list of descriptors generated in preceding sessions (alcohol, buttery, caramel, fruity, grassy, chemical, green leaves, oily, pine, sweaty, pungent/vinegar, sweet, unknown). After 2 and 4 months, they also sniffed a dummy sample to determine the noise level of the group of assessors. Detection frequencies were calculated for the different samples and subjected to Friedman two factor ranked analysis of variance to determine significant differences (O'Mahony, 1986). The significance level was $p < 0.05$.

2.2. Influence of sampling method

2.2.1. Materials

Commercially dried diced vegetables (red bell peppers and French beans) were supplied by Top Foods b.v. (Elburg, the Netherlands). The vegetables were packed in glass jars and stored at 4 °C in the absence of light until sampling.

2.2.2. Isolation of Volatile Compounds

The vegetables (1.2 g) were rehydrated by adding 10 ml of distilled water, followed by heating in a water bath at 100 °C for 10 min and then cooling to 25 °C for 4 min. The rehydrated vegetables were transferred into a sample flask and 4 ml of artificial saliva were added (van Ruth *et al.*, 1997). The isolation of volatiles from rehydrated bell peppers was compared in a dynamic headspace (DH), in a purge-and-trap (PT) and in a model mouth system (MMS) with identical dimensions. Dynamic headspace was performed by flushing the headspace of the sample with purified nitrogen gas at a flow rate of 20 ml/min for 1 h. For purge-and-trap, the inlet of the nitrogen gas was placed below the liquid surface of the sample and nitrogen gas passed through the bell pepper-saliva mixture at the same rate and for the same period. For model mouth measurements a plunger was added to the instrument, which made up-and-down screwing movements to mimic mastication. The headspace was flushed with nitrogen gas at the same rate and for the same period as for the other methods. For each of the three methods, volatile compounds were swept towards a glass trap filled with Tenax TA 35/60 (Alltech Nederland b.v., Zwijndrecht, the Netherlands).

In the second part of the studies on sampling methods, the time of isolation was varied (1, 1.5, 3 and 12 min), while volatile compounds from rehydrated French beans were released in the model mouth system.

2.2.3. Instrumental Analysis

For GC-O, desorption of the volatile compounds from the Tenax was performed by a thermal desorption (210 °C, 5 min)/cold trap (-120 °C) device (Carlo Erba TDAS 5000, Interscience b.v., breda, the Netherlands). Gas chromatography was carried out on a Carlo Erba MEGA 5300 (Interscience b.v., Breda, the Netherlands) equipped with a Supelcowax 10 capillary column, 60 m length, 0.25 mm i.d., 0.25 μm film thickness, and an FID at 275 °C. The effluent was split 1:2:2 for FID, sniff port 1, and sniff port 2, respectively. An initial oven temperature of 40 °C for 4 min was used, followed by a rate of 2 °C/min to 92 °C, and then by 6 °C/min to 272 °C. A panel of 12 assessors (aged 20-50) was selected and trained for GC-O analysis. Assessors used the same software as described for the training experiments. Detection frequencies and average FID peak areas were calculated and subjected to Friedman two factor ranked analysis of variance (O'Mahony, 1986). The significance level was $P < 0.05$.

The odor active compounds of the vegetables were identified using combined GC (Pye 204, Unicam Ltd, Cambridge, UK) and mass spectrometry (MS; VG MM 7070 F, Fisons Instruments, Weesp, the Netherlands). The volatile compounds were thermally desorbed (Chrompack TCT injector 16200, Chrompack, Middelburg, the Netherlands) and the column and oven temperature conditions were identical to those used in GC-O. Mass spectra were recorded in the electron impact mode at an ionization voltage of 70 eV and scanned from m/z=300 to 25 with a cycle time of 1.8 s.

3. RESULTS AND DISCUSSION

3.1. Influence of Training of Assessors

A mixture of eight odor active compounds was assessed by the GC-O panel after 8, 16, and 24 sessions, which corresponded to 1, 2, and 4 months of training/experience. The selection of the compounds was based on their physico-chemical properties and their sensory importance for food systems. Assessors attributed odor descriptors to the eight compounds and the retention indices of the compounds were calculated (Table 1). The compounds were detected separately, except that 2-butanone and diacetyl eluted very closely. FID detection showed some minor overlapping of the two compounds.

Table 1. Odor active compounds selected for training experiments, their retention indices and odor descriptors

Retention index[a]	Volatile compound	Odor descriptors[b]
653	2-Butanone	Buttery, caramel, fruity, sweet
655	Diacetyl	Caramel, buttery, sweet
663	Ethyl acetate	Caramel
759	3-Methyl-1-butanol	Sweaty, buttery, caramel, unknown
812	Ethyl butyrate	Fruity, sweet
817	Hexanal	Green leaves, grassy, sweet
906	2-Heptanone	Unknown, chemical, sweaty
952	α-Pinene	Pine, alcohol, fruity, green leaves

[a] Retention indices calculated according to the formula proposed by Van den Dool and Kratz (1963).
[b] In order of frequency.

The detection frequencies of the eight compounds in the mixture were calculated for the three sessions, which reflect the increased GC-O experience of the assessors (Figure 1). The detection frequency of the compounds varied slightly over the months, but did not differ significantly (Friedman two factor ranked analysis of variance, $P < 0.05$). Cumulative detection frequencies of the compounds per assessment were quite consistent (Session 8: 40; Session 16: 40; Session 24: 38). The response to 'noise' of the GC-O panel was calculated as the total number of perceptions minus those for the eight added compounds. Figure 1 shows the large decrease in 'noise' perceptions during the experiment. The detection frequency of 'noise' decreased from 161 in Session 8 to 50 and 30 for Sessions 16 and 24, respectively. In addition, dummy samples were assessed close to the Sessions 16 and 24. The analysis resulted in 43 and 24 'noise' detections, which are slightly lower than those for the real sample mixture. It is remarkable that the number of signal detections did not change significantly during the training period, which demonstrates the robustness of the detection frequency method. Nevertheless, the noise level reduced with training, assessors are obviously improving their skills in distinction of signals from noise due to ongoing training. The reduced noise level increases the sensitivity of the method. Pollien *et al.* (1997) suggested that the detection frequency method does not require any training. The present study shows that, although the signal detection does not change, the sensitivity is improved by training. The suggestion of Pollien *et al.* (1997) is, therefore, not entirely justified.

3.2 Influence of Sampling Method

Volatile compounds of rehydrated diced red bell peppers were isolated in DH, PT and MMS. The compounds were analyzed by gas chromatography combined with olfactometry detection (GC-O) and flame ionization detection (FID). The odor active compounds were subsequently identified by GC-MS and by their retention indices. Furthermore, the odor active compounds were characterized by their detection frequencies and their FID peak areas (Table 2). In terms of presence in the different isolates determined by GC-MS, the qualitative composition of the odor active compounds was similar for the isolates obtained with the three methods. However, the compounds showing odor activity in GC-O differed among the three methods. Fewer odor active compounds were determined in DH (6 odor active regions); PT isolation had 11, and MMS isolation had 10 odor active regions. Cumulative detection frequencies showed the same tendency, with a total of 36 perceptions for DH, 93 for PT and 90 for MMS isolation. Friedman two factor ranked analysis of variance showed a significant difference in detection frequencies between the sampling procedures (P < 0.05). These results show that the sampling method affects GC-O data and even attribution of odor activity to volatile compounds.

The volatile compounds isolated with the three sampling methods differed also significantly in FID response (Friedman two factor ranked analysis of variance, P < 0.05). However, the difference in FID response between the methods is considerably larger than the difference in detection frequency. This phenomenon is related to the log linear relationship between sensory intensity and physical concentration, which is known as Fechner's law (Meilgaard *et al.*, 1991). Consequently, a difference in concentration does not necessarily result in a difference in detection frequency or attribution of odor activity to compounds.

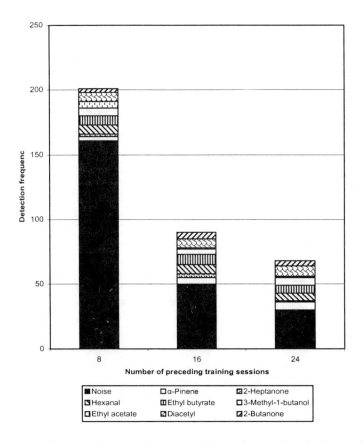

Figure 1. Detection frequency of odor active compounds in a mixture and 'noise' in gas chromatography-olfactometry analysis after 8, 16, and 24 sessions, which took place after 1, 2, and 4 months, respectively.

Table 2. Detection frequency and flame ionization detector response (FID) in gas chromatography-olfactometry analysis of odor active compounds isolated from rehydrated diced red bell peppers using a dynamic headspace technique (DH), a purge-and-trap technique (PT) and a model mouth system (MMS)

Compounds	Detection frequency [a]			FID response [V.s]		
	DH	PT	MMS	DH	PT	MMS
2-Methylpropanal	4	8	7	1.15	33.06	12.58
2-Methylbutanal	8	12	12	0.47	19.99	8.24
3-Methylbutanal				1.08	48.63	19.32
Diacetyl	7	9	11	0.01	---	0.16
1-Penten-3-one	---[b]	8	8	0.02	0.72	0.39
Hexanal	---	11	10	2.98	166.04	54.04
Heptanal	---	5	6	0.04	4.26	1.25
Unknown	---	9	9	---	0.40	0.09
1-Octen-3-one	5	8	9	---	0.73	0.24
Dimethyl trisulfide	4	12	7	---	0.31	0.02
2-Methoxy-3-isobutylpyrazine	8	11	11	---	---	---
β-Cyclocitral	---	5	---	---	0.21	0.08

[a] 2- and 3-methylbutanal could not be detected separately by the panel.

[b] Below detection level, detection level is 3 assessors for detection frequency, and 0.01 V.s for FID.

It is, therefore, even more remarkable that DH showed a smaller selection of odor active compounds, which suggests that the concentrations are considerably out of range compared to those isolated under mouth conditions.

In the second part of the study, the influence of different times of volatile sampling of rehydrated diced French beans on the attribution of odor active compounds and the detection frequency was studied. The odor active compounds and their detection frequencies for the various sampling times are presented in Figure 2. GC-O of dummy samples showed that a detection frequency of one could be considered as noise. Consequently, compounds with a detection frequency of two or higher are considered odor active. Despite the fact that the sampling time varied between 1 and 12 min, GC-O generated an identical selection of odor active compounds (Friedman two factor ranked analysis of variance, $P < 0.05$). Nevertheless, the various sampling times resulted in accordingly varying detection frequencies for the odor active compounds. Longer sampling times resulted in increased detection frequencies. However, as shown in the experiments on sampling method comparison, the sampling time, which was linearly related to the concentration in the isolate, was not linearly related to the detection frequency. Again, a logarithmic relationship between physical concentration and detection frequency is more likely. As a consequence, small changes in isolation time, which could be due to experimental errors, will hardly affect the outcome of the detection frequency data in GC-O analysis. This, once more, shows the robustness of the method.

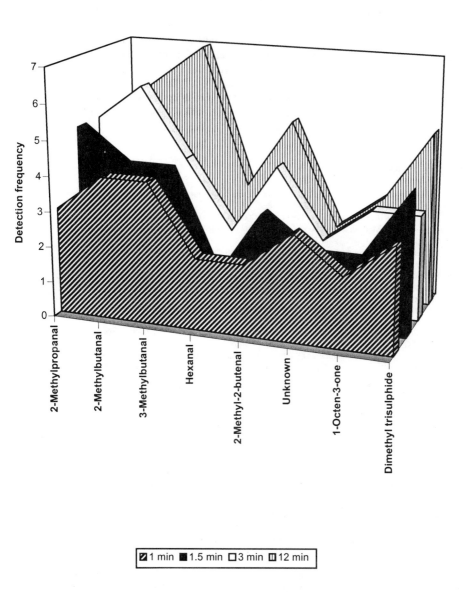

Figure 2. Detection frequency of odor active compounds of rehydrated diced French beans in gas chromatography-olfactometry analysis after isolation for different time periods in the model mouth system.

4. CONCLUSIONS

GC-O analysis is an important technique for attribution of odor activity to volatile compounds released from food products. The detection frequency method is a robust method for GC-O data collection and processing, since attribution of odor activity is hardly affected by the training of assessors and by volatile sampling time. Nevertheless, training does reduce noise levels and, therefore, increases sensitivity of the method. Dynamic headspace sampling results in volatile isolates, which are very different from the concentrations isolated under mouth conditions for the samples used. Consequently and despite the insensibility of the detection frequency method for small and moderate variations in concentration, it also affects the outcome of the GC-O analysis.

5. REFERENCES

Acree, T. E., and Barnard, J., Cunningham, D., 1984, A procedure for the sensory analysis of gas chromatographic effluents, *Food Chem.*, 14:273-286.
Acree, T. E., and Barnard, J., 1994, Gas chromatography-olfactometry and CharmAnalysis, in: *Trends in Flavour Research*, H. Maarse, and D. G. van der Heij, eds, Elsevier, Amsterdam, pp. 211-220.
Casimir, D. J., and Whitfield, F. B., 1978, Flavour impact values: a new concept for assigning numerical values for potency of individual flavour components and their contribution to overall flavour profile, *Ber. Int. Fruchtsaftunion,* 15:325-345.
de Roos, K. B., and Wolswinkel, K., 1994, Non-equilibrium partition model for predicting flavour release in the mouth, in: *Trends in Flavour Research*, H. Maarse, and D. G. van der Heij, eds, Elsevier, Amsterdam, pp. 15-32
Fuller, G. H., Steltenkamp, R., and Tisserand, G. A., 1964, The gas chromatograph with human sensor: perfumer model, *Ann. N. Y. Acad. Sci.,* 116:711-724.
Grosch, W., 1993, Detection of potent odorants in foods by aroma extract dilution analysis, *Trends Food Sci. Technol..* 4:68-73.
Guth, H., and Grosch, W., 1999, Evaluation of important odorants in foods by dilution techniques, in: *Flavor Chemistry*, R. Teranishi, E.L. Wick, and I. Hornstein, eds, Kluwer Academic/Plenum Publishers, New York, pp. 377-386.
Linssen, J. P. H., Janssens, J. L. G. M., Roozen, J. P., and Posthumus, M. A., 1993, Combined gas chromatography and sniffing port analysis of volatile compounds of mineral water packed in laminated packages, *Food Chem.,* 46:367-371.
Meilgaard, M., Civille, G. V., and Carr, B. T., 1991, *Sensory Evaluation Techniques*, CRC Press, Boca Raton.
O'Mahony, M., 1996, *Sensory Evalution of Food. Statistical Methods and Procedures*, Marcel Dekker, New York.
Pollien, P., Ott, A., Montignon, F., Baumgartner, M., Munoz-Box, R., and Chaintreau, A., 1997, Hyphenated headspace-gas chromatography-sniffing technique: screening of impact odorants and quantitative aromagram comparisons, *J. Agric. Food Chem.,* 45:2630-2637.
Reineccius, G., 1993, Biases in analytical flavor profiles introduced by isolation method, in: *Flavor Measurement*, C.-T. Ho, and C. H. Manley, eds., Marcel Dekker, New York, pp. 61-76.
Sanchez, N. B., Ledere, C. L., Nickerson, G. B., Libbey, L. M., and McDaniel, M. R., 1992, Sensory analytical evaluation of beers brewed with three varieties of hops and an unhopped beer, in: *Proceedings of the 6th International Flavor Conference*, Rethymnon, Crete, G. Charalambous. ed., Elsevier, Amsterdam, pp. 403-426.
Teranishi, R., 1998, Challenges in flavor chemistry: an overview, in: *Flavor Analysis. Developments in Isolation and Characterization*, C. J. Mussinan, and M. J. Morello, eds., American Chemical Society, Washington, DC, pp. 1-6.
Ullrich, F., and Grosch, W., 1987, Identification of the most intense volatile flavor compounds formed during autoxidation of linoleic acid, *Z. Lebensm.-Unters, Forsch.,* 184:277-282.
Van Den Dool, H., and Kratz, P., 1963, A generalization of the retention index system including linear temperature programmed gas-liquid partition chromatography, *J. Chromatogr.,* 11:463-471.
van Ruth, S. M., 2000, Aroma measurement, in: *Focus on Biotechnology VII*, M. Hofman, ed., Elsevier, Amsterdam, The Netherlands.

van Ruth, S.M., and Roozen, J.P., 1994, Gas chromatography/sniffing port analysis and sensory evaluation of commercially dried bell peppers (*Capsicum annuum*) after rehydration, *Food Chem.,* **51**:165-170.

van Ruth, S.M., Roozen, J.P., and Legger-Huysman, A., 1997, Relationship between instrumental and sensory time-intensity measurements of imitation chocolate, in: *Flavour Perception. Aroma Evaluation*, H.-P. Kruse, and M. Rothe, eds, Universität Potsdam, Potsdam, pp. 143-151.

SCREENING FOR SENSORY QUALITY IN FOODS USING SOLID PHASE MICRO-EXTRACTION TANDEM MASS SPECTROMETRY

Casey C. Grimm[*], Mary An Godshall, Terry J. Braggins and Steven W. Lloyd

1. INTRODUCTION

Tandem mass spectrometry has proven a useful technique for the elucidation of chemical structures. However, there are no applications of tandem mass spectrometry for routine applications. This is in part due to the limitation of samples being introduced into the ion source in the presence of a solvent. The relative large amount of solvent molecules overwhelms the presence of the analyte and the signal, if any, is lost in the chemical noise. With the introduction of solid phase microextraction (SPME), the analyte can now be placed in the mass spectrometric source in a solvent free environment (Belardi and Pawliszyn, 1989; Eisert and Levsen, 1996; Kataoka et al., 2000). The major contributors to the muddy/musty off-flavor in drinking water systems are 2-methylisoborneol (2-MIB) and geosmin (Lovell, 1983). These compounds are generally placed in the water column by blue-green algae, but are also produced by bacteria and fungi. Associated with algae blooms in late summer, the compounds are perceptible to the human nose at the low parts per trillion range (Persson, 1980). They plague drinking water systems and are particularly problematic in the warm water aquaculture production of farm-raised catfish. The current method for analysis of 2-MIB and geosmin is closed loop stripping (McGuire et al., 1981) and purge and trap (Johnsen and Lloyd, 1992) with GC/MS analysis. However, a technique employing SPME-GC/MS has recently been reported for the analysis of these compounds at the parts per trillion level (Lloyd et al., 1998; Watson et al., 2000). The SPME technique has proven to be an excellent method for the concentration of volatile compounds, and in combination with gas chromatography for separation and mass spectrometry for detection, provides a state-of-the-art analytical tool.

---

[*] Casey C. Grimm, USDA-ARS-SRRC, 1100 Robert E. Lee Blvd, New Orleans, LA 70124, Ph. (504) 286-4293; FAX (504) 286-4419; E-mail cgrimm@srrc.ars.usda.gov.

Quality of Fresh and Processed Foods, edited by Shahidi et al.
Kluwer Academic/Plenum Publishers, 2004.

The rate-limiting step in the process is the time required to perform a GC separation. Elimination of the gas chromatograph would permit maximum sample throughput. In certain cases, the gas chromatograph or separation step can be performed using tandem mass spectrometry (Johnson *et al.,* 1990; Braggins *et al.,* 1999). All compounds are introduced directly into the source of the mass spectrometer. Only a specific ion indicative of the targeted compound is collected. Normally this is the molecular ion. The selected or parent ion is then fragmented to produce progeny ions. The abundance of selected progeny ions is then measured to provide quantitative information. Hence, the mass spectrometer is used as a separation device and as a detector. The initial isolation of the parent ion serves as the separation step, and the second mass spectrometric determination serves as the detection step. Problems arise when isotopic compounds are present which are similar in structure to the compound of interest and produce similar progeny ions. This research explores the feasibility of performing semi-quantitative analysis to screen food for trace levels of selected compounds that have a high impact on the sensory impact.

2. EXPERIMENTAL

SPME fibers were obtained from Supelco (Ringoes, NJ). Geosmin and 2-methylisoborneol (2-MIB) were obtained from Waco Chemical (Osaka, Japan). Stock solutions of 1 part per thousand were made up in ethanol, with subsequent dilution's in Milli-Q water. For the detection limit determinations, a 100 μm film of polydimethylsiloxane was employed. Sample volumes of 8 ml were placed in 12 ml vials and sufficient NaCl (3g) was added to saturate the solution. The sample was stirred using a magnetic stirring bar and the sample vial placed in a water bath held at 40°C. The SPME fiber was exposed to the headspace of the solution for 10 minutes. The fiber was then immediately placed in the GC injection port and desorbed.

A Finnigan GCQ ion trap (Palo Alto, CA) was used as the tandem mass spectrometer. For SPME/MS/MS, a 25 cm piece of a DB-5 cm, 0.05 I.D. Column (J & W Scientific, Folsom, CA) was used between the injection port of the GC and the mass spectrometer (Figure 1). The GC injection port contained a 0.7mm I.D. injection liner and was operated in splitless mode at 250°C. Helium was used as the carrier gas and held at an initial pressure of 20 psi. A surge pressure of 60 psi was used from 0.1 to 1.0 minutes. The GC oven was held isothermally at 200°C during the analysis and heated to 250°C between analyses. The transfer line was held at 275°C. The source of the ion trap was held at 150°C and the offset between the trap and the source was 10 V. Methane was used as the reagent gas for chemical ionization and only positive ions were monitored. The ion trap was operated at a q=0.225, with a collision energy of 0.5 eV, a parent collection time of 2 ms, and a collision time of 30 ms. Data were collected over a total run time of 1 to 2 minutes.

For rice, 12 ml vials were filled half way with 5g of milled rice (~175 kernels). Analysis of sugar was conducted on 0.75 g samples employing a 2ml vial. Extraction of volatile compounds from the headspace was accomplished using a carboxen/DVB/PDMS SPME fiber at 65C for both samples. GC & MS conditions were the same as above. Election impact ionization at 70eV was used for the first MS stage.

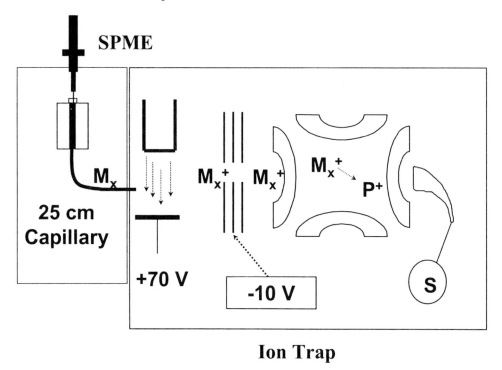

Figure 1. Block diagram of SPME/GC/MS/MS, where M_x is a molecule, which becomes ionized and then undergoes CID to give a progeny ion (P_x).

3. RESULTS AND DISCUSSION

3.1. Off-odors in Water

The ability to monitor a given compound is dependent upon a variety of factors including its mass spectral fragmentation pattern, the presence of interfering ions and the relative amount of analyte in the sample. A large difference in concentration of at least one order of magnitude and preferably two orders of magnitude is needed in order to successfully screen for the desired traits. Compounds that fragment little upon electron ionization and posses rather unique molecular ions are most suited for this type of screening.

Injections of a one ppm solution, employing standard GC using a 30 m DB5 ms column (J & W Scientific, Folsom, CA) and electron ionization MS were initially used for method development. The molecular ion for 2-MIB was observed at m/z 168 at an abundance of 2% relative to the base peak at m/z 95. The molecular ion for geosmin was less than 2% at m/z 182, relative to the base peak at m/z 112. To enhance the relative abundance of the molecular ions, chemical ionization was employed using methane as the reagent gas. The $[M+H]^+$ ion remained small, as the addition of a proton resulted in the loss of water. The pseudo molecular ion of $[M+H-H2O]^+$ at m/z 151 gave a relative abundance of 40% relative to the base peak of m/z 95. Optimum chemical ionization to produce the molecular ion or pseudo molecular ion as the base peak could not be achieved with the GCQ instrument.

The standard 30 m GC capillary column was removed and replaced with length of column only sufficient to connect the injection port with the source of the mass spectrometer. The SPME fibers incubated for 10 minutes in the headspace above water samples were thermally desorbed in the GC injection port. The desorbed compounds were swept into the external source of the ion trap within a few seconds, where they underwent chemical ionization. The ions were then pulsed into the ion trap. Potentials were applied to the trap to eject all ions with the exception of those falling in a 2 Dalton range centered on m/z 151. Potentials were then applied on the end caps to collisionally induce dissociation of the parent ion. The progeny ions were then determined by sequentially ejecting the ions, with subsequent detection at the electron multiplier. A two Dalton window centered on the progeny ions of m/z 81, 95, and 109 were monitored for 2-MIB. For geosmin, chemical ionization gave a pseudo molecular ion at m/z 163, $[M+H-H_2O-H_2]^+$, believed to result from the addition of a proton and the subsequent loss of a water molecule and molecular hydrogen. No progeny ion was observed at m/z 112, the base peak observed under normal electron ionization. However, progeny ions were observed and monitored at m/z's 95, 109, and 135.

Efficiency of the collisionally induced dissociation (CID) process for 2-MIB was analyzed by monitoring two progeny ions and the intact m/z 151 ions. In this manner the parameters of the ion trap, collection time, ionization energy, q value and reaction time were optimized to produce the greatest abundance of progeny ions. A maximum signal was obtained by setting the collection time of the parent ion at the minimum value allowed by the software, 2ms. The maximum allowable setting of 30 ms for the reaction time produced the highest number of progeny ions relative to the parent. The optimal ionization energy was found to be 0.5 eV. The allowable q values of 0.225, 0.300, and 0.450 were investigated with various ionization energies, collection times and reaction times. In all cases a q value of 0.225 gave the maximum signal.

Using the pressure surge of the GC injector increased the detection limits by producing a sharper peak at the front (Figure 2). An aqueous solution containing concentrations of 2-MIB of 0, 1, 5, 10, and 20 ppb were analyzed by SPME/MS/MS. Little or no difference was observed between the blank and the 1 ppb solution. The 5 ppb solution produced a peak maxim of 442 counts at ca. 0.1 minute (the beginning of the surge). As the baseline was at 40 counts, this gave a S/N ratio of 10:1. The limit of detection thus was between 1 and 5 ppb for the technique as described for 2-MIB.

3.2. Odors in Rice

Two compounds that impact the sensory quality of rice are indole and 2-acetyl-1-pyrroline (2-AP). An off-odor, resembling a barnyard type smell, occurs when indole is present at high concentration levels in rice. The presence of high levels of 2-AP results in a popcorn aroma and is generally considered a desired trait. Aromatic rice contains high levels of 2-AP and often exceeds concentration levels of 1 ppm (Buttery *et al.*, 1982; 1988). Although 2-AP can be found in all rice samples in non-aromatic (control) rice the concentration is several orders of magnitude lower (~1 ppb). Both compounds contain a single nitrogen and give rather unique molecular ions due to the incorporation of a single nitrogen atom in their chemical formulas (indole m/z 117, 2-AP m/z 111). For SPME

Table 1. Comparison of the MS/MS peak heights with GC/MS peak areas for a collection of rice samples. For 2-AP the parent ion was m/z 111 and the progeny ion was m/z 83. For indole the parent ion was m/z 117 and the progeny ion was m/z 90.

Rice		2-Acetyl-1-Pyrroline		Indole	GCMS
		MS/MS	GC/MS	MS/MS	
Della	Aromatic	271	1,123,440	1020	2,506,107
Dellmont	Aromatic	229	453,497	973	374,671
Goolara	Aromatic	256	1,142,973	483	252,413
Bengal	Barnyard	51	12,563	2354	13,279,032
NATO	Barnyard	167	36,295	2883	18,653,813
V4716	Barnyard	85	7,549	1722	11,607,319
Braz	Control	130	37,387	1970	3,548,549
Calrose 76	Control	129	14,980	476	33,456
Kosanbare	Control	102	44,518	1073	1,051,659
Koshihikari	Control	126	11,552	260	91,498
Leah	Control	51	9,581	403	62,520
Rexmont	Control	95	18,632	492	27,225

MS/MS, milled rice was analyzed as whole kernels requiring little or no sample preparation. Progeny ions of m/z 90 and 83 were monitored for indole and 2-AP, respectively. The maximum peak height observed for indole and 2-AP was compared with the peak area observed using GC/MS in Table 1. Due to the longer cycle times associated with MS/MS and the format of the MS/MS data, comparison of peak areas to peak areas was not possible. A correlation was observed for both the 2-AP and the indole based upon their concentration in the various rice samples. If the Braz and Kosenbare rice samples were eliminated, due to possible contamination as evidenced by the GC/MS data, there was a four to five fold increase in peak height for indole between the barnyard flavored rice and the control rice. The 2-AP data was less dramatic, which may be a result of the smaller relative abundance of the parent ion. Little fragmentation was observed in indole, the parent ion is the base peak and molecular ion at m/z 117. For 2-AP, the base peak was at 83 and the parent ion m/z 111 was less than 50%. Additional fragment ions were observed in the progeny spectrum of 2-AP, an indication of an isobaric contaminant with the parent ion (m/z 111). This may affect the total number of parent ions captured and thus affect the sensitivity of the analysis. Further optimization of the MS/MS parameters and/or different selection of the parent/progeny ion pairs may yield better results.

3.3. Odors in Sugar

Raw sugar and white sugars often contain odors which make them unacceptable for consumption. Refining of the sugar greatly reduces the problem, however, some odors may not be removed by the refining process. There are various compounds that can render an off-odor in sugar. One family of compounds is the short chain fatty acids. With the exception of propanoic acid, electron ionization gave a base peak at m/z 60 resulting

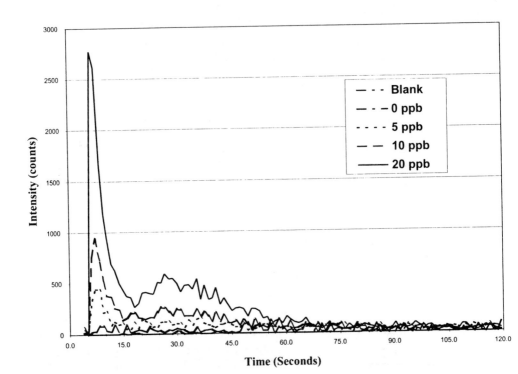

Figure 2. Total ion chromatograms of the *m/z* 81, 95 and 109 progeny ions of 2-MIB at concentrations of 0, 5, 10, 50,100 ppb.

Table 2. Comparison of GC/MS/MS with MS/MS peak areas for the progeny ion *m/z* 42.

Type	GC/MS/MS		MS/MS
	Sugar	(*m/z* 60→42)	(*m/z* 60→42)
Unacceptable	1	18723	11420
	2	8302	7943
	3	185332	45263
	4	178234	45245
Acceptable	5	39	269
	6	36	370
	7	24	159
	8	32	208

from β cleavage of the carbon-carbon bond along the backbone of the molecule. The propanoic acid molecular ion undergoes the McLafferty rearrangement and does not give a base peak at *m/z* 60 as observed in other short chain fatty acids. However, a relative abundance of a few % was observed at *m/z* 60 for propanoic acid and the base peak observed at *m/z* 74.

Initial sample profiling by SPME/GC/MS, showed the presence of ethanoic, butanoic and hexanoic acids, as well as propanoic, isopentanoic, pentanoic, heptanoic and octanoic acids. SPME/GC/MS/MS was then performed on four unacceptable sugars and four control sugar samples. The GC column was then removed and the 25 cm column installed. Employing m/z 60 as the parent ion and m/z 42 as the progeny ion, a comparison was made between the combined peak areas of all acids from the SPME/GC/MS/MS analysis and the SPME/MS/MS analysis. In this case only a single CID experiment was conducted (m/z 60 - m/z 42) and a direct comparison can be made of the peak areas. The SPME/MS/MS peak area was generated over a two minute period. The results of the comparison are presented in Table 2. In this case, where the analytes were present at the high ppb range, the SPME/MS/MS screening technique worked quite well with the peak areas of nearly two orders of magnitude, separating the acceptable sugars from the unacceptable sugars.

4. CONCLUSIONS

This research illustrated the practical aspects of employing SPME/MS/MS for the screening of trace levels of high odor impact compounds in foods. Two of the unique capabilities of the SPME technique, the ability to concentrate trace levels of volatile and semi-volatile compounds and to do so without solvent were exploited to provide a rapid, sensitive analytical screening technique. Instrumental analysis times of two minutes or less were typical. The method as outlined was adequate to detect ppb levels of off-flavor compounds in water, ppm levels of 2-AP and indole in rice and ppb to ppm levels of short chain fatty acids in sugars. These compounds can be screened for with minimal sample preparation and require very little analysis time. SPME/MS/MS involves the optimization of a large number of variables which interact to effect the total sensitivity of the analysis. Further optimization of SPME and/or MS/MS parameters could easily result in an increased sensitivity approaching an order of magnitude.

5. REFERENCES

Belardi, R. and Pawliszyn, J., 1989, The application of chemically modified fused silica fibers in the extraction of organics from water matrix samples and their rapid transfer to capillary columns, *Wat. Pollut. Res. J. Canada*, 24(1):179-191.
Braggins, T. J., Grimm, C. C., and Visser, F. R., 1999, Analysis of Food Volatiles Using SPME, in: *Applications of Solid Phase Microextraction*, J. Pawliszyn, ed., The Royal Society of Chemistry, Cambridge, UK. Chapter 31, pp. 407-422.
Buttery, R., Ling, L., and Juliano, B., 1982, 2-Acetyl-1-Pyrroline: An important aroma compound in cooked rice, *Chem. Ind.* (London), 4 Dec., 958.
Buttery, R., Turnbaugh, J., and Ling, L., 1988, Contributions of volatiles to rice aroma, *J. Agric. Food Chem.*, 36:1006-1009.
Eisert, R. and Levsen, K., 1996, Solid-phase microextraction coupled to gas chromatography: A new method for the analysis of organics in water, *J. Chromatogr. A.*, 733(1/2):143-157.
Johnsen, P. B., and Lloyd, S. W., 1992, Influence of fat content on uptake and depuration of the off-flavor 2-methylisoborneol by channel catfish (*Ictalurus punctatus*), *Can. J. Fish. Aqua. Sci.*, 49:2406-241.
Johnson, J. V., Yost, R. A., Kelley, P. E., and Bradford, D. C., 1990, Tandem-in-space and tandem in-time mass spectrometry: Triple quadrupoles and quadrupole ion traps, *Anal. Chem.*, 62:2162-2172.
Kataoka, H., Lord, H. L., and Pawliszyn, J., 2000, Applications of solid-phase microextraction in food analysis. *J. Chromatogr. A.*, 880:35-62.

Lloyd, S. W., Lea, J. M., Zimba, P. V., and Grimm, C. C., 1998, Rapid Analysis of Geosmin and 2-Methylisoborenol in Water using Solid-Phase Micro-Extraction Procedures, *Water Res.*, **32**(7):2140-2146.

Lovell, R. T., 1983, Off-flavors in pond-cultured channel catfish, *Water Sci. Technol.*, **15**:67-73.

McGuire, M. J., Krasner, S. W., Hwang, C. J., and Izaguirre, G., 1981, Closed-loop stripping analysis as a tool for solving taste and odor problems, *J. Am. Water Works Assoc.*, **73**:530-537.

Persson, P. E., 1980, Sensory properties and analysis of two muddy odour compounds, geosmin and 2-methylisoborneol, in water and fish, *Water Res.*, **14**:1113-1118.

Watson, S. B., Brownlee, B., Satchwill, T., and Hargesheimer, E., 2000, Quantitative Analysis of Trace Levels of Geosmin and MIB in Source and Drinking Water Using Headspace SPME, *Water Res.*, **34**:2818-2828.

MAILLARD REACTION-BASED GLYCOSYLATION OF LYSOZYME

Alex N. Yousif[1], Shuryo Nakai, and Christine H. Scaman[*]

1. INTRODUCTION

Modification of the functional properties of selected proteins can be achieved through deliberate addition of carbohydrates via the Maillard reaction, as first shown by Kato *et al.* (1990). This type of chemical modification is recognized as being advantageous over other chemical modifications (i.e. addition of carbohydrates via cyanogen bromide activation) in terms of safety and acceptability since only food ingredients are used (Kato, 1996). The reactants are a carbonyl group, as found at the reducing end of an oligo- or polysaccharide and a free amino group. The reaction has typically been carried out with protein and carbohydrate powders in the dry state, under controlled conditions of humidity (60 - 79%) and temperature (50-60°C) for up to several weeks. The rate of reaction varies with the type and size of carbohydrate, and the nature of the protein. Proteins from a variety of sources, including milk (Hattori et al., 2000; Dickinson and Galazka, 1991), fish (Wahyuni *et al.*, 1999, Tanaka, *et al.,* 1999, Matsudomi *et al.*, 1994), plant (Babiker and Kato, 1998; Babiker *et al.*, 1998; Kato *et al.*, 1991), and egg (Nakamura *et al.*, 1998; Kato *et al.*, 1993; Kato *et al.*, 1990; Nakamura *et al.*, 1992) have been derivatized.

The most commonly used carbohydrates for these conjugation reactions are dextrans and galactomannans, which vary in size from 10 to over 100 kD. Both of these polysaccharides readily react. Other carbohydrates, including xyloglucan (Shu *et al.*, 1996), alginate (Hwang *et al.,* 1997), and anionic and cationic derivatives such as dextran sulfate (Dickinson and Galazka, 1991), glucose-6-phosphate (Wahyuni *et al.,* 1999), glucosamine, and chitopentaose (Hattori *et al.,* 2000) have also been used.

---

[1] Alex N. Yousif, Shuryo Nakai, Food, Nutrition, and Health,University of British Columbia, 6650 North West Marine Drive, Vancouver, B.C., Canada, V6T 1Z4. *To whom correspondence should be addressed. Tel: 604-822-1804. Fax: 604-822-3959. E-mail: cscaman@interchange.ubc.ca

Quality of Fresh and Processed Foods, edited by Shahidi et al.
Kluwer Academic/Plenum Publishers, 2004.

The influence of molecular weight of the carbohydrate on extent of the reaction and the properties of the conjugate has been examined. Small molecular weight carbohydrates react more quickly than larger ones, for example xyloglucan hydrolysate of approximately 1.4 kD reacted with chicken egg white lysozyme in 3 days (Nakamura *et al.*, 2000) while a reaction carried out with lysozyme and dextran (60 kD) required approximately 2 weeks (Nakamura *et al.*, 1991). However, protein structure also has a strong influence on the rate of reaction. Casein was conjugated with dextran after 1 day at 60°C and 79% relative humidity (Kato *et al.*, 1992) while as noted above, a more structured protein, lysozyme, required several weeks for derivatization. A minimum size of polysaccharide of 10 kD has been suggested by Kato (1996) to be required to enhance the heat stability and emulsification properties of the derivatized protein. Results with mono- and disaccharides have been less consistent. Kato *et al.* (1990) reported that mono- and disaccharides conjugated to proteins resulted in insoluble aggregates with poor surface activity; however, some improvement in emulsification was reported for porcine globulin derivatized with various monosaccharides (Miyaguchi *et al.*, 1999).

Improvements in the functional properties of proteins associated with derivatization include increased solubility, emulsification, thermal stability, antioxidant activity, and antimicrobial activity. As well, a reduction in undesirable properties such as antigenicity has been achieved. For example, emulsification improvement has been shown with curdlan derivatized with phosvitin and alpha S-casein (Nakamura *et al.*, 2000). Beta-lactoglobulin emulsification improved and aggregation, antigenicity and immunogenicity were reduced by conjugation with chitopentaose or chitosan (Hattori *et al.*, 2000). In some cases, conjugates have been found to have dual functional properties. Ovalbumin conjugated with dextran or galactomannan has been shown to exhibit both antioxidant and emulsifying properties (Nakamura *et al.*, 1992). Salmine (Tanaka *et al.*, 1999), and polylysine (Yu *et al.*, 2000) conjugated with dextrans were reported to exhibit both antimicrobial activity and excellent emulsification properties.

Lysozyme has frequently been used in carbohydrate derivatization experiments. Lysozyme is well accepted as an antimicrobial agent in food, cosmetic and medical applications, but it would be even more useful if it were effective against pathogenic Gram-negative bacteria. The antimicrobial mechanism of lysozyme has traditionally been attributed to hydrolytic action against 1,4 beta linkages between N-acetylmuramic acid and N-acetyl-D-glucosamine residues in the bacterial cell wall peptidoglycan. The resistance of Gram-negative bacteria to native lysozyme is attributed to the inability of lysozyme to penetrate the outer lipopolysaccharide membrane. However, lysozyme, modified by physical or chemical treatments, may exhibit antimicrobial activity distinct from the lytic activity, possibly through binding to the bacterial membrane with perturbation of membrane function (Ibrahim, 1996; During *et al.*, 1999; Nakamura *et al.*, 1991).

To create a more surface active protein, hen egg white lysozyme (HEWL) has been conjugated with various carbohydrate polymers. These conjugates are reported to be effective against both Gram-positive and Gram-negative bacteria, as well as exhibiting enhanced thermal stability and improved emulsification activity (Nakamura *et al.*, 1991; Nakamura *et al.*, 1994; Shu *et al.*, 1996). In this work, lysozyme was conjugated to dextrans of varying molecular weight (10 - 188 kD) to systematically evaluate the effects of dextran size on conjugate characteristics. A facile method of separating the unreacted polysaccharide, unreacted lysozyme, and lysozyme-dextran conjugate was developed using ion exchange chromatography and ultrafiltration. The conjugates were characterized in terms of residual lytic activity, antimicrobial properties and thermal stability.

2. MATERIALS AND METHODS

2.1 Materials

Dextrans (average molecular weight 10, 37.5, 65.5, 188 kD) were obtained from Sigma Chemicals. Hen egg-white lysozyme (HEWL) was donated by Canadian Inovatech, Inc. (Abbottsford, BC). All other chemicals were of reagent grade and were used without any further purification.

2.2 Preparation of lysozyme-dextran conjugates

Lysozyme and the various dextrans, at the weight ratio of 1:5 were dissolved in a minimum amount of distilled deionized water, lyophilized, and incubated at 60°C under a relative humidity of 79% for 3 weeks. Relative humidity was maintained by placing containers with powdered lysozyme and dextran in a dessicator containing saturated KBr. The dessicator was then placed in an incubator to obtain the reaction temperature.

2.3 Conjugate purification and characterization

Conjugates were isolated using FPLC with a Pharmacia Hi Trap SP column, equilibrated with 20 mM phosphate buffer, pH 7.0, or 30 mM ammonium carbonate buffer, pH 7.7. Elution was carried out with a gradient of the same buffer containing 1 M NaCl. The protein content in each fraction was detected at 280 nm, and the carbohydrate content was determined with the phenol-sulfuric acid reaction (Dubois *et al.*, 1956). All fractions containing lysozyme-dextran conjugate were pooled and subjected to ultrafiltration (cut off 30 kD), and lyophilized. The free amino groups of control and derivatized lysozyme were determined using fluorescamine (Ford *et al.*, 1982). The number of modified lysine residues was calculated, considering the mole content of amino groups in native lysozyme is seven.

The lytic activity of native lysozyme and lysozyme conjugates was determined by monitoring the rate of lysis of *Micrococcus lysodeikticus*. A suspension of *M. lysodeikticus* cells was rehydrated in 100 mM sodium phosphate (pH 7.0), and the concentration of cells was adjusted to yield an absorbance between 0.6 and 1.0. After the cell suspension was equilibrated to 25 °C, an aliquot of lysozyme or conjugate was added and the drop in turbidity was monitored. One unit of activity is defined as the amount of enzyme which caused a decrease in turbidity of 0.001 per min at 450 nm at 25 °C.

Sodium dodecyl sulfate-polyacrylamide gel electrophoresis (SDS-PAGE) was carried out using 10% acrylamide separating gel, and 3% stacking gel (Laemmli, 1970). Gels were stained for proteins and carbohydrates with Coomassie blue G-250 and Alcian blue/silver staining (Moller and Poulsen, 1996), respectively.

The thermal stability of the conjugates was estimated by measuring the developed turbidity of a 0.02% protein aqueous solution of native lysozyme or lysozyme-dextran conjugates when heated in a water bath to 95°C. One milliliter aliquots were immediately chilled and turbidity measured at 500 nm.

2.4 Determination of antimicrobial effects

Gram-negative *Escherichia coli* strain IM294 and Gram-positive *Leuconostoc mestenteroids* strain IM006 were both from Canadian Inovatech Inc. The cultures were maintained on brain heart infusion agar and MRS agar, respectively. The two organisms were streaked onto the agar medium and incubated at room temperature until visible growth was observed. Bacterial cells were then harvested in phosphate-buffered saline and washed twice by centrifugation (10,000 x g, 10 min at 5 °C). Microorganisms were then resuspended in PBS to an $O.D_{540 \ nm}=1.0$. One milliliter aliquots of the bacterial suspension were then pelleted by centrifugation and the supernatant was discarded. Lysozyme or lysozyme-dextran conjugate were then added to the bacterial pellets to give a final lysozyme concentration of 0.05 %. Bacterial cells were kept in suspension by gentle mixing for 1 h at room temperature. Following the incubation period, 1 ml of the reaction mixture was diluted in PBS, plated, and incubated at room temperature. Colonies were counted after 48 h.

3.   RESULTS AND DISCUSSION

Separation of unreacted dextran and lysozyme from conjugate was achieved using cation exchange chromatography with both a phosphate buffer (Figure 1) and with a volatile buffer, ammonium bicarbonate (Figure 2). Use of the volatile buffer allowed unreacted components to be lyophilized and re-incubated to improve the efficiency of the reaction, with minimal increase in ionic strength to the reaction, and was used for all subsequent runs.

The conjugate was eluted at 0.2 M NaCl, and free lysozyme was eluted in a sharp peak by increasing the salt concentration to 1.0 M. SDS-PAGE of fractions obtained from the cation exchange column are shown in Figure 3.With all samples, a small fraction of protein was eluted early in the chromatographic run, along with the free dextran, and showed no affinity for the cation exchange column. Dialysis of samples prior to loading on the column had no effect on retention of this fraction. Less than one free amino group per lysozyme molecule was detected in this fraction. Therefore, it is likely that this protein fraction is highly derivatized lysozyme, and that the dextran chains effectively coat the surface of the protein, eliminating the affinity of lysozyme for the negatively charged resin.

On average, just over 2 of the 7 potential amino groups of lysozyme were derivatized under the conditions used (Table 1), similar to the extent of derivatization reported by Nakamura *et al.* (1991). Molecular weight was not related to the number of derivatized amino groups. Approximately 40% of lysozyme was derivatized with the 10 kD dextran during the three week incubation period. The amount of derivatization decreased with increasing molecular weight of the dextran, with approximately 35, 30 and 18% of lysozyme forming a conjugate with the 37.5, 65.5 and 188 kD dextrans, respectively. The amount of browning observed in the samples decreased with increasing molecular weight. SDS-PAGE of the 0.2 M NaCl fractions of each dextran obtained from Hi-Trap SP chromatography of conjugates is shown in Figure 4. The 10 kD dextran derivatized fraction was then subject to ultrafiltration to remove residual free lysozyme, which increased the amount of conjugate in the fraction to 75%. Similar increases in the percentage of conjugate were noted for derivatives of the other molecular weight dextrans after ultrafiltration.

Table 1. Chemical Characterization of Lysozyme-Dextran Conjugates.

Sample	Amino groups derivatized per lysozyme molecule [1]	% Lysozyme derivatized [2]
HEWL + 10	2.2	40
HEWL + 37.5	1.3	35
HEWL + 65.5	3.0	30
HEWL + 188	2.5	18

[1] Out of a total of 7 possible amino groups

[2] Estimated from FPLC chromatograms prior to ultrafiltration, after 3 weeks incubation

Table 2. Lytic and Antibacterial Activity of Native Lysozyme and Lysozyme-Dextran Conjugates

Sample	Lytic activity in U/mg protein[1] (% HEWL)	Survival of L. mesenteroides (%)[2]	Survival of E. coli (%)[3]
Lysozyme	17,160 (100%)	2	96
Lysozyme + 10	1888 (11%)	2	100
Lysozyme + 37.5	2928 (17%)	9	75
Lysozyme + 65.5	2368 (14%)	2	79
Lysozyme + 188	848 (5%)	0.1	79
65.5 Dextran	0 (0%)	92	83

[1] Determined against *M. lysodeikticus*, 1 U = amount of enzyme which will cause a decrease in absorbance of 0.001/min.

[2] Initial CFU/ml was 2.6×10^9

[3] Initial CFU/ml was 2.4×10^7

Although all conjugates retained less than 20% of the enzymatic (lytic) activity against *M. lysodeikticus* compared to native lysozyme, conjugates were just as effective as native lysozyme against *L. mesenteroides* (Table 2). This supports other studies that have found antimicrobial activity of lysozyme to be independent of enzymatic function (During *et al.*, 1999). However, there is an optimum level of conjugation for antimicrobial activity. Assays carried out with the highly conjugated lysozyme eluted first during the chromatography indicated that this fraction was at least 25 time less effective than the fraction eluted at 0.2 M NaCl against *L. mesenteroides*. Neither native

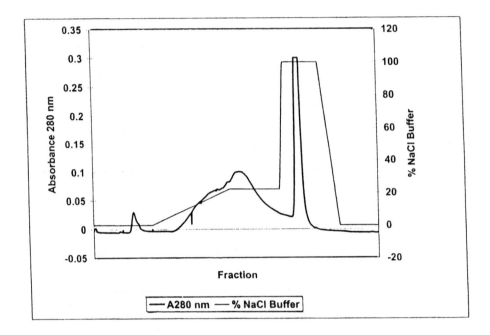

Figure 1. FPLC elution profile of lysozyme and 10 kD dextran reaction mixture from Hi-Trap SP column using 20 mM phosphate buffer, pH 7.0, and buffer containing 1 M NaCl. Insert Legend: Dashed line is carbohydrate as determined by the phenol sulphuric assay and solid line shows free amino groups per molecule of lysozyme.

lysozyme nor the conjugates were effective against *E. coli* under the assay conditions used. Lysozyme-polysaccharide conjugates have been shown to be effective against Gram-negative bacteria only when combined with a thermal treatment, which is likely to contribute to the destabilization and solubilization of the lipopolysaccharide (LPS) portion of the cell wall. Recently, *E. coli* has been shown to produce a potent protein inhibitor of type C lysozymes (Monchois *et al.*, 2001). Therefore, the inability of native lysozyme to lyse *E. coli* may be partially a result of inactivation through the binding of this inhibitor, and not solely due to an inability to penetrate the LPS layer, as commonly reported. However, for lysozyme conjugates where the antibacterial effect is related to surface activity rather than lytic activity, bacteria such as *E. coli* may be more susceptible under certain environmental conditions.

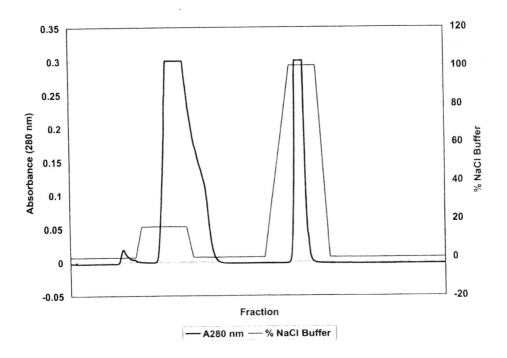

Figure 2. FPLC elution profile of lysozyme and 10 kD dextran reaction mixture from Hi-Trap SP column using 30 mM ammonium carbonate buffer, pH 7.7 and buffer containing 1 M NaCl.

All conjugates showed much greater thermal stability compared to free lysozyme (Figure 5) and may be beneficial for use in foods subject to thermal treatments. There was no difference in the thermal stability for lower molecular weight dextran (10 and 37.5 kD) conjugates. Shu *et al.* (1996) also reported that the heat stability of lysozyme

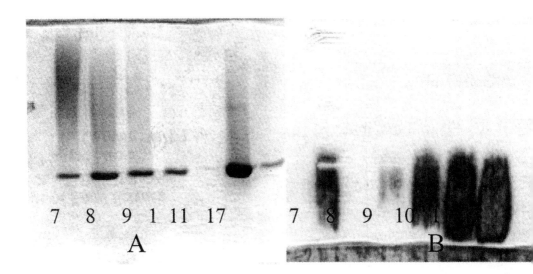

Figure 3. Sodium dodecyl sulfate-polyacrylamide gel electrophoresis of Hi-Trap SP column fractions (7-11, 17) of lysozyme and 10 kD dextran reaction mixture. A. Coomassie blue stain for protein. B. Alcian / Silver stain for carbohydrate.

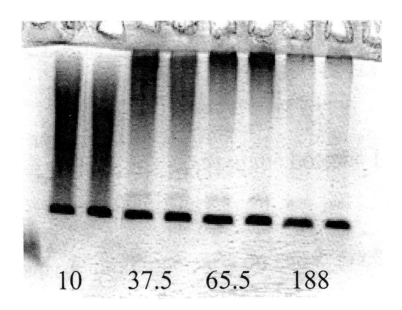

Figure 4. Sodium dodecyl sulfate-polyacrylamide gel electrophoresis of lysozyme-dextran conjugates after Hi-Trap SP chromatography (pooled fractions). Duplicate lanes of conjugates of each molecular weight dextran are shown.

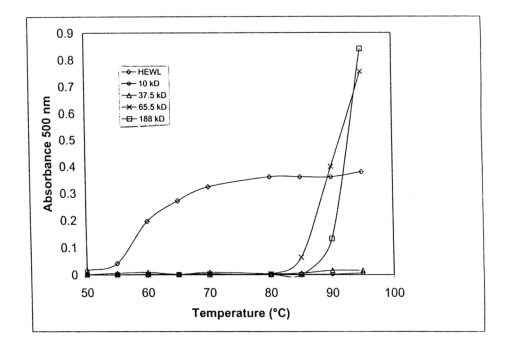

Figure 5. Thermal stability of lysozyme and lysozyme-dextran conjugates.

conjugates was similar for carbohydrates of molecular weight 3.5 to 24 kD. However, conjugates of high molecular weight (65.5 and 188 kD) dextrans were less stable at temperatures greater than 80°C. There is no advantage in using dextrans of molecular weight over 37.5 kD for derivatization, in terms of antimicrobial activity and heat stability. However, conjugation with carbohydrates may be useful in reducing allergenicity of lysozyme; the effect of the molecular weight of the dextran on this attribute of lysozyme has not been determined.

4. REFERENCES

Babiker, E. E., Kato, A., 1998, Improvement of the functional properties of sorghum protein by protein-polysaccharide and protein-protein complexes, *Nahrung,* **42**:286-289.

Babiker, E. E., Hiroyuki, A., Matsudomi, N., Iwata, H., Ogawa, T., Bando, N., Kato A., 1998, Effect of polysaccharide conjugation or transglutaminase treatment on the allergenicity and functional properties of soy protein, *J. Agric. Food Chem.,* **46**:866-871.

Dickinson, E., Galazka, V. B., 1991, Emulsion stabilization by ionic and covalent complexes of beta-lactoglobulin with polysaccharides, *Food Hydrocolloids,* **5**:281-296.

Dubois, M., Gilles, K. A., Hamilton, J. K., Rebers, P. A., Smith, F., 1956, Colorimetric method for determination of sugars and related substances, *Anal. Chem.,* **28**:350–356.

During, K., Porsch, P., Mahn, A., Brinkmann, O., Gieffers, W., 1999, The non-enzymatic microbicidal activity of lysozymes, *FEBS Lett.,* **449**:93-100.

Ford, T.F., Hermon-Taylor, J., Grant, D.A.W., 1982, A sensitive fluorometric assay for the simultaneous estimation of pepsin and pepsinogen in gastric mucosa, *Clinica Chimica Acta,* **126**:17-23.

Hattori, N., Numamoto, K., Kobayashi, K., Takahashi, K., 2000, Functional changes in beta-lactoglobulin by conjugation with cationic saccharides, *J. Agric. Food Chem.,* **48**:2050–2056.

Hwang, J. K., Choi, M. J., Kim, C. T. Emulsion properties of casein-alginate mixtures, *J. Kor. Soc. Food Sci. Nutr.,* **26**:1102–1108.

Ibrahim, H., Higashiguchi, S., Juneja, L. R., Kim, M., Yamamoto, T., 1996, A structural phase of heat-denatured lysozyme with novel antimicrobial action, *J. Agric. Food Chem.,***44**:1416-1423.

Kato, A., 1996, Functional protein-polysaccharide conjugates, *Comm. Agric. Food Chem.,* **3**:139-153.

Kato, A., Sasaki, Y., Furuta, R., Kobayashi, K., 1990, Functional protein-polysaccharide conjugate prepared by controlled dry heating of ovalbumin-dextran mixture, *Agric. Biol. Chem.,* **54**:107–112.

Kato, A., Sasaki, Y., Furuta, R., Kobayashi, K., 1991, Improvement of the functional properties of insoluble gluten by pronase digestions followed by dextran conjugation, *J. Agric. Food Chem.,* **39**, 1053-1056.

Kato, A., Mifuru, R., Matsudomi, N., Kobayashi, K., 1992, Functional casein-polysaccharide conjugates prepared by controlled dry heating, *Biosci. Biotech. Biochem.,* **56**:567-571.

Kato, A., Minaki, K., Kobayashi, K., 1993, Improvement of emulsifying properties of egg white proteins by the attachment of polysaccharide through Maillard reaction in a dry state, *J. Agric. Food Chem.,* **41**:540–543.

Laemmli, U.K., 1970, Cleavage of structural proteins during the assembly of the head of bacteriophage T4, *Nature,* **227**:680-685.

Matsudomi, N., Tsujimoto, T., Kato, A., Kobayashi, K., 1994, Emulsifying and bactericidal properties of a protamine-galactomannan conjugate prepared by dry heating, *J. Food Sci.,* **59**:428-431.

Miyaguchi, Y., Tsutsumi, M., Nagayama, K., 1999, Properties of glycated globin prepared through the Maillard reaction. *J. Japan. Soc. Food Sci. Technol.,* **46**:514-520.

Moller, H.J., Poulsen, J.H., 1995, Staining of glycoproteins/proteoglycans on SDS-Gels, in: *The Protein Protocols Handbook,* J. M. Walker (ed.)., Humana Press Inc., Totwa, NJ, pp. 627-631.

Monchois, V., Abergel, C., Sturgis, J., Jeudy, S., Claverie, J. M., 2001, *Escherichia coli ykfE* ORFan gene encoded a potent inhibitor of c-type lysozyme, *J. Biol. Chem.,* **276**:18437-18441.

Nakamura, S., Kato, A., Kobayashi, K., 1991, New antimicrobial characteristics of lysozyme-dextran conjugates, *J. Agric. Food Chem.,* **39**:647-650.

Nakamura, S., Kato, A., Kobayashi, K., 1992, Enhanced antioxidative effect of ovalbumin due to covalent binding of polysaccharides, *J. Agric. Food Chem.,* **40**:2033-2037.

Nakamura, S., Kobayashi, K., Kato, A., 1994, Role of positive charge of lysozyme in the excellent emulsifying properties of Maillard-type lysozyme-polysaccharide conjugate, *J. Agric. Food Chem.,* **42**:2688-2691.

Nakamura, S., Ogawa, M., Nakai, S., Kato, A., Kitts, D. D., 1998, Antioxidant activity of a Maillard-type phosvitin-galactomannan conjugate with emulsifying properties and heat stability, *J. Agric. Food Chem.,* **46**:3958-3963.

Nakamura, S., Ogawa, M., Saeki, H., Saito, M., Miyasaka, S., Hata, J., Adachi, N., Hwang, J. K., 2000, Improving the surface functionality of curdlan by conjugation with unfolding protein through naturally occurring Maillard reaction, *J. Food Sci. Nut.,* **5**:200–204.

Shu, Y., Sahara, S., Nakamura, S., Kato, A., 1996, Effects of the length of polysaccharide chains on the functional properties of the Maillard-type lysozyme-polysaccharide conjugate, *J. Agric. Food Chem.,* **44**:2544-2548.

Tanaka, M., Kunisaki, N., Ishizaki, S., 1999, Improvement of emulsifying and antibacterial properties of salmine by the Maillard reaction with dextrans, *Fish. Sci.,* **65**:623-628.

Wahyuni, M., Ishizaki, S., Tanaka, M., 1999, Improvement in the functional properties of fish water soluble proteins with glucose-6-phosphate through the Maillard reaction, *Fish. Sci.*, **65**:618-622.

Yu, T. H., Ishizaki, S., Tanaka, M., 2000, Improving emulsifying activity of epislon-polylysine by conjugation with dextran through the Maillard reaction, *Food Chem.*, **68**:449-455.

QUALITY MODIFICATION OF FOOD BY EXTRUSION PROCESSING

Christopher M. Gregson and Tung-Ching Lee*

1. INTRODUCTION

Extrusion is a production method widely used industrially to cook and texturize food products. The extruder applies a temperature, pressure and shear regime influenced by many factors including moisture content, barrel and die temperatures, screw and die geometry, screw speed and feed rate. These in turn determine the physical and chemical changes which occur in the extruder. Carbohydrate and/or protein macromolecules constitute the major structural components in most extruded foods and can undergo gelatinization, denaturation, depolymerization and formation reactions. Also, the moisture content in the final product largely determines their physical state (glass, rubber, crystalline). Other important changes can occur affecting lipids, colors, flavors, vitamins, minerals, antinutritional factors and enzymes, which can influence nutritional, organoleptic, functional and storage qualities. This chapter provides an updated review of our understanding of these changes focusing particularly on those that affect organoleptic quality.

2. EXTRUSION PROCESSING

Extrusion is a relatively new processing method that has rapidly gained popularity in the food industry. It is used to manufacture an ever-increasing variety of products, including textured vegetable proteins, ready to eat (RTE) breakfast cereals, expanded snacks and pasta products. Essentially, it involves conveying a feed material along a heated barrel using an Archimedes-type screw and then forcing it through a narrow slit or die. The advantages of using extrusion include the ability to process food continuously, energy efficiency, consolidation of multiple operations into a single process and high quality through high temperature short time (HTST) regimes. The problem, however, is that although the method is apparently simple, understanding extrusion processes fundamentally is extremely complex due to the large number of variable parameters (Table 1).

* Department of Food Science, the Center for Advanced Food Technology, and NJ-ECO COMPLEX Life Support Center, Rutgers University, 65 Dudley Road, New Brunswick, New Jersey 08901, U.S.A.

Quality of Fresh and Processed Foods, edited by Shahidi et al.
Kluwer Academic/Plenum Publishers, 2004.

Table 1. Major parameters involved in extrusion processing

Controllable (input)	Measurable/ calculable	Controlling	Quality (output)
Screw, barrel & die geometry	Residence time distribution	Substrates & concentration	Storage stability
Moisture content	Melt viscosity	Temperature	Color
Feed rate	Power	Shear	Taste
Screw speed	Strain (WATS)	Pressure	Aroma
Barrel and die temperature	Energy (SME)	Time	Microbiology
Feed material	Spot pressures	Mixing	Toxicology
	Spot temperatures	Water activity	Nutritional
	Shear rate		Functional
	Flow rate		Texture
	Torque		

Extrudate quality is determined by the starting material and the reactions/ rearrangements that take place due to the application of thermal energy, pressure, shear and mixing. Conditions within the extruder vary both axially and radially resulting in a distributed (rather than uniform) set of processing conditions. The same starch granule could remain intact or become fully gelatinized depending on the path it is forced to take through the barrel and die.

The conditions that determine the rate of reactions are difficult to measure and cannot be manipulated directly. Instead, control is achieved through a number of input parameters, such as screw speed and feed rate. The relationship between the controllable and the controlling parameters is often complex. Useful information can be gained from simple measurements and calculations that serve to bridge the gap.

3. MAJOR DIFFERENCES: SHEAR, PRESSURE AND SHORT RESIDENCE TIMES

The major difference between extrusion and other thermal processes is that extrusion is a multiple unit operation process that applies not only thermal energy, but also mechanical energy and high pressure. The application of mechanical energy, the excellent thermal contact between the product and extruder and the high temperatures possible at high pressure allow the rate of many important cooking processes to become much faster and residence times shorter. Additionally, extrusion performs mixing and shaping.

High pressures within an extruder are the combined result of the screw forcing the feed material into a small aperture (the smallest commonly being the die), the heating of volatile materials, such as water, to high temperatures and the high viscosity preventing flow. The maximum pressure within an extruder is normally less than 100MPa (Noguchi, 1989). According to Wang *et al.* (1992), the effect of normal pressure on internal energy is nil for incompressible materials. Pressure should, therefore, have little effect on reaction kinetics although the assumed incompressibility may not be valid for some materials especially at higher pressures. A major effect of high pressure is that it allows volatile compounds to remain liquid at high temperatures. This enables high inputs of thermal energy but the retention of moisture and other volatile components within the food product. When the extrudate exits the die into normal atmospheric conditions, these compounds volatilize rapidly leading to an expansion of the structure and a loss of volatiles.

Noguchi (1989) conducted experiments on defatted soy protein systems using a thermally controlled high-pressure cell, thereby removing the effects of shear and enabling experiments to be conducted on small samples. It was concluded that during high temperature treatment, pressures of up to 50 MPa (the maximum experimental value) had no effect on protein reactions, although reports cited in the literature suggest that pressure could theoretically influence protein reactions. Bates *et al.* (1994) also used a reaction cell to predict color development in a starch-glucose-lysine system as a result of twin screw extrusion. Unfortunately, although 'equivalent' pressures to those found at the die were applied, pressure was not used as a variable.

The other major difference between conventional thermal processes and extrusion cooking is that mechanical energy is applied to the food material. As many cooking reactions proceed much quicker when shear energy is applied, residence times within the extruder can be significantly shorter. It has been suggested by a number of authors that the sensitivity of a molecule to shear is related to its size, therefore, large polymers such as amylose are very susceptible to shear conversion.

Bulut *et al.* (1999) found that shear stress correlated strongly with bacterial destruction during low temperature extrusion experiments on a thermostable microorganism. Zheng and Wang (1994) showed that shear activation energy is 2-3 orders of magnitude lower than the thermal activation energy. In a subsequent study, it was found that tribological shear (a result of powdery friction) resulted in a lower activation energy than rheological shear (Wang and Zheng, 1995). At the lower end of the size scale, Van Den Hout *et al.* (1998) found that shear had negligible effect on the inactivation of trypsin inhibitor proteins in soy flour. Interestingly, Cho *et al.* (1995) showed how flavor compounds were more efficiently produced from starch-methionine and starch-cysteine mixtures using cold extrusion than either hot extrusion or thermal cooking.

The shear conditions within an extruder are complex. Shear is influenced by the screw speed, screw and die conformation and the rheology of the food material. Typically, shear is greatest at the die and reverse screw elements (Bulut *et al.*, 1999). Shear energy is converted to thermal energy within the extruder, thereby causing both local and bulk temperature changes further complicating the study of thermal versus shear reactions. Using rheometers, experiments can be conducted at constant, controllable shear and temperature conditions to determine the thermomechanical energy required for changes to occur. Both rotational (Bulut *et al.*, 1999) and capillary rheometers (Zheng and Wang, 1994) have been used for this purpose, however capillary rheometers are more commonly capable of reproducing the high shear conditions found within an extruder. Another approach used by a number of authors (for example Van Den Hout *et al.*, 1998) to determine the effect of shear is to calculate the theoretical thermal effect from zero shear experiments and compare this to the total effect.

4. TEXTURE

For many products, extrusion is primarily a texturization process as other factors, such as flavor and color, can be altered through the addition of ingredients post-extrusion. Texture is a measure of mechanical properties made in the mouth during eating and is therefore a function of macrostructure. The most important structural components in extruded food are usually starch and/ or protein (Smith, 1976). This section will concentrate on the influence of extrusion on starch and protein and the resulting texture.

Extrusion can be used to texturize a variety of protein-rich plant materials to yield meat analogs or extenders (Stanley, 1989). The goal of protein texturization is to produce a chewy, fibrous structure, the quality of which is determined by the proximity of its texture to meat. Flavor and color are often added either pre- or post-extrusion to yield a product used to replace or extend meat.

Commonly, defatted soy meal, which contains approximately 50% protein, is used as the base ingredient (Sevatson and Huber, 2000). High temperatures, pressures, and shear result in the dissociation of protein subunits and the formation of a continuous plastic melt (Stanley, 1989). Due to laminar flow conditions the protein molecules align parallel to the screw and then aggregate to form structured fibers. Release of steam at the die exit results in product expansion and the formation of a porous protein-rich structure containing carbohydrate inclusions. Both protein quality and quantity are important during texturization as well as insoluble carbohydrates due to their influence on the formation of air cells.

Protein solubility reduces as a function of temperature due to cross-linking. Cross-linking occurs due to ionic, hydrogen, hydrophobic, disulfide and perhaps other covalent bonds, however, the mechanism of texturization is not totally known (Stanley, 1989). Major aggregation of the dissociated subunits are hydrophobic and ionic bonds, however, the texture is probably the result of a relatively small number of bonds. Up to a point, increasing extrusion temperature produces a more textured product, the process is also influenced by pH and ionic strength. The process of defatting soy involves heating and can result in protein denaturation and a loss of functionality and extrusion performance. Frazier and Cranshaw (1984) showed that defatted soy protein with a low nitrogen solubility index texturized less. Rhee (1981) found that protein with a high nitrogen solubility index produced more fibrous and expanded structures.

Protein texturization is usually conducted at low moisture conditions (10-30%) using single-screw extruders. The result is slightly expanded, textured product with a low moisture content. However, structures much closer to fresh meat can be produced using high moisture extrusion with twin screw extruders as reviewed by Cheftel *et al.* (1992) and Noguchi (1989). The improved mixing, kneading, transportation capabilities, and controllable operations of the twin-screw extruder make it well suited to wet extrusion cooking. The extra water results in a low viscosity melt within the barrel and so cooling at the die is necessary to produce a self-supporting structure. A multiple orifice die (breaker plate) is often used to align the protein molecules and form a fibrous texture. Using methods borrowed from the polymer extrusion industry, Noguchi (1989) used injection molding to form the product into the desired shape.

Extruders are also used to texturize starch and flour towards the manufacture of various products including snack foods, RTE cereals, flat breads, and functional polysaccharides as reviewed by Colonna *et al.* (1989). The basic texture required for each product is different and can depend upon the macroscopic expanded structure, cohesiveness, shape, mouthfeel, thickening behavior and susceptibility to amylolysis. Second generation snack products, for example, normally have a crisp expanded structure, whereas, RTE breakfast cereals require thick cell walls, a high density and a slow rate of hydration for extended 'bowl-life'. The quality of many extruded products is dependent on their crispness, which correlates with a high elastic modulus and low rupture strength (Launay and Lisch, 1983).

Extrusion has no effect on the monosaccharide subunits comprising starch or on the total starch concentration (Colonna et al., 1989). The size of high molecular weight structures is reduced through the action of thermal and mechanical forces, whereas low molecular weight polymers are largely unaffected. Amylopectin and amylose, and 1-6 and 1-4 bonds, have approximately the same susceptibility to breakdown. Starch modification is reduced when lipids are present as they act as lubricants by reducing friction within the extruder. Conversely, by adding a thermostable alpha amylase during extrusion, starch can be made to undergo rapid liquefaction and thereby produce low molecular weight maltodextrin syrups. Using glucoamylase, these syrups can be further depolymerized to yield high glucose syrups with DE's of up to 88% (Chouvel et al., 1983).

Native starch does not absorb water at room temperature and has a very low viscosity, whereas extruded starch absorbs water rapidly to form a paste. Granular and crystalline structures usually disappear during extrusion cooking, however, they may be able to endure high moisture or low shear conditions (Colonna et al., 1989). Extruding starch results in a structured material comprising of a continuous amorphous phase, often containing regions of crystalline amylose-lipid complexes that are formed at the cooling stage.

High pressure and temperature at the die exit results in product expansion, which is essential to the texture of many products e.g. second-generation snacks (Colonna et al., 1989). Air comprises 85-92% of the total volume of the open foam structure. The structure of extrusion-expanded products depends on starch gelatinization and the starch melt at the die. At the die exit, water rapidly vaporizes causing the structure to expand. The product becomes rigid as the temperature falls below Tg during cooling and thereby retains its shape. The degree of expansion may be related to starch content. Starch damage prior to extrusion results in products with smaller pores, softer textures, greater solubility and a sticky character when eaten (Launay and Lisch, 1983). Bran can result in reduced expansion through cell walls rupture (Guy and Horne, 1988).

Colonna et al. (1989) reviewed the effect of added ingredients and process variables on extruded starch. Expansion increases up to 5% fat, after which it drops rapidly. Concentrations of sucrose greater than 15% increase expansion but it decreases with increasing gluten content. Expansion generally decreases rapidly with increasing moisture content and reaches a maximum at 170°C. Mechanical strength is governed by the extent of heat treatment and the degree of starch transformation before extrusion. A low moisture content and high temperature results in increased crispness although an amylose concentration of 5-20% is usually required to afford an acceptable texture.

5. FLAVOR

One of the factors determining the palatability of an extruded product is flavor: a parameter incorporating aroma, taste, texture and some aspects of mouthfeel such as astringency. In real food products, flavor is rarely determined by a single compound but rather by a combination of tens or hundreds of both volatile and non-volatile ingredients. It should be remembered that it is the balance of flavors released at the time of eating rather than concentrations within the product that determine consumer acceptance and that both storage and extrusion can have significant effects on the flavor profile.

The basic ingredients used in extrusion (e.g. wheat, maize, soy) are often bland. Unlike many conventional cooking procedures, such as baking, very little flavor is produced from these materials during extrusion despite high temperatures (Izzo *et al.*, 1994). This is due to retention times being insufficient for polymer degradation and subsequent flavor production. The normal approach used to produce a desirable product is to minimize flavor production during extrusion and to add the desired compounds post-extrusion (Villota and Hawkes, 1994). This allows the manufacturer to have full control over the outcome and the ability to manufacture multiple flavors of the same product with very little difficulty. The main issue to consider in this case is the degradation and production of flavors during storage.

An alternative approach is to add flavors prior to or during extrusion, which is preferable to the addition of flavors post extrusion in a number of ways. The flavor is dispersed evenly within the product rather than coating the surface, thereby improving consumer acceptance. As the compounds are less exposed to their surroundings, some degree of protection is provided against oxidative damage during storage. Additionally, manufacturing is simplified as a processing operation is removed.

Low-temperature low-pressure extrusion provides few problems as flavor retention is high. When the process is more severe, various effects must be considered including thermal degradation, oxidation, polymerization, reaction with the matrix and volatilization. Palkert and Fagerson (1980) reported that only 4 to 22% of flavor volatiles were retained when added to textured soy protein prior to extrusion. Sadafian and Crouzet (1988) showed that by adding flavors as free volatiles losses of over 90% can occur.

Several solutions have evolved that increase flavor retention. By incorporating flavor close to the end of the extrusion process, process severity (as experienced by the flavor compound) is reduced yet the extrusion process is unaffected (Kollengode and Hanna, 1997b). A number of flavors with increased heat stability have been developed although with limited success and a restricted range of flavors (Maga, 1989). Sadafian and Crouzet (1988) showed that single layer encapsulation with a polysaccharide or hydrocolloid resulted in a reduction in losses from 90 to 70% whereas multilayer encapsulation resulted in just a 10% loss. Kollengode and Hanna (1997a and b) showed that complexing a flavor volatile with cyclodextrin resulted in 70 to 100% recovery for a variety of flavor compounds. Concern has been raised over the ability of very stable forms of volatile to be released upon eating.

In some cases volatilization has been found to be more important than chemical reaction in determining the degree of flavor retention (Villota and Hawkes, 1994). High pressures and temperatures are often found within the extruder. At the die exit there is a sudden reduction in pressure resulting in the volatilization of various components. This leads not only to the expansion of the product, but also the flashing off of flavor volatiles.

This can be detrimental when desirable flavors are lost or beneficial by driving away off-flavors e.g. in soy beans during the manufacture of textured protein products (Kinsella, 1978).

The degree to which a compound is lost through volatilization depends upon its volatility (Villota and Hawkes, 1994). Kim and Maga (1994) found that retention increased with increasing chain length, was greater for alcohols than acids and even lower for aldehydes. Chen (1986) produced two models to predict volatile loss; one based on a thermodynamic approach and the other using the volatility of a compound relative to water. Nair *et al.* (1994) and Vodovotz *et al.* (2000) described an apparatus designed to collect the volatiles produced at the extruder die so that further analysis could be conducted using GCMS. Nair *et al.* (1984) found a range of compounds including Maillard and carotenoid decomposition products and discovered that the balance of volatiles released at the die were significantly different from those remaining in the extrudate. Hwang *et al.* (1997) found that adding cysteine to wheat flour before extrusion resulted in an increased release in volatiles released at the die, especially sulfur containing species.

Flavors can also be produced through reactions in the barrel although as already mentioned retention times are too low for sufficient production from the basic ingredients used in extrusion. Bredie *et al.* (1998) found that when high temperatures and low moisture contents were used to extrude maize flour, the majority of the resulting flavor compounds were lipid degradation products. The same sample extruded at a high temperature and low moisture content resulted in flavor compounds being predominantly Maillard reaction derived. Katz (1994) stated that Maillard products produced during extrusion are less acceptable than those produced using conventional cooking processes.

By adding flavor precursors to the mix, the Maillard reaction can be tailored to produce flavors within the extruder of the desired species and concentration. Tanaka (1994) showed that glycoside could be used as a flavor precursor. Hwang *et al.* (1997) added cysteine to wheat flour resulting in the production of increased concentrations of sulfur containing Maillard products. Izzo *et al.* (1994) showed that ammonium bicarbonate and pyruvaldehyde enhanced the production of heterocyclic pyrazines and therefore an increased toasted aroma. Bailey *et al.* (1994) showed that the addition of whey protein concentrate to cornmeal prior to extrusion resulted in the formation of lipid oxidation and Maillard products. Cho *et al.* (1995) showed that cold extrusion of starch-methionine and starch-cysteine mixtures resulted in higher levels of flavor volatiles and a more pleasant profile than when hot extrusion was used.

6. COLOR

Color is an important parameter determining the aesthetic quality of food is usually determined in a product by a small number of compounds or a group of compounds with similar properties. Generally, the feed materials used for extrusion are light in color (Berset, 1989). Moreover, HTST processes do not provide the conditions required for, or are tailored to prevent, significant color formation. Without the addition of extra ingredients, extrudates remain similar in color to their feed material. Generally, colored compounds are not volatile and so are not affected by pressure drops at the die exit. Although normally associated with chemical degradation, decreases in product color can also occur as a result of increased expansion (reduction in concentration per unit area) (Maga and Cohen, 1978) and reduced gelatinization (change in the way light is reflected, refracted and absorbed).

Color is often controlled by adding pigments before, during or after extrusion and interesting products can be made through the use of more than one color in conjunction with innovative processing, such as co-extrusion (Berset, 1989). When colors are sprayed on post-extrusion only the outer surface of the product becomes colored, however, colors that are not extrusion stable can be used in this way. Adding colors before or during extrusion allows for even distribution. The colors must be highly stable as severe processing conditions and long shelf-life cause degradation, however, a range of natural and artificial colors are available suitable for use in extruded products.

Excessive heating or local superheating can cause degradation. Some colors are relatively heat stable, such as "Tartrazine" and "Green S" (Berset, 1989). Reducing agents (e.g. ascorbic acid, reducing sugars, metal ions) can increase the effect of temperature. Light can also cause color loss, indigo carmine and erythrosine being the most affected, however, this is a storage issue that can be largely solved through correct packaging.

There is a trend in the food industry towards the use of natural colors. With the exception of some carotenoids and caramel, most 'natural' colors are unsuitable for extrusion cooking due to the high temperatures involved (Berset, 1989). However, by adding color towards the end of the process more heat labile compounds can be used.

Carotenoids (e.g. annatto, β-carotene, canthaxanthin) give colors from yellow to red and are quite temperature stable. Marty (1986) showed that extrusion cooking of β-carotene at 180°C was more destructive than simple heating at the same temperature for 2 h. Extrusion was found to enhance cis-trans isomerization and the initiation of sequential reactions that lead to oxidative degradation. Increased degradation is perhaps due to the combined effects of temperature, local superheating, pressure and shearing although conventional wisdom suggests that β-carotene is non-volatile and too small to be affected by shear. The form in which a carotenoid is added also seems important in determining its thermal stability. Berset (1989) reported 75% loss of β-carotene when added pure, yet less than 10% loss when incorporated into a matrix of gelatin and maltodextrin.

Browning can occur through the Maillard or caramellization reactions. The Maillard reaction occurs in high temperature, low moisture conditions when reducing sugars and free amine groups are present (Ilo and Berghofer, 1999). Starch hydrolysis can result in the liberation of reducing sugars although this process is often too slow for sufficient color formation to occur. In the same way as for flavor production, precursors can be added to the raw ingredients to promote the Maillard reaction. In some products, for example baked goods, browning is required for product acceptance. Too much, however, can result in quality reduction through the formation of dark colors and bitter tastes (Berset, 1989). Enzymes can also cause the browning of foods, however, they are usually inactivated during extrusion.

7. NUTRITION

As food is designed primarily to supply the consumer with nutrients, the level and bioavailability of these compounds is an important quality determining factor. The nutritional quality of extruded products is becoming a more important issue due to the increasing number of products manufactured using this technique. Nutrition is especially important when the product is eaten as the sole food, for example infant formula. Nutrients vary considerably in their chemical structure and therefore in their ability to withstand processing. A number of reviews cover the subject of nutrition in relation to extruded products and the effects of extrusion processing, for example Bjorck and Asp (1983), de la Guerivier *et al.* (1985), Cheftel (1986), Camire *et al.* (1990), Killiet (1994) and Camire (2000).

Minerals are largely unaffected by extrusion as they are very stable and unlikely to be lost through volatilization. The bioavailablility of minerals can be a problem as binding can occur with constituents such as fiber and phytates (Camire, 1990). Extrusion, however, has been found to reduce phytate content in wheat (Fairweather-Tait *et al.*, 1989) although it remains unaffected in legumes (Lombardi-Boccia *et al.*, 1991). The concentration of trace metals can actually rise during extrusion due to screw and barrel wear, which increases with fiber content and extrusion temperature. Fairweather-Tait *et al.* (1987 and 1989) showed that screw wear iron and zinc are well utilized by the body.

Products are often fortified with minerals to increase their nutritional quality. Kapanidis and Lee (1986), however, showed that heavy fortification of rice with iron containing phenolic compounds can result in the formation of dark compounds although color formation can be reduced through the use of ferrous sulfate heptahydrate. Matinez-Bustos *et al.* (1998) showed that when calcium hydroxide was incorporated into cornmeal, expansion decreased and the product became lighter in color.

Killiet (1994) provided an excellent review of vitamin retention during extrusion cooking. He concluded that in general vitamin destruction increases with increasing temperature, shear stress, energy and decreasing moisture content, die cross sectional area and throughput. In summary then, factors that increase temperature and shear stress increase destruction.

Vitamins vary a lot in their chemical structure as well as their stability to processing (Camire, 2000). Another issue is the stability of natural versus added vitamins. Vitamins D and K are relatively stable. Vitamins A, C and E and related compounds are sensitive to both oxygen and high temperature. Gusman-Tello and Cheftel (1990) found that β-carotene (a precursor of vitamin A) was sensitive to both temperature and oxygen. Of the B vitamins, thiamine is most sensitive to high temperature.

The vitamin content of extruded foods can be increasing through post extrusion application, which is the best method economically as no processing losses occur and palatability is largely unaffected. Unless surface addition is used to add flavors or colors, however, post extrusion application complicates processing by introducing an extra step. The other option is to add vitamins before or during extrusion. The forms of the vitamin with the greatest stability and bioavailability should be chosen. Killiet (1994) stated that thiamine hydrochloride is more stable than thiamine mononitrate and L-ascorbyl-2 polyphosphate is more stable than ascorbic acid during extrusion processing. Lee *et al.* (1978) found that retinal acetate and retinal alcohol (forms of Vitamin A) were relatively stable during extrusion when compared to provitamin A carotenoids such as β-carotene or canthaxanthin. Putnam (1986) showed that vitamin A was readily oxidized during extrusion, but to a lesser extent when added in bead form.

Starch, the most important source of energy in the human diet, becomes more digestible as a result of extrusion or other cooking processes and digestibility is related to the degree of gelatinization. High digestibility is required for infant formula and weaning foods whereas low digestibility, perhaps through the production of enzyme resistant starch, may be desirable in reduced calorie products (Camire, 2000). Chiu et al. (1994) patented a procedure to produce 30% resistant starch by extruding high amylose starch with pullulanase.

Protein digestibility is generally improved by extrusion (Asp and Bjorck, 1989). Denaturation can lead to exposure of enzyme susceptible sites and a resulting increase in digestibility. The concentration of lysine (and therefore the quality of the protein) may reduce due to the Maillard reaction.

The concentration of lipids in extruded products is generally low (Camire, 2000). Levels as low as 5 % can reduce extruder performance by reducing shear due to slip and expansion due to lower pressures. A widely researched topic is the ability of amylose and protein to complex with lipids during extrusion, although as complexed lipids are digestible nutritional quality is unaffected.

8. MICROBIOLOGY

The concentration and type of microbes within a product is a quality-determining factor as some organisms can cause food borne illness or spoilage. The water activity of most extruded products is low, in which case they can be regarded microbiologically stable (Likimani et al., 1990). Extrusion is often a high-temperature short-time (HTST) process, which reduces microbial numbers whilst maintaining other quality attributes such as color. Extrusion has been shown effective at reducing total microbial count (e.g. Gry et al., 1984), non-sporulating bacteria (e.g. Queguiner et al., 1989), bacterial spores (e.g. Li et al., 1993) as well as yeast (Kauffman and Hatch, 1977). The degree of reduction can vary tremendously depending on the extrusion conditions, substrate and type of microorganism. Likimani and Sofos (1990) showed that spores not destroyed by extrusion processing may still be damaged, evidenced by the fact that a richer medium is required for growth to occur.

Applications have been developed which use extrusion as a method of reducing microbial load. Kelley and Walker (1999) showed that a single screw, dry extrusion process was able to reduce the concentration of potentially pathogenic bacteria in food waste during the production of animal feed. Using a dry extrusion process, Gry et al. (1984) reduced the microbial count of various spices by several orders of magnitude whilst maintaining other quality attributes. The high lethality of medium temperature extrusion used in conjunction with low-dose irradiation also shows the potential of combined techniques to control microbial populations (Garcia Zapeda et al., 1997).

As with other thermal processes, increased residence time and mass temperature leads to increased destruction of microorganisms. Mild extrusion conditions result in low lethality. Walsh and Funke (1975), for example, showed that the number of non-sporulating bacteria S. aureus was reduced by less than one log cycle during the production of spaghetti (35-50°C, 12-30rpm). Very high temperatures, however, can be effective at sterilizing products. Bouveresse et al. (1982), for example, showed that extrusion at 182°C reduced a population of *B. stearothermophillis* spores by 107 in a biscuit dough mixture. In order to predict the destruction of microbial spores during extrusion cooking, a method of calculating D and z values was developed by Likimani *et al.* (1990) based on the average residence time and mass temperature.

Likimani *et al.* (1990) stated that 'the effect of shear on the destruction of bacterial spores is believed to be negligble due to their small size' as illustrated by experiments with *B. globigii*. Reduction in the number of vegetative cells resulting from low temperature extrusion shows that a factor other than thermal energy, perhaps shear energy or pressure, could bring about the destruction of non-sporulating microorganisms. Kauffman and Hatch (1977) found a decrease in yeast viability in semi-dry dog food extruded at 40°C as screw speed was increased suggesting that shear energy was a factor in determining microbial destruction. Using low-temperature high-shear experiments on the non-sporulating organism *Microbacterium lacticum* with gelatin as the carrier, Bulut *et al.* (1999) clearly showed that shear energy can kill bacteria. Extruding at a shear stress of 409kPa for 49-58 secs (Tmax=73°C, H20=65%) resulted in effective sterilization. Additionally, it was shown that shearing the sample in a rotational rheometer at 2.8kPa for 4 minutes at a similar temperature resulted in a 1.4 log cycle reduction.

9. CONCLUSION

The quality of extruded materials and the changes undergone as a result of extrusion processing can be measured after extrusion using the whole range of laboratory techniques available to the food scientist. For the majority of products, structure is the most important parameter as extrusion is largely used as a texturization process. Normally, insufficient color and flavor is provided by the basic ingredients or produced during processing. Additional materials are often added either prior to, during or post-extrusion but effects such as degradation and volatilization can be a problem. Microbiological safety is usually assured through HTST high shear regimes and low final moisture content. The nutritional quality is becoming of great importance as extrusion is now used to create a wide range of products and can form a significant proportion of the diet.

10. REFERENCES

Asp, N. G. and Bjorck, I., 1989, Nutritional properties of extruded foods, in: *Extrusion Cooking*, C. Mercier, P. Linko, and J. M. Harper, eds., AACC, St. Paul, MN, pp. 399-434.

Bailey, M. E., Gutheil, R. A., Hsieh, F. H., Cheng, C. W., and Gerhardt, K. O., 1994, Maillard reaction volatile compounds and color quality of a whey protein concentrate - corn meal extruded product, in: *Thermally Generated Flavors*, T. H. Parliment, M. J. Morello and R. J. McGorrin, eds., American Chemical Society, pp. 317-327.

Bates, L., Ames, J. M., and MacDougall, D. B., 1994, The use of a reaction cell to model the development and control in extrusion cooked foods, *Lebens.-Wiss. und Technol.*, **27**:375-379.

Berset, C., 1989, Color, in *Extrusion Cooking*, C. Mercier, P. Linko and J. M. Harper, eds., AACC, St. Paul, MN, pp. 371-386.

Bjorck, I. and Asp, N. G., 1983, *The effects of extrusion cooking on nutritional value – a literature review, J. Food Eng.*, 2:281-308.

Bouveresse, J. A., Chouvel, H., Pina, M., Graille, J., and Cheftel, J. C., 1982, Influence of extrusion-cooking on the thermal-destruction of bacillus-thermophillus spores in a starch-protein-sucrose mix. *Lebens.-Wiss. und Technol.*, **15**:135-138.

Bredie, W. L. P., Mottram, D. S., and Guy, R. C. E., 1998, Aroma volatiles generated during extrusion cooking of maize flour, *J. Agric. Food Chem.*, **46**:1479-1487.

Bulut, S., Waites, W. M., and Mitchell, J. R., 1999, Effects of combined shear and thermal forces on destruction of microbacterium lacticum, *Appl. Exper. Microbiol.*, **65**:4464-4469.

Camire, M. E., Camire, A., and Krumhar, K., 1990, Chemical and nutritional changes in foods during extrusion, *Crit. Rev. Food Sci. Nutr.*, **29**:35-57.

Camire M. E., 2000, Chemical and nutritional changes in food during extrusion, in: *Extruders In Food Applications*, M. N. Riaz, ed., Technomic Publishing Company, Lancaster, pp. 127-147.

Cheftel, J. C., 1986, Nutritional effects of extrusion-cooking, *Food Chem.*, **20**:263-283.

Cheftel, J. C., Kitagawa, M. and Queguiner, C., 1992, New protein texturization processes by extrusion cooking at high moisture levels, *Food Rev. Int.*, **8**:235-275.

Chen, J., Reineccius, G. A., and Labuza, T. P., 1986, Prediction and measurement of volatile retention during extrusion processing, *J. Food Technol.*, **21**:365-383.

Chiu, C. W., Henley, M., and Altieri, P., 1994, Process for making amylase resistant starch from high amylose starch. United States Patent US05281276.

Cho, M. H., Zheng, X., Wang, S. S., Kim, Y., and Ho, C. T., 1995, Production of natural flavors using a cold extrusion process, in: *Flavor Technology: Physical Chemistry, Modification, and Process.* American Chemical Society, Washington, DC, pp. 120-126.

Chouvel, H., Chay, P. B., and Cheftel, J. C., 1983, Enzymatic hydrolysis of starch and cereal flours at intermediate moisture contents in a continuous extrusion reactor, *Lebens.-Wiss. und Technol.*, **16**:346-353.

Colonna, P., Tayeb, J., and Mercier, C., 1989, Extrusion cooking of starch and starchy products, in: *Extrusion Cooking*, C. Mercier, P. Linko, and J. M. Harper, eds., AACC, St. Paul, MN, pp. 247-319.

De la Guerivier, J. F., Mercier, C. and Baudet, L., 1985, Incidences of extrusion-cooking on certain nutritional parameters of food products, especially cereals, *Cah. Nutr. Diet.*, **20**:201-214.

Fairweather-Tait, S. J., Portwood, D. E., Symss, L. L., Eagles, J., and Minski, M. J., 1989, Iron and zinc absorption in human subjects from a mixed meal of extruded and nonextruded wheat bran and flour, *Am. J. Clin. Nut.*, **49**:151-155.

Fairweather-Tait, S. J., Symss, L. S., Smith, A. C. and Johnson, I. T., 1987, The effect of extrusion cooking on iron absorption from maize and potato, *J. Sci. Food Agric.*, **39**:341-348.

Frazier, P. J., and Crawshaw, A., 1984, Relationship between die-viscosity, ultrastructure and texture of extruded soya proteins, in:*Thermal Processing and Quality of Foods*, P. Zeuthen *et al* eds., Elsevier Applied Science Publishers, New York, pp. 89-95.

Garcia Zapeda, C. M., Kastner, C. L., Wolf, J. R., Boyer, J. R., Kropf, D. H., Hunt, M. C., and Setser, C. S., 1997, Extrusion and low-dose irradiation effects of destruction of clostridium sporogenes spores in a beef-based product, *J. Food Protec.*, **60**:777-785.

Gry, P., Holm, F., and Kristensen, K. H., 1984, Degermination of spices in an extruder, in: *Thermal Processing and Quality of Foods*, P. Zeuthen *et al* eds., Elsevier Applied Science Publishers, New York, pp. 185-188.

Guy, R. C. E. and Horne, A. W., 1988, Extrusion and co-extrusion of cereals. In: *Food Structure – Its Creation and Evaluation.* J. M. V. Blanshard and J. R. Mitchell, eds., Butterworths, London, pp. 331-349.

Guzman-Tello, R. and Cheftel, J. C., 1990, Colour loss during extrusion cooking of beta-carotene- wheat flour mixes as an indicator of the intensity of thermal and oxidative processing, *Int. J. Food Sci. Technol.*, **25**:420-434.

Hwang, C. F., Riha, W., Jin, B., Karwe, M. V., Hartman, T. G., Daun, H., and Ho, C. T., 1997, Effect of cysteine addition on the volatile compounds released at the die during twin-screw extrusion of wheat flour, *Lebens.-Wiss. und Technol.*, **30**:411-416.

Ilo, S. and Berghofer, E., 1999, Kinetics of color changes during extrusion cooking of maize grits. *J. Food Eng.*, **39**:73-80,

Izzo, H. V., Hartman, T. G., and Ho, C. T., 1994, Ammonium bicarbonate and pyruvaldehyde as flavor precursors in extruded food systems, in: *Thermally Generated Flavors*, T.H. Parliment, M.J. Morello and R.J. McGorrin, eds., American Chemical Society, pp. 328-333.

Kapanidis, A. N. and Lee, T. C., 1996, Novel method for the production of color-compatible ferrous sulfate-fortified simulated rice through extrusion, *J. Agric. Food Chem.*, **44**:522.

Katz, I., 1994, Maillard, microwave, and extrusion cooking. Generation of aromas, in: *Thermally Generated Flavors* T. H. Parliment, M. J. Morello and R. J. McGorrin, eds., American Chemical Society, pp. 2-6.

Kauffman, R. E. and Hatch, R. T., 1977, Shear effects on cell wall viability during extrusion of semi-moist foods. *J. Food Sci.*, **42**:906-910.

Kelley, T. R. and Walker, P. M., 1999, Bacterial concentration reduction of food waste amended animal feed using a single-screw dry-extrusion process, *Bioresource Technol.*, **67**:247-253.

Killeit, U., 1994, Vitamin retention in extrusion cooking, *Food Chem.*, **49**:149-155.

Kim, C. H. and Maga, J. A., 1994, Chain length and functional group impact on flavor retention during extrusion, in: *Thermally Generated Flavors*, T. H. Parliment, M. J. Morello and R. J. McGorrin, eds., American Chemical Society, Washington, DC, pp. 380-384.

Kinsella, J. E., 1978, Texturised proteins: fabrication, flavoring and nutrition, *Crit. Rev. Food Sci. Nutr.*, 11:147-207.

Kollengode, A. N. R. and Hanna, M. A., 1997, Cyclodextrin complexed flavor retention in extruded starches, *J. Food Sci.*, 62:1057-1060.

Kollengode, A. N. R. and Hanna, M. A., 1997, Flavor retention in pregelatinized and internally flavored starch extrudates, *Cereal Chem.*, 74:396-399.

Launay, B. and Lisch, J. M., 1983, Twin screw extrusion cooking of starches: behavior of starch pastes, expansion and mechanical properties of extrudates, *J. Food Eng.*, 2:259-280.

Lee, T. C., Chen, T., Alid, G., and Chichester, C. O., 1978, Stability of vitamin A and provitamin A (carotenoids) in extrusion cooking processing, *AIChE Symposium Series*, 74(172):192-201.

Li, Y., Hsieh, F, Fields, M. L., Huff, H. E., and Badding, S. L., 1993, Thermal inactivation and injury of clostridium sporogenes spores during extrusion of mechanically deboned turkey mixed with white corn flour, *J. Food Process. Preserv.*, 17:391-403.

Likimani, T. A., Alvarez-Martinez, L, and Sofos, J. N, 1990, The effect of feed moisture and shear strain on destruction of bacillus globigii spores during extrusion cooking, *Food Microbiology*, 7:3.

Likimani, T. A. and Sofos, J. N, 1990, Bacterial spore injury during extrusion cooking of corn soybean mixtures, *Int. J. Food Microbiol.*, 11:243-249.

Likimani, T. A., Sofos, J. N, Maga, J.A., and Harper, J. M, 1990, Methodology to determine destruction of bacterial spores during extrusion cooking, *J. Food Sci.*, 55:1388-605.

Lombardi-Boccia, G., DiLullo, G., and Carnovale, E., 1991, In vitro iron dialysability from legumes: influence of phytate and extrusion cooking, *J. Sci. Food Agric.*, 55:599-605.

Maga, J. A., 1989, Flavor formation and retention during extrusion, in: *Extrusion Cooking*, C. Mercier, P. Linko, and J. M. Harper, eds., AACC, St. Paul, pp. 387-398.

Maga, J. A. and Cohen, M. R., 1978, Effect of extrusion parameters on certain sensory, physical and nutritional properties of potato flakes, *Lebens.-Wiss. und Technol.*, 11:195.

Matinez-Bustos, F., Chang, Y. K., Bannwart, A. C., Rodriguez, M. E., Guedes, P. A., and Gaiotti E. R., 1998, Effects of calcium hydroxide and processing conditions on corn meal extrudates, *Cereal Chem.*, 75:796-801.

Nair, M., Shi, Z., Karwe, M. V., Ho, C. T., and Daun, H., 1994, Collection and characterization of volatile compounds released at the die during twin screw extrusion of corn flour, in: *Thermally Generated Flavors*,T. H. Parliment, M. J. Morello and R.J . McGorrin, eds., American Chemical Society, pp., 334-347.

Noguchi, A., 1989, Extrusion cooking of high-moisture protein foods, in: *Extrusion Cooking*, C. Mercier, P. Linko, and J. M. Harper, eds., AACC, St. Paul, MN, pp. 343-370.

Palkert, P. E. and Fagerson, I. S., 1980, Determination of flavor retention in pre-extrusion flavored textured soy protein, *J. Food Sci.*, 45:526-528, 533.

Queguiner, C., Dumay, E., Cavalieri, R. L., and Cheftel, J. C., 1989, Reduction of streptococcus thermophilus in a whey protein isolate by low moisture extrusion cooking without loss of functional properties, *Int. J. Food Sci. Technol.*, 24:601-612.

Rhee, K. C., Kuo, C. K., and Lusas E. W. (1981) Texturisation, in: *Protein Functionality in Foods*, J. P.Cherry, ed., American Chemical Society, Washington, DC, pp. 51-88.

Sadafian, A. and Crouzet, J., 1988, Aroma compounds retention during extrusion-cooking. in: *Frontiers of Flavor: Developments in Food Science 17*, G. Charalambous, ed., Elsevier Science Publishers: Amsterdam, The Netherlands.

Sevatson, E. and Huber, G. R., 2000, Extruders in the food industry, in: Extruder in Food Applications, M.N. Riaz, ed., Technomic Publishing, Lancaster, PA, pp. 167-204.

Smith, O. B., 1974, Extrusion cooking, in: *New Protein Foods* Vol.2, A. M. Altschul, ed., London Academic Press, pp. 86-121.

Stanley, D. W., 1989, Protein reactions during extrusion cooking, in: *Extrusion Cooking*, C. Mercier, P. Linko, and J. M. Harper, eds., AACC, St. Paul, MN, pp. 321-342.

Tanaka, S., Karwe, M. V., and Ho, C. T., 1994, Glycoside as a flavor precursor during extrusion, in glycoside as a flavor precursor during extrusion, *Thermally Generated Flavors*, T. H. Parliment, M. J. Morello and R. J. McGorrin, eds., American Chemical Society, Washington, DC, pp. 370-379.

van den Hout, R., Jonkers, J., van Vliet, T., van Zuilichem, D. J., and van 't Riet, K., 1998, Influence of extrusion shear forces on the inactivation of trypsin inhibitors in soy flour, *Trans IChemE Part C*, 76:155.

Villota, R. and Hawkes, J. G., 1994, Flavoring in extrusion: an overview, in: *Thermally Generated Flavors*, T. H. Parliment, M. J. Morello and R. J. McGorrin, eds., American Chemical Society, Washington, DC, pp. 280-295.

Vodovotz, Y., Zasypkin, D. V., Lertsiriyothin, W., Lee, T. C., and Bourland, C. T., 2000, Quantification and characterization of volatiles evolved during extrusion of rice and soy flours, *Biotechnol. Progress*, **16**:299-301.

Walsh, D. E. and Funke, B. R., 1975, The influence of spaghetti extruding, drying and storage on survival of Staphylococcus aureus, *J. Food Sci.*, **40**:714.

Wang, S. S., Chiang, W. C., Zheng, X., Zhao, B., and Yeh, A. I., 1992, Application of an energy equivalent concept to the study of the kinetics of starch conversion during extrusion, in: *Food Extrusion Science and Technology*, J. L Kokini, C. T. Ho, and M. V. Karwe, eds., Marcel Dekker, New York, pp. 165-176.

Wang, S. S. and Zheng, X., 1995, Tribological shear conversion of starch, *J. Food Sci.*, **60**:520.

Zheng, X. and Wang, S. S., 1994, Shear induced starch conversion during extrusion, *J. Food Sci.*, **59**:1137-1143.

11. ACKNOWLEDGEMENT

This research was supported by NJ-NSCORT (New Jersey NASA Specialized Center of Research and Training), and NJ-ECO Complex Life Support Center, and their support is greatly appreciated.

STABILITY OF ASEPTIC FLAVORED MILK BEVERAGES

Ranjit S. Kadan[*]

1. INTRODUCTION

The dairy industry has promoted milk as nature's near perfect drink as well as an important beverage for U.S. consumers. Milk is high in calcium and therefore, a major source of dietary calcium, especially for persons of Caucasian origin. However, during the last half century, the per capita consumption of milk as a beverage has been consistently decreasing, whereas the consumption of soft drinks and other novel beverages has been steadily increasing. There are many reasons for these changes in beverage consumption patterns. For example, many consumers consider milk high in calories, fat and cholesterol. Other consumers are lactose intolerant and thus avoid milk and milk products. Perhaps the primary explanation for the decline in milk consumption is that most U.S. consumers, especially the younger generation, do not consider milk as refreshing. They find milk bland in taste, with an objectionable mouth coating after taste. Many adolescents simply consider milk as an old-fashioned beverage (Sloan, 1995; Harnack *et al.*, 1999). The dairy industry has sold pasteurized flavored milks such as chocolate and strawberry for a long time. Mixtures of fruit juices and milk have also been developed (Shenkenberg *et al.*, 1971; Luck and Rudd, 1972) since consumers consider both fruit juices and milk very healthy foods. Research development of other novel dairy beverages has recently been discussed (Kadan and Champagne, 1997). Pasteurized milk beverages require refrigeration and therefore, have very limited shelf life at room temperature. Soft drinks, on the other hand, are stable at room temperature and enjoy the advantage and convenience of storage under such conditions. Aseptic processing allows milk beverages to be stored at room temperature with long shelf life. In the U.S., the Food and Drug Administration (FDA) requires the precise temperature/time conditions for aseptic processing of dairy beverages and process (Dunkley and process (sterilized in the container)). Continuous flow sterilization is

[*] Southern Regional Research Center, P.O. Box 19687, New Orleans, Louisiana 70179

Quality of Fresh and Processed Foods, edited by Shahidi et al.
Kluwer Academic/Plenum Publishers, 2004.

achieved in heat exchangers, followed by immediate cooling and aseptic packaging. Sterilization temperatures range from 130-150°C with a holding period of 2-5 s for continuous sterilization ((referred to as ultra-high-temperature UHT) or high temperature short time (HTST)). With the case of in-container sterilization (retort), the beverage is filled in the metal or glass containers and then autoclaved at 117-120°C for 10-20 min. The pasteurization or ultra-pasteurization requirements are controlled by specific standards for time/temperature requirements by the Pasteurized Milk Ordinance (FDA, 1999).

Pasteurized milk has been flavored with chocolate, strawberry and other flavors, primarily for children for a long time and presumably could also be UHT processed. However, for this presentation, only the milk flavored with either natural orange juice or coffee will be discussed, since it is believed that high amounts of organic acids present in citrus juices (Izquierdo and Sendra, 1993) and coffee (Clarke, 1993) would not allow them to be UHT processed without precipitating casein, the main protein of milk, which has an isoelectric point at pH 4.6.

2. ORANGE JUICE MILK BEVERAGES

Several pasteurized fruit juice-milk based beverages have been reported. A cherry-milk beverage, stabilized by addition of dibasic potassium phosphate, was developed by Schanderl and Hedrick (1969). Shenkenberg et al. (1971) reported a stable milk-orange juice beverage. In Shenkenberg's process, the milk protein casein was stabilized by the addition of carboxymethyl cellulose (CMC) before mixing with orange juice. Several other fruit juices have been successfully incorporated into fruit juice-milk beverages by using stabilizers (Luck and Rudd, 1972). In all cases, the casein in milk is stabilized by interacting it with either CMC or pectin before adding fruit juice to the milk. According to (Hercules/Anon, 1983a) the supplier of both CMC and Genu pectin stabilizers, these gums form complexes with casein, which allow the milk to stay in suspension. The complex formed between casein and CMC is stable in the pH range of 3.0-5.5, with an optimum around pH 4.7. The Genu pectin-casein complex is stable between pH 3.7 and 4.3. Unfortunately, both stabilizers also increase the viscosity of the fruit juice-milk mixtures.

A process to produce aseptic orange juice-milk (OJM) beverage by using UHT was developed at the Southern Regional Research Center (SRRC) in New Orleans, Louisiana. The composition of the beverage was:

Skim milk	45.7%
Orange juice concentrate 60 Bx	9.1%
Water	36.2%
Corn Sweetner (high fructose corn syrup)	8.5%
Genu Pectin (stabilizer)	0.5%

The stabilizer was mixed without any lumps in water (1 part to 10 parts) and added to the milk cooled to 2-4°C. The mixture was stirred for about 15-30 min without letting it foam. The remaining water was stirred moderately into the orange concentrate and mixed with stabilized milk for another 15 min. HFCS was added to the mixture while it was being stirred. The resultant pH of the mixture was 4.5. The mixture was continuously homogenized at 500 psi first stage and 50 psi second stage and UHT processed at 110°C for 22 s, using a coiled tubular aseptic (CTA) system. The OJM thus produced was cooled to 5°C and aseptically packaged in tetrapack packages and in stainless steel containers. The tetrapack package liner was composed of polyethylene-primacor-polyethylene foil. The stainless steel cans were heated to 260°C, and cooled to 60°C just before final packaging in a closed system.

Physical stability of the OJM is seen in Table 1. OJM viscosity is about four times that of milk. The pH of the product increased slightly with increased storage time and temperature, suggesting the casein-pectin complex is not stable during storage. The quantity of pelletable solids (g/100 ml at 921 xg) and viscosity decreased with the increase in storage temperature and time, due to the unstable nature of the casein-pectin complex. This effect appears to be more pronounced in a tetrapak package than in stainless steel container suggesting that some components of OJM react with the tetrapak package liner and thus decreases the stability of the beverage. Mannheim et al. (1987) have shown such as interaction of the orange juice components with the polyethylene liner of the container. There is little information available, regarding the stability and biochemical characteristics of the casein-pectin stabilizer complex on comparable products. There is a need to characterize the nature of this complex, and particularly the effect of storage and temperature on its integrity.

Orange juice is a mixture of water soluble and suspended components (floc) from the orange fruit. The juice normally separates during storage and is shaken before pouring for consumption. The OJM in the present study, exhibits suspension stability comparable to regular orange juice. Refrigerated and room temperature stored OJM beverage, when shaken before pouring, show no detectable settling for about 30 min.

Organoleptic evaluation of OJM indicates that both tetrapak and stainless steel packaged beverages have a slight, but noticeable caramelized flavor. The characteristic flavor of orange juice is very susceptible to changes during processing and storage. Moshonas and Shaw (1989) reported that the flavor scores of commercial orange juice packaged aseptically, were significantly lower than the control even after 1-2 weeks of storage. Both skim milk and orange juice have large amounts of reducing sugars. These sugars are expected to undergo nonenzymatic browning and related reactions during UHT processing and storage, particularly at high temperatures. Nevertheless, the over all flavor of OJM, particularly in tetrapak containers was acceptable for about three months even in samples stored at room temperature. OJM stored in a stainless steel container developed a slightly scorched and caramelized flavor after canning, which tended to have a slight metallic and bitter taste after two months storage at either room or at high temperature. It is suggested that the objectionable flavor described above is developed due to the cooled OJM coming in contact with hot stainless steel surfaces during filling.

The effect of package and storage temperatures on the terpenoid concentrations of OJM is shown in Table 2. The content of these volatile components was determined by purge and trap method (Kadan et al., 1993). The concentrations of alpha-pinene, beta-myrcene and d-limonene in stainless steel containers were about twice the concentration of that in tetrapak packages, suggesting that these and other volatile compounds had reacted with the polyethylene-primacor-polyethylene foil. Mannheim et al. (1987) also observed a decrease in d-limonene in aseptically-packed citrus juices in polyethylene lined containers. Since the entire product is blended in the same batch and packaged either in stainless steel cans or tetrapack containers, the indication is that the hydrophilic

Table 1. Physical stability of Orange Juice Milk (OJM) during storage

Storage Time (Months)	Package	pH Change			Stability - Centrifuge solids at 921 xg (g/ml)			Viscosity (centipose)		
		5°C	24°C	51°C	5°C	24°C	51°C	5°C	24°C	51°C
0	Stainless steel	4.5	-	-	5.8	-	-	4.31	-	-
1	Stainless steel	4.4	4.5	4.5	5.9	5.5	4.8	3.71	3.59	2.92
2	Stainless steel	4.5	4.5	4.5	5.8	5.8	2.6	2.08	2.92	2.25
3	Stainless steel	4.6	4.6	4.6	5.0	5.2	2.7	4.28	3.71	3.32
0	Tetrapak	4.5	-	-	5.1	-	-	4.17	-	-
1	Tetrapak	4.5	4.5	4.5	5.0	4.7	4.1	3.64	3.30	2.69
2	Tetrapak	4.5	4.5	4.5	4.2	2.8	2.6	3.81	2.92	2.25
3	Tetrapak	4.7	4.7	4.6	4.6	2.5	1.7	3.64	3.41	2.41

Table 2. Effects of storage on terpenoid in an Orange Juice-Milk beverage (ppm)

1st Month Compound	Stainless steel can			Tetrapak		
	5°C	24°C	51°C	5°C	24°C	51°C
α-Pinene	2.02	1.94	0.67	0.15	0.22	0.00
β-Myrcene	8.94	8.86	4.07	0.34	0.41	0.29
d-Limonene	186.63	169.92	86.75	42.34	24.55	23.12
Sabinene	0.19	0.61	0.00	0.00	0.00	0.00
γ-Terpinene	1.05	1.09	0.40	0.00	0.00	0.00
Linalool	0.26	0.26	0.02	1.39	0.52	0.00
α-Terpineol	0.003	0.03	0.00	0.91	0.00	0.00
Valencene	0.94	1.38	0.02	0.05	0.00	0.00

2nd Month Compound	Stainless steel can			Tetrapak		
	5°C	24°C	51°C	5°C	24°C	51°C
α-Pinene	1.26	1.24	1.66	0.31	0.01	0.10
β-Myrcene	2.70	1.66	1.62	1.01	0.96	0.44
d-Limonene	162.15	128.99	110.32	60.77	60.71	25.78
Sabinene	0.42	0.00	0.00	0.00	0.00	0.00
γ-Terpinene	0.00	0.28	0.31	0.09	0.12	0.08
Linalool	0.13	0.25	0.05	0.07	0.17	0.00
α-Terpineol	0.00	0.00	0.00	0.00	0.00	0.00
Valencene	0.00	0.00	0.00	0.00	0.00	0.00

Table 2. Continued...

3rd Month	Compound	Stainless steel can			Tetrapak		
		5°C	24°C	51°C	5°C	24°C	51°C
	α-Pinene	4.70	1.30	0.28	1.04	0.78	0.18
	β-Myrcene	13.78	4.83	4.62	4.07	3.12	0.60
	d-Limonene	114.90	63.84	66.14	41.96	36.04	15.67
	Sabinene	0.76	0.47	0.10	0.35	0.38	0.00
	γ-Terpinene	0.54	0.21	0.30	0.13	0.11	0.00
	Linalool	0.03	0.00	0.00	0.00	0.00	0.00
	α-Terpineol	0.01	0.00	0.00	0.00	0.00	0.00
	Valencene	0.01	0.00	0.00	0.00	0.00	0.00

polyethylene liner adsorbs the orange flavor volatile components.   In spite of the pronounced scorched taste in stainless steel canned OJM, no additional major volatile compounds were found.   The changes in volatile compounds during storage at various temperatures show that concentrations of α-limonene and a-pinene decrease with time in both containers.   With an increase in storage temperature and time, the number of terpenes and sesquiterpenes increased, suggesting isomerization of the indigenous compounds.   None of the major volatile compounds isolated suggested browning or Maillard reaction in high temperature stored samples, even though flavor evaluation indicated pronounced browning reactions.   It therefore appears likely that most of the browning reaction occured in the orange juice components, and skim milk contributed little or no off flavor in the OJM.

It can be concluded that an acceptable aseptic OJM that will have about two months shelf life at room temperature can be developed.   The product does have about four times the viscosity of milk and may develop some scorched/objectionable taste.   It may be desirable to use less than aseptic processing conditions (e.g., ultra-pasteurization) and storing the product at refrigeration temperature, as showed by a recent U.S. patent by Yang *et al.* (1997).   The product will still be too viscous for beverage refreshing properties and probably will be diluted with water to lower its viscosity.   At the present there is no OJM like beverage in the U.S. market.

3.   COFFEE MILK BEVERAGES

3.1. Milk Flavored Coffee Beverages

These include a number of coffee/tea drinks, such as Blue Mountain Gold Ice Coffee (Tigers Foods, Taiwan), Nestle Coffee (Nestle Foods), P'nosh Iced Coffee and P'nosh Iced Cappuccino (P'nosh Beverages, Inc., Brooklyn, N.Y.) and are sold in cans. Essentially, these products are hot packed or are retorted products; the stability of the suspended mixture is achieved by addition of disodium phosphate.   These products contain about 5-10% skim milk products and require shaking before consumption.   Shelf life is typically six months or more for these products.

3.2. Coffee Milk Beverage (CMB)

This is essentially skim milk flavored with real coffee extracts.   For a long time food scientists assumed that producing an aseptic coffee milk beverage was not technically possible because the acidity present in the coffee (Clarke, 1993) would speed milk casein during UHT processing.   It is well known that during the UHT process, milk undergoes profound physicochemical change that results in a decrease in pH, a decrease in soluble calcium-phosphate, protein denaturation, and an increase in the size of calcium-casein-phosphate particles that affect the stability of milk (Tumerman and Byron, 1965).

An aseptic coffee milk beverage was developed at SRRC.   It consisted of skim milk (60 parts), liquid coffee concentrate (LCC) (one part), and corn sweetener (HFCS) (five parts).   The components are mixed vigorously at 5°C for about 15 min and homogenized at 1000 psi in the first stage.   They were then homogenized at 500 psi in the second stage of preparation.   Processing continued in a commercial plant using a CTA system at

133°C for 21 s. The beverage was cooled immediately to 5°C, packaged aseptically and stored either in tetrapak packages or stainless steel containers. The tetrapak containers had a liner compressed of polyethylene-primacor-polyethylene foil. The stainless steel cans were heated to 260°C and subsequently cooled to 60°C prior to filling in a closed system. The packaged beverage was divided into three groups for storage at 5, 24, and 51°C. The stored samples were periodically tested for flavor, physical stability and changes in volatile contents. The physical stability data is shown in Table 3. The product packaged in both the stainless steel and tetrapak packages showed good stability even after two months of storage at all three temperatures. There were some noticeable trends. The pH, centrifuge solids and viscosities of the beverages tended to decrease slightly with the increase in storage temperature and time. However, the CMB packaged in both containers had excellent stability and flavor for six months or longer at room temperature.

Green coffee has large amounts of organic acids such as chlorogenic acid (Clarke, 1993). Furthermore, the other carbohydrates in the coffee beans can also undergo thermal decomposition to carboxylic acids, and can therefore increase the acidity during roasting. The small decrease in pH to about 6.5 in CMB from pH 6.8 of milk, is probably due to the large buffering capacity of milk proteins that are most likely responsible for product stability during UHT processing without the addition of stabilizers. The LCC and the CMB were not analyzed for their organic acids' contents. However, the appearance of oxalic acid in 51°C and its absence at lower temperature stored tetrapak CMB, suggests that high storage temperature probably promotes thermal decomposition similar to that seen in roasting.

The volatile compounds found in LCC and CMB stored at 5, 24, and 51°C for one month are given in Table 4. A comparison of the LCC and selected coffee volatiles (Sivetz, 1963) showed that among the many compounds reported, only 2-methylbutanal and acetic acid were present in LCC. However, the LCC must have most of the flavor precursors present in the extract, as the volatiles from the coffee milk beverages had developed acetone, methylfuran, methylacetone, 2-butanone, 2-methylbutanal, dimethylfuran, diacetyle, dimethylsulfide, furancarboxyaldehyde and acetic acid. In fact, all the volatiles identified from the present CMB study have previously been reported in coffee beverage (Flament, 1989).

Flavor evaluation of a stored samples indicated that samples stored at lower temperature stored had better flavor than samples stored at higher temperature. Similarly, samples stored in stainless steel cans had better flavor than those stored in tetrapak containers. Tetrapak storage material, having polyethylene liners, is not completely resistant to the diffusion of air (Burton, 1988). Our data suggests that some oxygen must enter the tetrapak container since both the 24 and 51°C stored samples had large amounts of Nonanal. Nonanal is a common oxidation product and was not found in the samples stored at high temperature. Furthermore, the data suggest that significant amounts of browning reaction occurs during storage at high temperatures as seen by the presence of 2-butanone, 2-methylbutanal, 3-methylbutanal and dimethylfuran. These browning reaction products were present in higher amounts in tetrapak packaged CMB than in canned products at both 24 and 51°C, indicating clearly the unsuitability of polyethylene lined packages. Nevertheless, when stored for 6-12 months at room temperature, CMB packaged in both types of containers had an acceptable flavor.

Table 3. Physical stability of Coffee Milk during storage

Storage Time (Months)	Package	pH Change			Stability - Centrifuge solids at 921 xg force (g/100)			Viscosity (centipose)		
		51°C	5°C	24°C	5°C	24°C	51°C	5°C	24°C	51°C
0	Stainless steel	6.6	-	-	0.9	-	-	2.3	-	-
1	Stainless steel	6.5	6.4	6.1	0.7	0.7	0.8	1.9	1.8	1.7
2	Stainless steel	6.6	6.5	6.0	0.6	0.8	0.7.	1.7	1.6	1.5
0	Tetrapak	6.5	-	-	0.7	-	-	2.0	-	-
1	Tetrapak	6.5	6.4	6.0	0.9	0.8	0.8	1.8	1.7	1.6
2	Tetrapak	6.5	6.5	5.9	0.7	0.7	0.7	1.8	1.7	1.6

Table 4. FID integrator count/1000 of votatile compounds from Liquid Coffee Concentrate (LCC) and Coffee Milk stored for 1 month

RT[1]	Compound	5°C			24°C		51°C	
		LCC	Tetra	Can	Tetra	Can	Tetra	Can
3.64	Acetone	0	274	7	115	40	81	102
5.79	MeFuran	0	0	8	0	2	139	54
6.40	MeAcetate	0	0	0	0	0	0	0
7.06	2-Butanone	0	19	9	40	11	49	32
7.44	2-MeButanal	7	76	2	96	7	65	3
7.69	3-MeButanal	0	74	0	38	7	30	6
8.08	Benzene	2	0	14	0	21	0	15
9.07	DiMeFuran	0	0	1	0	1	19	13
10.79	Diacetyl	0	84	1	0	4	0	0
12.99	Toulene	0	0	4	45	4	11	7
14.87	DiMeSulfide	0	27	11	19	4	14	24
17.42	Xylene	0	0	3	0	2	0	7
20.25	Limoene	11	23	2	0	0	0	0
27.86	Nonanal	0	0	7	102	0	27	0
33.74	Benzaldehyde	0	25	0	28	0	0	2
34.06	FuranCarboxaldehyde	0	67	0	0	0	0	0
37.21	Acetic Acid	237	16	40	61	92	56	25
37.13	Butanediol	0	0	0	0	23	0	0
38.35	2-Furanmethanol	56	0	3	29	21	29	10
39.87	Oxalic Acid	0	0	0	0	0	6	0

The technology to produce CMB was commercialized under a cooperative research and development agreement (CRADA) in collaboration with National Fruit Flavor Co. (New Orleans, LA). The CRADA resulted in the introduction and marketing of the Caffe Fantastico brand of coffee milk beverages (3 flavors) in 1991. About two years later Piacere (Glendale, CA), introduced Piacere Cappuccino, a ready to use CMB. Then Caffe Del Mar (Solana, CA) introduced Mocha Milano, a ready to drink coffee beverage. During the last 2-3 years, Starbucks, a North American Coffee Partnership (Somers, N.Y.) introduced Frappuccino coffee drink. Frappuccino has small amounts of maltodextrin and pectin added, probably to give a better mouthfeel rather than improve the stability of the mixture. All these CMB products have a shelf life of six months or better at room temperature. However, these products require shaking before consumption, suggesting that some components do settle during the long storage period.

According to Anon. (2000b) Beverage Isle, a trade magazine, there is evidence that consumption of flavored milk products is increasing among middle-aged childless couples possibly due not only to the products' flavor also because of this group's perception of the need for calcium supplementation to their daily diet. The total ready to drink (RTD) coffee beverage market was estimated at 59.1 units or about 65 million dollars. This compares with about 20 billion market for milk and about $50 billion for the soft drink market. However, for the last several years, RTD's market has been increasing at the rate of 12.5% per year. The affluent U.S. consumer will probably continue to demand nutritious, fun, and novel beverages. Milk enjoys a very healthy image because of its high calcium and nutrients content. Therefore, milk based, novel beverages, such as RTD, CMB appear to have a great potential.

4. REFERENCES

Anonymous, Down the aisle, Beverage Aisle, 2000b, 9(10), 10.
Anonymous, 1983a, Guide to Food and Pharmaceutical Uses for Genu Pectins and Gums for Food, Hercules, Inc., Wilmington, pp. 19894.
Burton, H., 1988, Ultra-High-Temperature Processing of Milk and Milk Products, Elsevier Applied Science, New York.
Clarke, R. J., 1993, Encyclopaedia of Food Science Food Technology and Nutrition, Academic Press, New York. pp. 1114-1120.
Dunkley, W. L., Stevenson, K. E., Ultra-high temperature processing and aseptic packaging of dairy products, 1987, J. Dairy Sci., 70:2192-2202.
Flament, I., 1989, Food-Rev.-Intern., 5:317-414.
Food and Drug Administration, 1999, Grade AA@ Pasteurized Milk Ordinance, U.S. Department of Health and Human Services, Washington, D.C.
Harnack, L., Stang, J., Story, M., 1999, Soft drink consumption among US children and adolescents: nutritional consequences, J. Am. Diet. Assoc., 99:436-441.
Izquierdo, L., Sendra, J. M., 1993, Encyclopaedia of Food Science Technology and Nutrition, Academic Press New York, pp. 999-1006.
Kadan, R. S., Ziegler, G. M., Grim, C. C., 1993, Unpublished data.
Kadan, R. S., Champagne, E. T., 1997, Chemistry of Novel Foods. Allured Publishing Co. Carol Stream, Il. pp. 217-231.
Luck, H., Rudd, S., 1972, S. Afr. J. Dairy Technol., 4:153-158.
Mannheim, C. H., Miltz, J., Letzter, A., 1987, Interaction between polyethylene laminated cartons and aseptically packed citrus juices, J. Food Sci., 52:737-740.
Moshonas, M.G., Shaw, P.E., 1989, Flavor evaluation and volatile flavor constituents of starch aseptically packaged orange juice, J. Food Sci., 54: 82-85.
Schanderl, S. H., Hedrick, T. I., 1968, Cherry-milk beverage stabilization and its measurement, Food Technol., 22(11): 95-98.

Shenkenberg, D. R., Chang, J. C., Edmondson, L. F., 1971, Develops milk-orange juice, *Food Eng.*, **43(4)**, 97-101.

Sivetz, M., Coffee Processing Technology, 1963, The AVI Publishing Company, Inc. Westport, pp. 62-84.

Sloan, A. E., 1995, Menus map the future, *Food Technol.*, **49(9)**, 26.

Tumerman, L., Bryon, H. W., 1965, Fundamental of Dairy Chemistry. The AVI Publishing Co., Inc., Westport, pp. 506-589.

Yang, D. K., Helsey, T., Nunes, V., 1997, Process for preparing chilled beverage products containing milk and acid, U.S. Patent #5,648,112.

SENSORY AND PEPTIDES CHARACTERISTICS OF SOY SAUCE FRACTIONS OBTAINED BY ULTRAFILTRATION

Anton Apriyantono , Dwi Setyaningsih, Purwiyatno Hariyadi, Lilis Nuraida *

1. INTRODUCTION

Soy sauce is a light brown to black liquid having umami and delicious tastes with an appetizing aroma. The most known type of soy sauce is the salty soy sauce. This soy sauce can be grouped into two types, i.e., a Japanese type (only found in Japan) and a Chinese type (produced in China and South East Asia). The Japanese soy sauce uses soybean and wheat (ratio 1:1) as the main raw material, whereas the Chinese type uses no or very little wheat besides soybean (Röling, 1995). In Indonesia, traditional soy sauce producers usually use black soybean, but some use yellow soybean, whereas modern producer use a mixture of wheat and defatted soybean. The main preparation of this soy sauce involves mold and brine fermentation to afford moromi which is then filtered and pasteurized to give soy sauce.

In Indonesia, two types of soy sauce are available in the market, i.e., salty and sweet soy sauce (kecap manis). The sweet soy sauce is more popular. It has a sweet taste in addition to *gurih* (tasty, umami) taste which is also the main taste of salty soy sauce. Our previous results showed that coconut sugar, which is one of the main raw materials for preparation of sweet soy sauce, contributes to the sweet taste of the soy sauce, whereas non-volatile components present in moromi are likely to be responsible for the *gurih* taste (Apriyantono *et al.*, 1996). Therefore, attempts have been made to characterize the components, peptides in particular, responsible for the *gurih* taste, which are present in moromi filtrate.

Several research groups have isolated peptides present in soy sauce (Kirimura *et al.*, 1969; Oka and Nagata, 1974a; Oka and Nagata, 1974b). However, none of the isolated peptides had umami taste and/or had direct contribution to the umami taste. On the other hand, Arai *et al.* (1972) investigated that some peptides, i.e., Glu-Asp, Glu-Gly-Ser, Glu-Ser, Glu-Glu, isolated from soybean hydrolyzed by pepsin, possessed a brothy taste. In addition, quite a number of peptides having umami taste were identified in fish protein hydrolysate (Noguchi *et al.*, 1975), gravy of beef meat (Yamasaki and Maekawa,

---

*Department of Food Technology and Human Nutrition, Bogor Agricultural University, Kampus IPB Darmaga, PO Box 220, Bogor 16002, Indonesia

Quality of Fresh and Processed Foods, edited by Shahidi et al.
Kluwer Academic/Plenum Publishers, 2004.

213

1978) and chicken protein hydrolysate (Maehashi *et al.*, 1999). Therefore, it is very likely that some peptides having umami taste are present in soy sauce, but they have not been identified as of yet.

2. EXPERIMENTAL

2.1. Materials

Materials for moromi preparation, i.e. yellow, black and defatted soybean, salt as well as *Aspergillus oryzae* starter, were obtained either from a local market in Bogor (Indonesia), or traditional *kecap manis* industries located in Bogor.

2.2. Moromi Preparation

Yellow and black soybean (22 kg) were each boiled in 66 l of water (1:3 w/v) for one hour, cooled and drained. Defatted soybean (22 kg) was moistened with 6.6 l of water and then steamed in an autoclave at 115°C for 45 min, cooled and drained. Each cooked soybean was inoculated with 0.5% (w/w) *Aspergillus oryzae* starter.

Each inoculated soybean was incubated at room temperature for three days to give koji. Each koji (26 kg) was soaked in 78 liters of 20% brine solution in a plastic drum and then incubated at room temperature for two months. The soaked soybean is called moromi. Each moromi was stirred once a day and then sampled at 0, 2, 4 and 8 weeks brine fermentation time. Preparation was done in duplicate for each type of soybean.

2.3. Moromi Filtrate Preparation

The moromi of each batch (consisted of 1 kg fermented soybean and 3 l the aqueous phase) was homogenized, put in cheesecloth and then pressed using a hydraulic press at an increasing pressure from 0 to 100 kg/cm^2 for 10 min. The obtained suspension was centrifuged at 3000 rpm for 20 min at 4°C, decanted and finally the filtrate was collected. Each moromi filtrate was kept in a plastic bag and stored in a freezer until ultrafiltration process and analysis.

2.4. Ultrafiltration Process

Each moromi filtrate (250 ml) was submitted to ultrafiltration in a stirred cell ultrafiltration with capacity of 50 ml (Amicon) connected with 1.5 liters reservoir at refrigeration temperature (5 - 10°C) under 2.5 - 3.0 bar N$_2$ pressure. Ultrafiltration was carried out using 0.45 μm porosity followed by YM 10 (molecular weight cut-off/MWCO 10,000 Da), YM 3 (MWCO 3000 Da) and YC 05 (MWCO 500 Da) membranes (Amicon) step wisely (see Fig. 1).

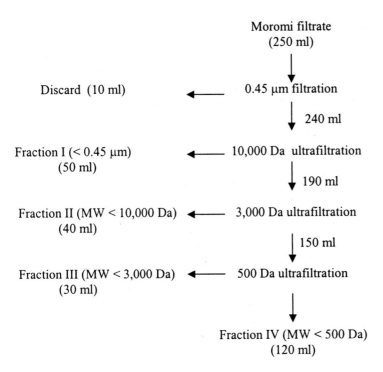

Figure 1. Scheme of the ultrafiltration process. Moromi filtrate was ultrafiltered on a 10,000 Da cut-off membrane leading to a retentate (Fraction I). The permeate was ultrafiltered through a 3,000 Da cut-off membrane to obtain a retentate (Fraction II) and finally the permeate was ultrafiltered using 500 Da cut-off membrane to obtain a retentate (Fraction III) and a permeate (Fraction IV).

2.5. Chemical Analysis

α-Amino nitrogen was analyzed by TNBS method (Adler Niesen, 1979). L-Glutamic acid was quantified using a Boehringer-Mannheim kit (Boehringer Mannheim, 1995). Total nitrogen and formol nitrogen were determined using Kjeldahl method and formol titration method (Judoamidjojo, 1986), respectively, whereas soluble peptides was analysed by Lowry method.

2.6. Peptide Profile

Peptides were analysed using a capillary electrophoresis (Hewlett-Packard) with a barefused silica (50 μm i.d. and 64.5 cm total length) as the capillary and CZE mode of separation using 50 mM borate buffer, pH 9.3. The separation took place for 35 min. The voltage used was 30 kV and the temperature was maintained at 25°C. Injection was done at 50 mBar pressure. Prior to injection, 100 μl of filtrate were added with 900 μl H_2O and then filtered through a 0.2μm PVDF membrane (Millipore). A photodiode array detector was used for detection.

2.7. Sensory Analysis

Six trained panelists performed sensory analysis for umami intensity and quantitative taste profiling. A series of standard for sensory taste description was used (Table 1) for both analyses (Apriyantono *et al.*, 1999). Sensory taste profiling (sweet, sour, salty and bitter) was performed for Fraction III and IV only.

Table 1. Standards used for sensory taste description

Description	Score	Standard
Sweet	50	7.50 % sugarcane solution
	20	2.50 % sugarcane solution
Salty	80	0.50 % sodium salt solution
	50	0.25 % sodium salt solution
Umami	150	0.20 % MSG solution
	50	0.05 % MSG solution
Bitter	50	0.05 % caffeine solution
	20	0.025 % caffeine solution
Sour	50	0.10 % citric acid solution
	20	0.05 % citric acid solution

3. RESULTS AND DISCUSSION

3.1. Moromi Filtrate

A large amount soybean, i.e. 22 kg for each batch, was used in preparation of moromi. This is to avoid batch-to-batch variation, since the moromi was sampled at 0, 2, 4 and 8 weeks. The sample was taken from each batch at specified times. If the moromi were prepared in small batch based on the specified fermentation time, batch-to-batch variation would be expected, since the process involved spontaneous fermentation.

In order to evaluate the rate of hydrolysis of soybean protein, total and formol nitrogen of moromi filtrate, as mentioned before, were monitored during brine fermentation. Total nitrogen of moromi filtrate increased up to 4 weeks brine fermentation time (especially for defatted soybean) and then leveled off (Fig. 2). This means the amount of soluble nitrogenous components, mostly soluble protein, increased during 4 weeks of brine fermentation. This paralleled the increase of formol nitrogen during brine fermentation up to 8 weeks of brine fermentation (Figure 3). A substantial increase took place during the first 2 weeks of brine fermentation followed by a slight increase up to 8 weeks. The increase in formol nitrogen is in agreement with the observations of Röling *et al. (*1994) during brine fermentation of traditional Indonesian soy sauce preparation.

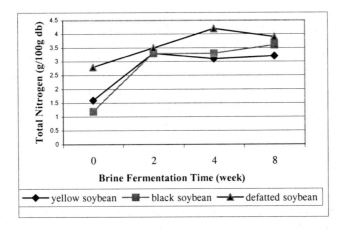

Figure 2. Change of total nitrogen of moromi filtrate during brine fermentation.

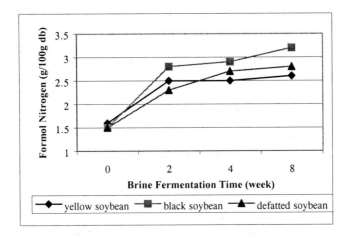

Figure 3. Change of formol nitrogen of moromi filtrate during brine fermentation time.

It should be noted that the higher amount of total nitrogen of defatted soybean as compared to the other two types of soybean is due to the higher protein content of defatted soybean. This is because lipid is removed in preparation of defatted soybean, therefore, at the same wet weight the protein content of defatted soybean would be higher than that of black or yellow soybean.

It is also interesting to note that the increase of formol nitrogen during brine fermentation of yellow soybean differed with that of black soybean. This may correlate with the composition of both soybeans, which are different; this eventually affects the fermentation process, especially affects the protease complex enzymes which work in degrading proteins and peptides during fermentation.

The formation of peptides and free amino acids during brine fermentation was expected. The increase of total nitrogen and formol nitrogen during fermentation (Fig. 3) indicates the increase of soluble protein and the formation of peptides as well as free amino acids. This expectation is also supported by the fact that from 1.50 to 1.65% (w/v) of total nitrogen of Japanese soy sauce, 45% of which were simple peptides (as reported by Fukushima, 1989). In order to characterize the peptides formed during soy sauce preparation, the peptides were fractionated and isolated. Therefore, moromi filtrate was fractionated by ultrafiltration using membranes with molecular weight cut-off of 10,000, 3,000 and 500 Da.

3.2. Fractionation Process

As can be seen from Fig. 1, membrane with 0.45 μm porosity was used to remove the insoluble particulates, the permeate was then ultrafiltered using a membrane with MW cut-off of 10,000 Da to give a retentate (Fraction I). The permeate was then ultrafiltered with a membrane with MW cut-off of 3,000 Da to afford a retentate (Fraction II) and a permeate which was ultrafiltered further using a MW cut-off of 500 Da membrane to give a retentate (Fraction III) and a permeate (Fraction IV). There was no attempt to purify the retentates, therefore Fraction I contained molecules with MW of more than 10,000 Da and less than 10,000 Da, Fraction II, III and IV contained molecules with MW of less than 10,000 Da, 3,000 Da and 500 Da, respectively. However, Fraction I contained molecules with MW of more than 10,000 Da 4.8 times more concentrate than those present in moromi filtrate. Fraction II contained molecules with MW of 10,000 Da 4.75 times more concentrate than those present in moromi filtrate and Fraction I. Fraction III contained molecules with MW of 3,000 Da 5 times more concentrate than those present in moromi filtrate, Fraction I and Fraction II. Fraction IV contained entirely molecules with MW of less than 500 Da. In addition, molecules with MW of less than 500 Da were theoretically present in all fractions at the same concentration.

The ultrafiltration process done was basically aimed at obtaining peptides with a MW of less than 500 Da, since many tasty (umami) peptides reported so far have MW of less than 500 Da (Maehashi *et al.*, 1999; Roudot-Algaron, 1996). The presence of these peptides in all fractions would be advantageous, since synergism of these peptides with those having MW of more than 500 Da in the formation of umami taste can be assessed. However, it is realized that molecules with MW more than 500 Da present in moromi filtrate were not only peptides, but also other molecules such as free amino acids, sugars, organic acids, salts and glycerol.

3.3. Chemical Characteristics of the Fractions

3.3.1. Soluble Peptides

Lowry's method was used for analysis of soluble peptides. This method actually measures the number of peptide bond; therefore all protein and peptides can be quantified. All fractions may contain soluble peptides, and only Fraction I may contain protein since it contained not only molecules with MW of less than 10,000 Da, but also those with MW of more than 10,000 Da.

Unlike total nitrogen of moromi filtrate, which leveled off after 2 or 4 weeks in brine fermentation, the concentration of soluble peptides present in all fractions increased up to 8 weeks of brine fermentation. A significant increase took place mainly during the first 4 weeks of brine fermentation (Fig. 4). This suggests that significant hydrolysis of insoluble protein to soluble protein took place up to 2 or 4 weeks brine fermentation, but hydrolysis of soluble protein to soluble and smaller peptides still took place up to 8 weeks of brine fermentation.

It is interesting to note that only a slight increase of soluble peptides happened after 4 weeks of brine fermentation of defatted soybean, whereas the increase was quite significant for the other two types of soybean (Fig. 4). This indicates defatted soybean protein is more prone to hydrolysis during the first 4 weeks of brine fermentation. This difference in the rate of hydrolysis may be due to different composition where defatted soybean contains less lipid, and this eventually affects the performance of the hydrolysis enzymes.

In general, the amount of soluble peptides (and protein for Fraction I) present in Fraction I was the highest during brine fermentation up to 8 weeks, the second highest was those present in Fraction II followed by Fraction III and those present in Fraction IV was the lowest. This means Fraction I contained the highest peptides and proteins mass and Fraction IV contained the lowest mass of peptides.

3.3.2. α-Amino Nitrogen and L-Glutamic Acid

The change of α-amino nitrogen during brine fermentation was similar with the trend of change of soluble peptides. The amount of α-amino nitrogen present in each fraction of 3 types of soybean increased with brine fermentation time (Fig. 5). This suggests the amount of peptides and/or free amino acids increased during brine fermentation up to 8 weeks. This was in line with the increase of the concentration of L-glutamic acid present in Fraction IV during brine fermentation (Table 2). Marked increase of α-amino nitrogen occurred in yellow and black soybeans, whereas in defatted soybean, the marked increase only occurred for the Fraction I (Fig. 5).

The ratio of α-amino nitrogen to soluble peptides reflects the relative number of peptides. Although Fraction IV contained lower amount of soluble peptides, it contained relative higher number of peptides, since the ratio of α-amino nitrogen to soluble peptides was higher than that of other fractions (Fig. 6). This is expected, since Fraction IV contained peptides and/or free amino acids all with MW of less than 500 Da.

3.3.3. Peptide Profile

Most amino acids lack a strong chromophore for detection by UV-Vis detector (Frazier *et al.*, 1999). The detector used for CE analysis was a photodiode array UV-Vis detector; therefore it is reasonable to expect that most components detected by this detector would be peptides. However, the presence of organic acids cannot be ruled out, since many organic acids can be detected at 215 nm. In addition, no attempt was made to identify each peptide, thus it is appropriate to expect the results of analysis as peptide profile. Also, only CE analysis for Fractions IV is reported, since the concern was peptides with MW of less than 500 Da.

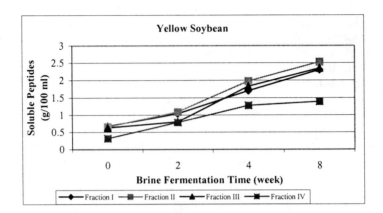

Figure 4. Change of soluble peptides of moromi filtrate fractions during brine fermentation time.

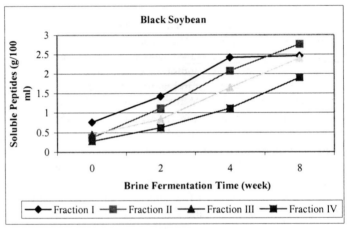

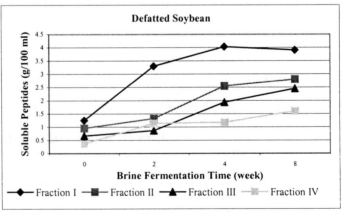

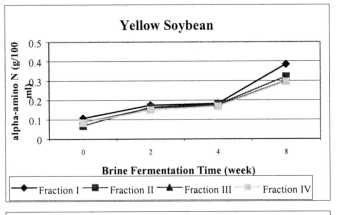

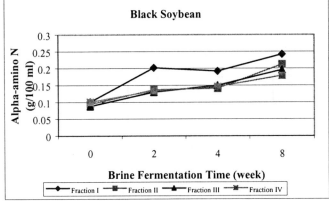

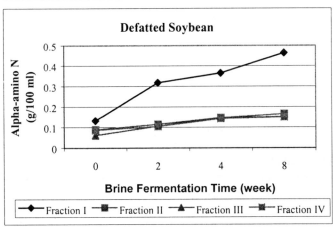

Figure 5. Change of α-amino N of moromi filtrate fractions during brine fermentation time.

Table 2. L-Glutamic acid content of moromi filtrate fractions of black soybean (g/100 ml)

Fraction	Brine fermentation time (week)			
	0	2	4	8
I	na[a]	0.24	0.33	0.34
II	na[a]	0.24	0.29	0.31
III	na[a]	0.26	0.31	0.35
IV	0.02	0.20	0.22	0.26

[a]na = not available

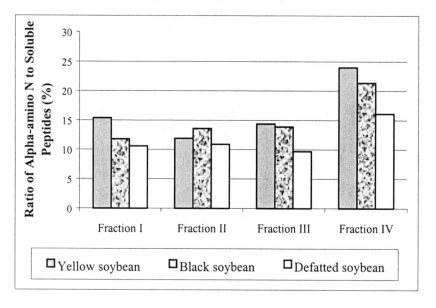

Figure 6. Ratio of α-amino nitrogen to soluble peptides of moromi filtrate fractions (the data of each fraction was the average of the data obtained from 0, 2, 4 and 8 weeks brine fermentation time).

Moromi filtrate obtained from 0 week brine fermentation was actually koji soaked in 20% brine solution. Koji was prepared by mold fermentation for 3 days. Apparently, part of the protein had already degraded to peptides during mold fermentation. This was evidenced by the fact that peptide profile of Fraction IV obtained from koji (0 week brine fermentation) was similar to that obtained from moromi filtrate prepared by 8 weeks of brine fermentation. In fact, peptide profile of Fraction IV prepared by 0, 2, 4 and 8 weeks of brine fermentation was similar (Fig. 7). This indicates that brine fermentation did not produce different types of peptides in a significant amount. Brine fermentation may play a role in increasing the amount of peptides, since the amount of α-amino nitrogen increased during brine fermentation. It should also be noted that peptide profile of Fraction IV of the three types of soybean was also similar (Fig. 7).

3.4. Sensory Characteristics of the Fractions

Intensity of umami taste of all fractions increased with brine fermentation time of up to 4 weeks, and then decreased (Table 3). This was concomitant with the increase of the amount of L-glutamic acid present in Fraction IV up to 4 weeks of brine fermentation. The amount of L-glutamic acid also decreased after 4 weeks of brine fermentation and this paralleled with the decrease of the umami taste intensity. As far as the umami taste is concerned, 4 weeks is likely to be the optimal brine fermentation time in preparation of soy sauce. This confirms our previous conclusion which showed one month brine fermentation was the optimal time for preparation of Indonesian soy sauce (Apriyantono et al., 1999; Röling et al., 1996).

Table 3. Umami taste intensity of fractions obtained by ultrafiltration of moromi filtrate

Brine fermentation time (week)	Type of soybean	Umami intensity score[a,b]			
		Fraction I	Fraction II	Fraction III	Fraction IV
0	Yellow	57.7	55.9	44.2	38.5
	Black	51.6	53.5	56.1	49.7
	Defatted	47.0	39.3	43.1	47.7
2	Yellow	70.4	74.1	76.9	60.7
	Black	56.2	67.0	65.4	53.0
	Defatted	55.7	61.8	67.9	74.6
4	Yellow	116.0	108.6	118.6	117.8
	Black	116.7	120.6	111.2	108.9
	Defatted	99.7	117.1	109.8	104.2
8	Yellow	69.4	73.6	68.7	77.6
	Black	67.0	81.6	72.6	69.5
	Defatted	68.4	76.0	88.4	91.4

[a]The score range was based on intensity of umami taste standard as presented in Table 1.
[b]The figures were obtained from the average of 6 trained panelists.

The decrease of umami intensity after 4 weeks of brine fermentation may correlate with the increase of bitter and sour taste intensity as occurred for Fractions III and IV (Table 4). It has been known that one taste attribute is affected by other taste attributes; therefore it is very likely that the increase of sour and bitter tastes lower the umami taste intensity.

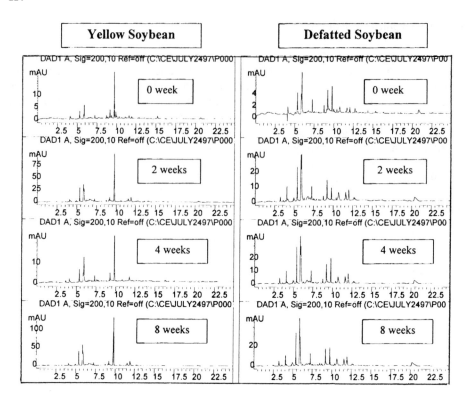

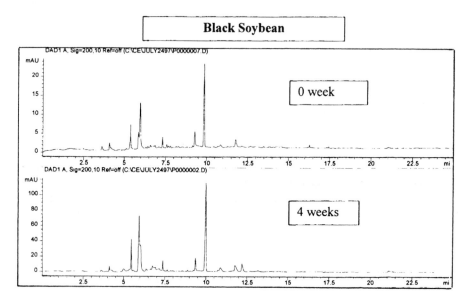

Figure 7. Electropherograms of fraction IV.

Interestingly, with a few exceptions, umami taste intensity of all fractions obtained from the same brine fermentation time was similar (Table 3). This suggests that molecules responsible for the umami taste of these fractions were mainly molecules with a MW of less than 500 Da, since these molecules were assumed to be present in all fractions at the same concentrations. The presence of molecules with a MW of more than 500 Da (as the case of Fraction II, III and IV) did not increase the umami intensity. In addition, Fraction IV of black soybean had the lowest L-glutamic acid as compared to other fractions (Table 2), although the umami intensity of Fraction IV was similar with that of other fractions. This indicates that the molecules responsible for the umami taste was not only L-glutamic acid, which has been known to possess umami taste, but also peptides with a MW of less than 500 Da.

Table 4. Taste profile of yellow soybean fractions obtained by ultrafiltration

Brine fermentation time (week)	Fraction	Intensity score[a,b]			
		Sweet	Sour	Salty	Bitter
0	III	22.6	34.7	66.0	na[c]
	IV	25.7	18.3	55.3	22.0
2	III	23.3	22.6	56.6	20.4
	IV	22.8	19.4	57.9	25.1
4	III	42.3	30.6	72.8	18.3
	IV	44.2	25.0	85.0	26.2
8	III	37.2	34.2	79.2	31.0
	IV	29.5	36.9	81.5	38.0

[a]The score was based on intensity of each taste description standard as presented in Table 1.
[b]The figures were obtained from the average of 6 trained panelists.
[c]na = not available

4. REFERENCES

Adler-Niesen, J., 1979, Determination of the degree of hydrolysis of food protein hydrolyzates by trinitro benzensulfonic acid, *J. Agric. Food Chem.*, **27**:1256-1262.

Apriyantono, A., Wiratma, E., Husain, H., Nurhayati, Lie, L., Judoamidjojo, M., Puspitasari-Nienaber, N. L., Budiyanto, S., and Sumaryanto, H., 1996, Analysis of volatiles of *kecap manis* (a typical Indonesian soy sauce), in: *Flavour Science. Recent Developments,* A.J. Taylor, and D.S. Mottram, eds., The Royal Society of Chemistry, Cambridge, pp. 62-65.

Apriyantono, A., Husain, H., Lie, L., Judoamidjojo, M., and Puspitasari-Nienaber, N. L., 1999, Flavor characteristics of Indonesian soy sauce (*kecap manis*), in: *Flavor Chemistry of Ethnic Foods*, F. Shahidi and C-T. Ho, Kluwer Academic, New York, pp. 15-32.

Arai, S., Yamashita, M., and Fujimaki, M., 1972, Glutamyl oligopeptides as factors responsible for tastes of a proteinase-modified soybean protein, *Agric. Biol. Chem.*, **36**:1253-1258.

Boehringer Mannheim, 1995, *Methods of Enzymatic BioAnalysis and Food Analysis*, Boehringer Mannheim GmbH Biochemicals, Mannheim, pp. 56-59.

Frazier, R. A., Ames, J. M., and Nursten, H. E., 1999, The development and application of capillary electrophoresis methods for food analysis, *Electrophoresis*, **20**:3156-3180.

Fukushima, D., 1989, Industrialization of fermented soy sauce production centering around Japanese shoyu, in: *Industrialization of Indigenous Fermented Foods*, K. Steinkraus, ed., Marcel Dekker, New York, pp. 1-88.

Judoamidjojo, M., 1986, *The Studies on the Kecap: Indigenous Seasoning of Indonesia*, PhD Thesis, The Tokyo University of Agriculture, Tokyo, Japan.

Kirimura, J., Shimizu, A., Kimizuka, A., Ninomiya, T., and Katsuya, N., 1969, The contribution of peptides and amino acids to the taste of foodstuffs, *J. Agric., Food Chem.,* **17**:689-695.

Maehashi, K., Matsuzaki, M., Yamamoto, Y., and Udaka, S., 1999, Isolation of peptides from an enzymatic hydrolysate of food proteins and characterization of their taste properties, *Biosci. Biotechnol. Biochem.,* **63**:555-560.

Noguchi, M., Arai, S., Yamashita, M., Kato, H., and Fujimaki, M., 1975, Isolation and identification of acidic oligopeptides occurring in a flavor potentiating fraction from a fish protein hydrolysate, *J. Agric. Food Chem.,* **23**:49-53.

Oka, S., and Nagata, K., 1974a, Isolation and characterization of neutral peptides in soy sauce, *Agric. Biol. Chem.,* **38**:1185-1191.

Oka, S., and Nagata, K., 1974b, Isolation and characterization of acidic peptides in soy sauce, *Agric. Biol. Chem.,* **38**:1195-1198.

Röling, W. F. M., 1995, *Traditional Indonesian Soy Sauce (Kecap): Microbiology of the Brine Fermentation,* PhD Thesis, Free University of Amsterdam, Amsterdam, The Netherlands.

Röling, W. F. M., Apriyantono, A., and van Verseveld, H. W., 1996, Comparison between traditional and industrial soy sauce (*kecap*) fermentation in Indonesia, *J. Ferment. Bioeng.,* **81**:275-278.

Roudot-Algaron, F., 1996, Le gout des acides amines, des peptides et des proteins: exemple de peptides sapides dans les hydrolysats de caséines, *Le Lait,* **76**:313-348.

Yamasaki, Y., and Maekawa, K., 1978, A peptide with delicious taste, *Agric. Biol. Chem.,* **42**:1761-1765.

5.   ACKNOWLEDGEMENT

The authors wish to thank the Directorate General of Higher Education, Ministry of National Education of Indonesia for the financial support for this project via a University Research Graduate Education (URGE) project. We are indebt to Dr. J.M. Ames of Reading University for providing facility for CE analysis. We also greatly appreciate Dr. L. Royle for performing CE analysis. We finally thank P.T. Givaudan Indonesia and P.T. Essence Indonesia for providing financial support to present this work in Pacifichem 2000 Congress.

QUALITY ASSESSMENT OF A LOW-SALT SOY SAUCE MADE OF A SALTY PEPTIDE OR ITS RELATED COMPOUNDS

Rie Kuramitsu[*]

1. INTRODUCTION

Health has been one of the most important matters for humans. Recently in Japan, excessive intake of salt is claimed to be one of the major causes of hypertension, arteriosclerosis, thrombosis and other life style-related diseases, and reduction of salt intake has been a serious problem. In 1998 Japanese consumed an average of 12.7 g of salt a day per head. The amount of salt intake recommended by the Ministry of Health and Welfare is 10.0 g (Ministry of Health, 1998). Thus the actual amount consumed was about 30 % more than the recommended amount. It is noted that considerable amount of salt is taken from soy sauce by Japanese and that 20 % or more of the total salt consumption is derived from foods containing soy sauce (Kawano, 1990).

The ordinary processes of reducing the salt content in soy sauce as practiced so far is as follows. The soy sauce produced by the conventional method is treated by ion exchange membrane, electrolysis or electrophoresis. This process certainly gives a low salt soy sauce, but while it contains decreased amount of sodium chloride but it also decreases the *umami* component, thus making the taste of soy sauce blur. An ordinary method employed for overcoming this defect is the addition of potassium chloride or another salt-substituting substance after desalting. In actual cases these processes are often employed. However, since potassium chloride has a bitter taste in addition to salty taste, it poses problem since the addition of more than 10 % of potassium chloride would influence the taste of foods to which soy sauce is added. Therefore, a process was devised by adding a salty peptide in place of potassium chloride.

---

[*] Rie Kuramitsu, Akashi National College of Technology, Uozumi, Akashi, 674-8501, Japan

Quality of Fresh and Processed Foods, edited by Shahidi et al.
Kluwer Academic/Plenum Publishers, 2004.

2. MAINTENANCE OF SALTINESS BY THE ADDITION OF SALT-SUBSTITUTING SUBSTANCES

As a "salt-substituting substance" that has a taste resembling salt is the "basic dipeptide" or ornityltaurine that was reported by Okai and others (Tada *et al.,* 1984; Seki *et al.,* 1990; Tamura *et al.,* 1984). However, the synthesis of 1 g of this substance requires higher than 1500 dollars. Therefore, we looked for a substance with simpler structure and thus compared the saltiness strength of known salt substitutes such as potassium chloride and glycine ethyl ester hydrochloride or lysine hydrochloride. Results are shown in Table 1. As represented by the number of + mark for sodium chloride, salty taste was obtained with potassium chloride and glycine ethyl ester hydrochloride. However, the taste of other substances was unacceptable.

Table 1. Sensory analysis of salt-substituting substances

| Compound | Saltiness, Concentration as NaCl (%) | | |
	1.0	0.5	0.25
NaCl	+++	++	+
KCl	+	+	+
MgCl$_2$	-	-	-
MnCl$_2$	-	-	-
Citric acid	-	-	-
Na Citrate	-	-	-
K Citrate	-	-	-
Gly-OEt HCl	+	+	+
Lys HCl	+	-	-
Orn HCl	+	-	-
Tau	-	-	-
Orn + Tau	-	-	-

3. PRODUCTION OF A LOW-SALT SOY SAUCE

The experiments were carried out by the cooperation of a food company that performs short-term fermentation of soy sauce. Soy bean was hydrolyzed for 24 h before it was submitted to aging and aging was completed in 1 month. Accordingly, the required amount of salt was less than 16% and set at 4.6%. This preparation is referred to as "pre-soy sauce." Except the low salt concentration the pre-soy sauce is not different from ordinary commercial soy sauce. This was used for the production of a low salt soy sauce by adding the peptidic salt substitute and so other ingredients of soy sauce were not affected.

The term of 72% replacement in Table 2 means that the shortfall of salt in the pre-soy sauce is all filled with a substituting substance to make the saltiness comparable to that of commercial soy sauce containing 16.2% salt. However, this could not secure the saltiness of the ordinary soy sauce. The term of quality means the so-called "soy sauce-likeness," and the above replacement poses a problem in this regard. Then the rate of replacement was gradually decreased to attain an appropriate taste. When the replacement rate was made 67%, sufficient saltiness was obtained, but the taste with glycine ethyl ester hydrochloride included an acidic taste and that with potassium chloride had a bitter taste. When the replacement rate was made 50% the salty taste was perfectly secured. Other unfavorable tastes were improved considerably and the production containing glycine ethyl ester hydrochloride fairly resembled the commercial soy sauce. The replacement rate was decreased to as low as 33%. Then, all of the 3 substitutes showed saltiness and soy sauce-likeness quite close to those of commercial soy sauce. Here, what the term soy sauce-likeness means is the good balance of salty and *umami* tastes. In these experiments the strength of *umami* taste changed although its content was not altered. This change was thought to be due to a phenomenon characteristic to ionic radicals.

Table 2. Sensory analysis of soy sauce in which NaCl was partially replaced with salt-substituting substances

Formulation	Saltiness			Quality		
	1.0	0.75	0.50	1.0	0.75	0.50
Commercial soy sauce	+++	++	+	+++	+++	+++
72% replacement						
Gly-OEt HCl	+	+	+	+	+	+
Lys HCl	+	+	-	+	+	+
KCl	+	+	-	+	+	+
50% replacement						
Gly-OEt HCl	+++	+++		++	++	
Lys HCl	++	++		+	++	
KCl	+++	+++		+	++	
33% replacement						
Gly-OEt HCl	+++	+++	++	++	+++	+++
Lys HCl	++	++	++	++	++	+++
KCl	+++	++	++	++	++	++

From these results it may be concluded that glycine ethyl ester hydrochloride and lysine hydrochloride are more effective than potassium chloride in that they are less affected by undesirable tastes. Then, reduction of salt by using ornithyltaurine that has an excellent salty taste was studied (Table 3). With 60% replacement rate the product from this peptide was similar to ordinary soy sauce in quality, and with 50% replacement the product was nearly equivalent to ordinary soy sauce (Table 3).

Table 3. Sensory analysis of soy sauce in which NaCl was partially replaced with Orn-Tau · 1.2HCl

Types		Saltiness, Concentration as NaCl(%)			Quality
		0.75	0.50	0.25	
Commercial soy sauce		+++	++	+	+++
Replacement	60%	+++	++	+	++
	50%	+++	++	+	+++
	30%	+++	++	+	+++

4. QUALITY PROBLEMS IN LOW-SALT SOY SAUCE

The first objective of securing sufficient salty taste in low salt soy sauce was thus accomplished by replacement. However, the addition of large amounts of salt-substituting compounds caused differences in taste between commercial soy sauce and the low salt soy sauces. The difference was particularly noticeable in *umami* taste. Although the content of sodium glutamate, which is the major *umami* component in soy sauce, was hardly changed, the added substituting compounds certainly influenced the *umami* taste. Then, the manner in which the *umami* taste in soy sauce was influenced by salt-substituting substances was studied. If the *umami* taste in soy sauce was influenced by the addition of salt-substituting compounds, it should be required to compare quantitatively the strength of saltiness and *umami* tastes. However, it is nearly impossible to estimate such different tastes as salty and *umami* in a similar manner. For convenience, commercial soy sauce was considered as having a perfect balance between salty and *umami* tastes. The salty taste of soy sauce is represented by sodium chloride and the *umami* taste by sodium glutamate. Quantification of salty taste was done as follows: A series of salt solutions, with different concentrations, were prepared. Soy sauce was diluted to the same extent as the salt solution in the above series, and the corresponding salt concentrations determined as stated above. From these results, a table of scores was obtained (Table 4). Strength of *umami* taste was evaluated by comparing the diluted soy sauce solutions, which were used for making the scores for salty taste, and a series of sodium glutamate solutions with different concentrations. Accordingly, the value of a score should show the strength of salty and *umami* tastes in correspondence to a given concentration of soy sauce. Thus, a score corresponds to an identical dilution of soy sauce.

By using Table 4 of scores, the strength of salty and *umami* tastes in commercial soy sauce was examined. The taste strengths were in good agreement with scores, and thus the validity of this Table was confirmed. Then, low salt soy sauces were prepared by replacing half an amount of salt of commercial soy sauce with substituting substances, and the tastes were compared with the original soy sauce. Results are shown in Figure 1. The product made with ornithyltaurine hydrochloride showed good balance between salty and *umami* tastes in all concentrations studied. With glycine ethyl ester hydrochloride salty taste was rather prominent in higher concentrations. With lysine hydrochloride, by contrast, *umami* taste was stronger in higher concentrations. With potassium chloride, which is a known and frequently used salt

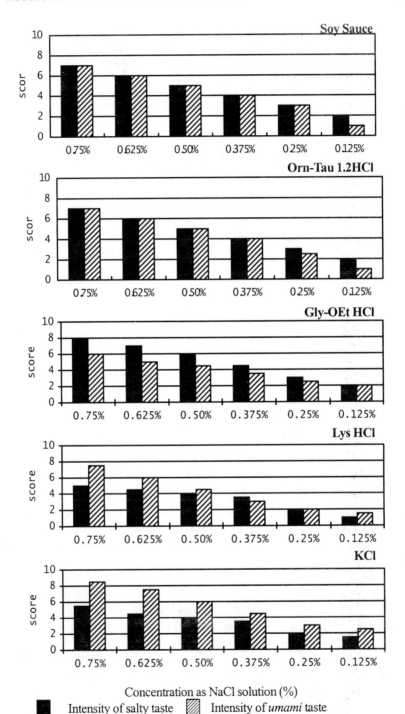

Figure 1. Comparison of intensity of salty and *umami* tastes in soy sauce and pre-soy sauce containing salt-substituting substances (50% replacement).

substitute, *umami* taste was stronger over all concentrations by 3 scores.   This means that the difference was 1.5-folds in quantity.   In other words, this substitution brought about one-third decrease in the effect of salty taste indicating a serious defect.   In this way various substituting substances reveal different effects on *umami* taste, and so reduction of salt concentration must be done with careful consideration of this problem.   The second important problem of making low salt soy sauce, besides securing the necessary saltiness, is to hold the good balance between salty and *umami* tastes.

Table 4.   Scores of intensities of salty and *umami* tastes in diluted soy sauce

Salty taste NaCl (%)	Score	*Umami* tasete MSG(%)
1.200	10	0.45
1.000	9	0.40
0.875	8	0.35
0.750	7	0.30
0.625	6	0.25
0.500	5	0.20
0.375	4	0.15
0.250	3	0.10
0.125	2	0.05
0.063	1	0.03

5.   THE THIRD FACTOR CONCERNING THE FORMATION OF SOY SAUCE FLAVOR

After securing salty taste and holding good balance between salty and *umami* tastes, the third problem to be considered is salt concentration (concentration of neutral salt or ion strength). Thus, amino acids were studied in addition to sodium chloride and sodium glutamate, which would have an influence on salt concentration.   Soy sauce contains many sorts of amino acids, but it is not known if all of these amino acids are necessary for taste?   Thus, a model system was considered and the correlation of the soy sauce in the model system to food materials was also studied.

Table 5 shows an example of analysis of the representative commercial soy sauce.   The taste components are classified roughly into salty, *umami* and other components.   The salty taste is represented by sodium chloride, and its content as inferred from the sensory test and the one obtained by analysis were in good agreement.   For *umami* taste, estimation of its content by the sensory test as sodium glutamate gave 142-1200 mM, that is a 10-fold variation depending on dilution.   Thus, a value of 390 mM which was obtained at the dilution in which the sodium chloride concentration was 0.5% was employed.   Sugars, alcohol and acids had no significant effects on the salty and *umami* tastes in soy sauce.

Table 5. Analysis of commercial soy sauce taste components

Salty taste	*Umami* taste	Other taste	
NaCl	MSG	Amino acids	
16.3%	390mM	Acidic group	166 mM
		Basic group	51 mM
		Hydrophobic group	132 mM
		Neutral group	198 mM
		Sugar	33%
		Alcohol	2.7%
		Acids	0.6%

A significant role of amino acids in soy sauce is thought to be a buffering action. Then the rates of ionization of amino acids were estimated on the basis of electric conductivity at the concentration at which salt-substituting substances show identical saltiness as sodium chloride (Table 6). While the values were considerably different among substituting substances, dividing the values by molar concentration of solution gave similar values. Upon this finding the total mole number of amino acids contained in soy sauce was kept constant

Table 6. Characteristics of NaCl and salt-substituting substances

Salt-substituing substance	NaCl	KCl	Gly-OEt□HCl	Lys□HCl	OTA□1.2HCl
Molecular weight	58.5	74.6	139.5	182.7	268.8
Concentration when salty intensity is 3*					
(%)	0.25	0.67	0.63	7.47	0.72
(mM)	42.8	89.5	45.0	409.1	30.0
Ionic equivalent conductivity					
	100	250	105	500	69
Ionic equivalent conductivity/mM					
	2.34	2.80	2.33	1.22	2.30

*   Score 3 is equal to the salty intensity of 0.25% NaCl solution.

The extent to which the amino acid composition might be simplified was also studied. Firstly, basic amino acids were examined, as shown in Table 7. When 3 basic amino acids lysine, histidine and arginine were replaced by lysine alone, the taste was mainly bitter and sweet, and the threshold value was 1.9 mM indicating that the taste of lysine alone was not greatly different from that of the above 3 amino acids in mixture. Then, hydrophobic amino acids were examined (Table 8). When 5 hyrophobic amino acids were replaced by phenylalanine alone, a solution with stronger bitter taste than the original mixture was obtained. For the sake of reducing the hydrophobicity a mixed solultion made of phenylalanine and leucine or β-alanine was prepared but the taste of this mixture was distinctly different from that of the mixed solution of the 5 amino acids.   Then, mixtures of phenylalanine and glycine were studied, and a solution of equimolar mixture of phenylalanine and glycine was found to give a solution close to the mixed solution of the 5 amino acids with regard to the kind of taste and the threshold value.   Subsequently, neutral amino acids were studied (Table 9).   A solution obtained by replacing neutral amino acids in a mixed solution with glycine alone gave a taste in which sweetness was overwhelming and taste was simpler than that in the original mixed solution.   The threshold value of the glycine solution was nearly identical with that of the original mixed solution.

Table 7.   Tastes and threshold values of mixed solutions of basic amino acids

Solution	Mixed		Lys alone	
Contents (mM)	Lys	25	Lys	51
	His	18		
	Arg	8		
TV (mM)	2.0		1.9	
Tastes	bitter, sweet > sour		sweet, bitter	
Similarity of Lys solution to mixed solution			++	

Table 8.   Tastes and threshold values of mixed solutions of hydrophobic amino acids

Solution	Mixed		Phe + Gly					
Contents(mM)	Leu	48						
	Vau	39						
	Phe	39	Phe	36	Phe	66	Phe	96
	Tyr	6						
	Ile	3						
			Gly	96	Gly	66	Gly	36
TV (mM)	22.5		25.0		22.5		20.0	
Taste	bitter >> sweet, *umami*		bitter, sweet		bitter >> sweet		bitter >> sweet	
Similarity of Phe + Gly solution to Mixed solution			++		++		++	

Table 9.   Tastes and threshold values of mixed solutions of neutral amino acids

Solution	Mixed		Gly		Gly + Pro		Gly + Ser	
Contents(mM)	Pro	53			Pro	53		
	Ala	45						
	Ser	42					Ser	42
	Gly	32	Gly	198	Gly	145	Gly	156
	Thr	26						
TV (mM)	21.0		18.6		10.5		12.5	
Taste	sweet		sweet		sweet		sweet, bitter	
Similarity of Gly, Gly +Pro, and Gly +Ser solution to Mixed solution			++		++		++	

From these results, it became clear that all the basic, hydrophobic and neutral amino acids that constitute soy sauce were not strictly required for its taste and that the amino acid composition of soy sauce can be simplified. Then the amino acids contained in soy sauce were replaced by lysine, phenylalanine and glycine. A small amount of methionine which is contained in soy sauce and exerts a characteristic flavor, was also added to the replacing amino acids. Results are shown in Table 10. In this way the 14 amino acids that constituted soy sauce were shown in column (A) and that of the 4 amino acids in column (B). Comparison of the two solutions showed similar taste characteristics, but the threshold value of the replaced amino acid solution (B) was lower than that of the original solution (A). Then sodium glutamate, the major *umami* component in commercial soy sauce, was added to the 14 amino acid solution (C) and to the replaced, 4 amino acid solution (D) in a concentration identical to that in soy sauce. Quality and strength of taste and other properties in the solutions (C) and (D) were more closely similar to each other than those in (A) and (B). Thus, limitation of not altering the ratio of the above groups of amino acids in commercial, the amino acids that were originally contained in soy sauce may be greatly removed and simplified. The experiments were successful by keeping the concentration constant. If test solutions were prepared on the basis of taste, data obtained would be confusing.

Table 10. Tastes and threshold values of mixed solutions of amino acids for the simplification of the amino acid composition in soy sauce

Solution	(A)		(B)		(C)		(D)	
Contents(mM)	Lys	25	Lys	51	Lys	25	Lys	51
	His	18			His	18		
(Amino acids	Arg	8			Arg	8		
total 389 mM)	Leu	48			Leu	48		
	Val	39			Val	39		
	Phe	36	Phe	66	Phe	36	Phe	66
	Tyr	6			Tyr	6		
	Ile	3			Ile	3		
	Met	8	Met	8	Met	8	Met	8
	Pro	53			Pro	53		
	Ala	45			Ala	45		
	Ser	42			Ser	42		
	Gly	32	Gly	66 + 198	Gly	32	Gly	66 + 198
	Thr	26			Thr	26		
					MSG	390	MSG	390
TV(mM)		24.3		17.2		2.2		2.0
Taste		bitter,sweet		bitter,sweet		*umami*,bitter,sweet		*umami*,bitter,sweet
Similarity of (B) to (A)				++	Similarity of (D) to (C)			+++

The amino acids composing soy sauce were, thus, simplified, and in a subsequent experiment, to this solution salt-substituting substances in amounts corresponding to 16.3% of sodium chloride were added.   When 50% of sodium chloride was replaced with a salt-substituting substance, the mixed solution secured desired saltiness but in regard to soy sauce-likeness it was considerably different from a solution in which only sodium chloride was added.   In the case of 33% replacement (Table 11), both glycine ethyl ester hydrochloride and potassium chloride showed taste characteristics identical with that of sodium chloride.   Lysine hydrochloride also showed a good similarity in quality with sodium chloride.   These results are in accordance with the experimental results on securing saltiness by using pre-soy sauce. Heating did not cause any distinct change in taste and flavor of the synthetic soy sauces.   From these results the synthetic soy sauce obtained by simplifying the amino acid composition from commercial soy sauce, referred to as "model soy sauce", was used in subsequent experiments.

Table 11.   Taste characteristics of synthetic soy sauce

	Synthetic soy sauce		Low-salt synthetic soy sauce with 33% replacement for NaCl			
	NaCl type		KCl type		Gly-OEt□HCl type	Lys□HCl type
Contents						
Salty taste	NaCl	16.3%	NaCl	10.9%	NaCl 10.9%	NaCl 10.9%
			KCl	5.4%	Gly-OEt□HCl 5.4%	Lys□HCl 5.4%
Umami taste			MSG			390mM
Amino acid			Lys			51mM
			Phe			66mM
			Met			8mM
			Gly			264mM
Taste intensity						
Salty/ *umami*	5.0 / 5.0		5.0/5.5		5.1/4.7	4.7/5.0
Taste						
	Salty, *umami*		salty, *umami*		salty, *umami*	salty, *umami*
			> bitter		>sour	> sweet
Similarity of commercial soy sauce						
	+++		+++		+++	++

Soy sauce is generally used for eating together with food materials. Naturally the taste and flavor of soy sauce is influenced by the properties of food materials. The extent to which taste was affected by food materials when the model low-salt soy sauce was added to various food materials was studied.   Results are shown in Table 12. With the model low salt soy sauce, sucrose increased salty taste while cellobiose, gelatin, oleic acid and linoleic acid by contrast decreased it. With regard to the *umami* taste, starch and gelatin increased it while oleic acid and linoleic acid decreased it.   In conclusion it was found that fats generally decreased both salty and *umami* tastes, but with other substances there was found no effect on the tastes of soy sauce. When these results are compared with commercial soy sauce, the effects of various foods on the tastes were quite similar in commercial soy sauce and the model low-salt soy sauce, except for the salty taste-deceasing effect of gelatin. The results thus demonstrate that both the ordinary and low salt synthetic soy sauces are similar in properties. When only salt-substituting substances were added to various food materials, no significant changes in the strength of salty taste were observed

Table 12. Taste characteristics of raw food materials after the addition of low-salt soy sauce

Food material	Taste intensity(salty/*umami*)					
	Synthetic Soy sauce	Low-salt synthetic soy sauce with 33% replacement for NaCl				Commercial soy sauce
(1%)	NaCl	KCl	Gly-OEt□HCl	Lys□HCl	OTA□1.2HCl	
Water	5.0/5.0	4.9/5.3	5.2/4.7	4.7/5.2	5.0/5.0	5.0/5.0
Cellobiose	4.9/5.0	4.7/4.6	4.6/4.8	4.8/5.3	4.7/4.8	4.7/5.0
Glucose	5.0/4.8	5.2/5.0	5.0/4.5	4.7/4.5	5.1/5.0	4.9/5.0
Sucrose	5.3/4.8	5.2/4.7	5.4/4.5	5.3/4.9	5.3/5.0	5.2/4.6
Starch	5.0/5.2	5.1/5.2	5.0/5.0	4.8/5.4	4.5/5.0	4.9/5.3
Glycogen	5.0/5.0	4.7/5.2	4.9/4.3	5.0/4.5	5.2/4.8	5.0/5.0
Gelatin	4.5/5.3	4.8/5.3	4.7/5.0	4.5/5.2	4.7/5.2	5.0/5.7
Albumin	5.0/4.5	4.7/5.5	4.5/4.7	4.4/4.5	4.5/4.7	4.5/5.3
Oleic acid	4.7/4.3	4.5/4.2	4.4/4.1	4.4/4.0	4.3/3.8	4.6/4.0
Linoleic acid	4.2/3.8	4.0/3.3	4.1/3.8	3.6/3.2	3.7/3.7	4.3/3.7

The model low salt soy sauces with 33% replacement were added to foods with more complicated compositions; results are shown in Table 13. Whole egg was found to augment salty taste while cucumber reduced it. Cucumber and salad oil reduced the *umami* taste. These results were in accordance with those with commercial soy sauce. However, as the system became complicated by increased number of food components, there occur many factors, such as distribution coefficient, adsorption (to protein) and other physical conditions, formation of taste of neutral salt and appearance of increasing or decreasing factors. These effects make change of taste strength complicated and so it is not possible to generalize the effects of foods on the taste of soy sauce. However, attempts were made to simplify the composition of soy sauce as far as possible is to change the composition of seasoning to meet desired objective of making specific blends of seasonings it. In model products, it is possible to both increasing and decreasing of components and analysis of foods becomes easier.

Table 13. Taste characteristics of raw food materials after the addition of low-salt soy sauce

Food material	Taste intensity (salty/*umami*)				
	Synthetic Soy sauce	Low-salt synthetic soy sauce with 33% replacement for NaCl			Commercial soy sauce
	NaCl	KCl	Gly-OEt HCl	Lys HCl	
Water	5.0/5.0	4.9/5.3	5.2/4.6	4.7/5.2	5.0/5.0
Cucumber (100%)	4.6/4.6	4.7/4.6	4.6/4.2	4.3/4.8	4.6/4.2
Starch (10%)	5.0/5.0	4.8/5.0	5.0/5.2	4.5/5.6	4.8/5.0
Tofu (100%)	4.7/5.3	5.1/5.0	4.7/5.0	4.7/5.6	4.6/4.2
Milk(100%)	4.8/5.3	5.3/5.4	5.2/5.8	4.8/5.6	5.2/5.0
Egg (100%)	5.5/5.4	5.1/5.4	5.5/5.8	5.2/5.6	5.2/5.0
Salad oil (100%)	4.1/4.0	4.2/3.5	4.8/4.0	4.2/3.5	4.8/4.0

6. CONCLUSIONS

The production of low salt soy sauce by employing peptide-type, salt-substituting substances was successful. Ornityltaurine hydrochloride showed 60%, glycine ethyl ester hydrochloride 50% and lysine hydrochloride 33% of salt-reducing effect. Thus, securing of saltiness of commercial soy sauce was made possible with these substances. The results were beyond the 24% salt reduction as proposed by the Ministry of Health and Welfare. Recently, soy sauce has been used in making gravies and dressings, among others, and thus usage of soy sauce has become versatile. Therefore, soy sauce may take a more important position as a food material in our daily eating habits.

7. ACKNOWLEDGMENTS

I thank Tenyo Takeda Co., Ltd for generous supply of commercial soy sauce and pre-soy sauce that were used in the present experiments.

8. REFERENCES

Kawano, T. Ed., Science of Soy Sauce; Association of Science, Technology and Education: Tokyo, 1990.

Ministry of Health and Welfare, ed; The Actual Situation in National Nutrition -1998-; Daiichi: Tokyo, 1998.

Seki, T., Kawasaki, Y., Tamura, M., Tada, M., and Okai, H., 1990, Role of Basic and Acidic Fragments in Delicious Peptides (Lys-Gly-Asp~Glu-Glu-Ser-Leu-Ala) and the Taste Behaviour of Sodium and Potassium Salts in Acidic Oligopeptides, *J. Agric. Food Chem.*, **38:** 25-29.

Tada, M., Shinoda, I., and Okai, H., 1984, L-Ornithyltaurine, A New Salty Peptide, *J. Agric. Food Chem.*, **32:** 992-996.

Tamura, M., Seki, T., Kawasaki, Y., Tada, M., Kikuchi, E., and Okai, H., 1989, An Enhancing Effect on the Saltiness of Sodium Chloride of Added Amino Acids and their Esters, *Agric. Biol. Chem.*, **53:** 1625-1633.

QUALITY CHARACTERISTICS
OF EDIBLE OILS

Fereidoon Shahidi[*]

1. INTRODUCTION

Edible oils provide a concentrated source of energy and essential fatty acids through daily dietary intake. Lipids also serve as an important constituent of cell walls and carrier of fat-soluble vitamins. In addition, lipids provide flavor, texture and mouthfeel to the food. Edible vegetable oils also serve as a heating medium and are important in the generation of aroma, some of which arise from direct interaction of lipids and/or their degradation products with food constituents.

Oilseeds and tropical fruits are a major source of food lipids. The edible oils from oilseeds may be produced by pressing, solvent extraction or their combination. The seeds may first be subjected to a pretreatment heating to deactivate enzymes present. The oils after extraction are subsequently subjected to further processing of degumming, refining, bleaching, deodorization, and if necessary, stabilization.

Edible oils from source materials are composed primarily of triacylglycerols (TAG). In addition, phospholipids, glycolipids, waxes, wax esters, hydrocarbons, tocopherols and tocotrienols, other phenolics, carotenoids, sterols and chlorophylls, and hydrocarbons among others, may be present as minor constituents and these are collectively referred to as unsaponifiable matter (Shahidi and Shukla, 1996). During processing, storage and use, edible oils undergo chemical and physical changes. Often, process-induced changes of lipids are necessary to manifest specific characters of food, however, such changes should not exceed a desirable limit. Both TAG and minor constituents of the oil exert a profound influence on quality characteristics of the oils and hence their effect on health promotion and disease prevention.

Following oil extraction, the left over meal may also serve as a source of fat-insoluble phytochemicals. Obviously, hulls, might be included in the meal if the seeds are

---

[*] Department of Biochemistry, Memorial University of Newfoundland, St. John's, NL A1B 3X9 Canada.
Tel.: (709) 737-8552; Fax: (709) 737-4000; E-mail: fshahidi@mun.ca

Quality of Fresh and Processed Foods, edited by Shahidi et al.
Kluwer Academic/Plenum Publishers, 2004.

not dehulled prior to oil extraction. The importance of bioactives in processing by-products of oilseeds may thus require attention.

The quality characteristics of edible oils, as noted earlier, depend on the composition of their fatty acids, positional distribution of fatty acids, non-triacylglycerol components, presence/absence of antioxidants, the system in which the oil is present such as bulk oil versus emulsion and low-moisture foods, as well as the storage conditions, as summarized below.

2. TRIACYLYGLYCEROLS

Neutral lipids, mainly triacylglycerols, usually account for over 95% of edible oils. The fatty acids present are either saturated or unsaturated (Figure 1). Although saturated lipids were generally condemned because of their perceived negative effect on cardiovascular disease, more recent studies have shown that C18:0 is fairly benign while C14:0 may possess adverse health effects. Furthermore, a better understanding of detrimental effects of trans fats has served as a catalyst for the return from using hydrogenated fats in place of naturally-occurring edible oils with a high degree of saturation. Hence, palm oil now serves as an important constituent of non-hydrogenated margarine formulations.

Of the unsaturated lipids, monounsaturated fatty acids, similar to saturated fatty acids, are non-essential as they could be synthesized in the body *de novo*. However, polyunsaturated fatty acids (PUFA; containing 2 or more double bonds) could not be made in the body and must be acquired through dietary sources. The parent compounds in this group are linoleic acid (LA, C18:2ω6) and linolenic acid (LNA, C18:3ω3). The symbols ω6 (or n-6) and ω3 (or n-3) refer to the position of the first double bond from the methyl end group as this dictates the biological activity of the fatty acid molecules involved (Figure 2) (Shahidi and Finley, 2001).

Dietary lipids are composed mainly (≈80%) of C18 fatty acids and it is recommended that the ratio of ω6 to ω3 fatty acids be at least 5:1 to 10:1, but the western diet has a ratio of 20:1 or less. Enzymes in our body convert both groups of PUFA through a series of desaturation and elongation steps to C20 and C22 products, some of which are quite important for health and general well-being. The C20 compounds may subsequently produce a series of hormone-like molecules known as eicosanoids which are essential for maintenance of health. Obviously, elongation of LA and alpha-linolenic acid (ALA) to other fatty acids may be restricted (to about 5%) by rate determining steps and lack of the required enzymes in the body (See Figure 1). Thus, pre-term infants and the elderly may not be able to effectively make even these limited transformations and that production of docosahexaenoic acid (DHA) from eicosapentaenoic acid (EPA) is rather inefficient. Furthermore, supplementation with gamma-linolenic acid (GLA) as a precursor to arachidonic acid (AA) might also be necessary.

Lipids are generally highly stable in their natural environment; even the most unsaturated lipids from oilseeds are resistant to oxidative deterioration prior to extraction and processing. Nature appears to be able to protect itself as higher level of antioxidants are generally found in highly unsaturated oils. It is believe that unsaturated oils generally co-exist with antioxidants in order to protect themselves from oxidation; of course the natural

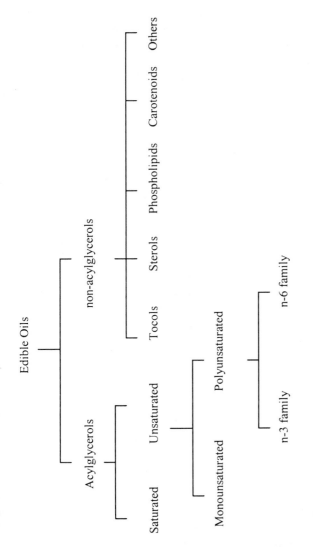

Figure 1. Edible oils and their fatty acid and non-triacylglycerol components.

capsule or seed coat provides a barrier to light and oxygen as well as compartmentalization of oil cells and inactivity of enzymes prior to crushing. However, upon crushing and oil extraction, the stability of edible oils is compromised and this is dictated by several factors, including the degree of unsaturation of fatty acid constituents. This topic will be discussed in further detail in a later section. In addition, the position of fatty acids in the triacylglycerol molecule (Sn-1, Sn-2 and Sn-3) would have a considerable effect on their assimilation into the body. Generally, fatty acids in the Sn-1 or Sn-3 position are hydrolyzed by pancreatic lipase and absorbed while those in the Sn-2 position are used for synthesis of new TAG and these might be deposited in the body.

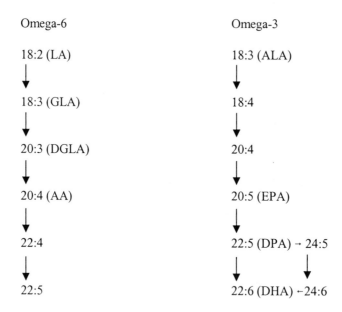

Figure 2. Essential fatty acids (EFA) of the ω6 and ω3 families.

3. STABILITY OF EDIBLE OILS

Stability of edible oils is affected by their constituents fatty acids and minor components as well as storage conditions. Thus as the degree of unsaturation of an oil increases, its susceptibility to oxidation, under similar conditions, increases. Furthermore, the condition in which the oil is present, e.g. bulk versus emulsion, has a profound effect on the stability of the oil. In addition, presence of antioxidants as well as metal ions, light, chlorophylls and other pigments might influence the stability of the oil. In our studies, we found that while crude green tea extracts behaved pro-oxidatively in a bulk oil model system, removal of the chlorophyll reversed the situation.

3.1. Degree of Unsaturation

As the number of double bonds in a fatty acid increases, its rate of oxidation increases. Thus, docosahexaenoic acid (DHA, C22:6ω3) is more prone to oxidation than eicosapentaenoic acid (EPA, C20:5ω3) followed by arachidonic acid (ARA C20:4ω6), alpha-linolenic acid (ALA, C18:3ω3), linoleic acid (LA, C18:2ω6) and oleic acid (OA, C18:1ω9). This trend is also reflected in the triacylyglycerols of different oils. Thus, as the iodine value (IV) of oils increases, their oxidation potential is also increased. Table 1 summarizes the induction period of several oils with different degrees of unsaturation as obtained using a Ranciment apparatus.

Table 1. Rancimat induction periods of selected oils at 100°C.

Oil	Induction period (h)
Palm	35.55
Palm (stripped)	17.38
Corn	20.79
Canola	15.59
Rice bran	12.07
Algal (DHASCO)	0.67
Seal blubber	0.17
Menhaden	0.13

3.2. Non-triacylyglycerols Constituents

The non-triacylyglycerol or unsaponifiables matter content varies from one oil to another. It appears that as the iodine value of oils increases, their content of tocols increases as well (see Table 2). Different classes of compounds belonging to this group and their application areas are summarized in Table 3. In most oils these constitute approximately 1%, but in others they may be present at 2-8% and in some cases even higher. Many of the unsaponifiable matter are recovered from oil during processing steps of degumming, refining, bleaching and deodorization. Thus, loss of sterols, tocopherols, carotenoids and related compounds during oil processing may range from 35 to 95% (Table 4). These material may be collected as distillates during the deodorization process. Distillates that are rich in certain components may be separated and marketed for use in nutraceutical applications. Thus, mixed tocopherols, tocotrienols, carotenoids, lecithin and other constituents may be separated from soybean, palm oil, rice bran oil and barley oil, depending on their prevalence.

Table 2. Correspondence of degree of unsaturated and tocol content of selected oils.

Oil	Iodine Value (g/100g oil)	Tocol content (mg/100g)
Coconut	9	2
Illipe	33	10
Mango	47	15
Tea seed	86	20
Rapeseed	107	65
Soybean	135	100

Table 3. Non-triacylglycerols of edible oils and their use

Constituent	Example	Application
Hydrocarbon	Squalene	Skin care
Phospholipids	Phosphatidylcholine & Phosphatidylserine (Lecithin)	Dietary supplement, OTC
Sterols	β-sitosterol	Functional food ingredient, OTC
Tocols	tocopherols & tocotrienols (Vitamin E)	Dietary supplement, OTC
Ubiquinone	Coenzyme Q10	Dietary supplement, OTC
Carotenoids	β-carotene & Xanthophylls	Dietary supplement, OTC
Phenolics	flavonoids, isoflavonoids	Dietary supplement

3.3. Tocols and Ubiquinones

Tocols include both tocopherols (T) and tocotrienols (T3) which are present in edible oils in different compositions and proportions. Eight different compounds exist; each series of T and T3 includes four components designated as α, β, γ and δ, depending on the number and position of methyl groups on a chromane ring. The α-isomer is 5, 7, 8-trimethyl; β-isomer, 5, 8-dimethyl; γ-isomer, 7,8-dimethyl; and δ is the 8-methyl isomer.

The occurrence of tocopherols in vegetable oils is diverse, but animal fats generally contain only α-tocopherol. However, absence of α-tocopherol in blood plasma and other body organs may not negate the importance of other tocopherols such as γ-tocopherol. In oilseed lipids, there appears to be a direct relationship between the degree of unsaturation as reflected in the IV, and the total content of tocols, mainly tocopherols (see Table 2). Most vegetable oils contain α-, γ- and δ-tocopherols, while β-tocopherol is less prevalent, except for wheat germ oil. Meanwhile, tocotrienols are present mainly in palm and rice bran oils. The antioxidant activity of tocotrienols generally exceeds that of their corresponding tocopherols. Meanwhile, the antioxidant activity of tocopherols is generally in the order of $\delta > \gamma > \beta > \alpha$. The content of tocopherol/tocotrienol in selected oils is summarized in Table 5. With respect to ubiquinones, also known as coenzyme Q, they occur as 6 to 10 isoprene unit compounds; that is Q_6 (UQ-6) to Q_{10} (UQ-10). Coenzymes Q_{10} (UQ-10) to a lesser extent Q_9 (UQ-9) are found in vegetable oils. Ubiquinone provides efficient protection *in-vivo* for mitochondria against oxidation, similar to vitamin E in the lipids and lipoproteins.

3.4. Phospholipids

Phospholipids possess fatty acids which are generally more unsaturated than their associated tricylglycerols and hence are more prone to oxidation. However, this situation does not necessarily hold for marine oils such as seal blubber oil whose phospholipids are less unsaturated than its triacylglycerols. The role of phospholipids as pro- or antioxidants is, however, complex because in addition to their lipid moiety, they contain phosphorous- and nitrogen-containing groups that dictate their overall effect in food systems.

Extensive studies have demonstrated that phospholipids may exert an antioxidant effect in vegetable oils and animal fats. The exact mechanism of action of phospholipids in stabilizing fats and oils remains speculative; however, evidence points out to the possibility of their synergisms with tocopherols, chelation of prooxidant metal ions as well as their role in the formation of Maillard-type reaction products. King *et al.* (1992) found a positive relationship between the presence and type of phospholipids and stability of salmon oil in the order given below.

Sphingomyelin ≈ lysophosphatidylcholine ≈ phosphatidylcholine ≈ phosphatidyl-ethanolamine > phosphatidylserine > phosphatidylinositol > phosphatidylglycerol.

3.5. Phytosterols

Edible oils generally contain a variety of sterols which exist in the free form, as sterol ester of fatty acids and sterol glycosides or esters of sterol glycosides. Sterols are heat-stable molecules with no flavor of their own and exhibit antipolymerization activity during frying. Sterols serve as a means for fingerprinting of vegetable oils and lend themselves for detection of adultration of oils. Among sterols, Δ^5-avenasterol, fucosterol and citrostadienol have been shown to exhibit antioxidant properties. Donation of a hydrogen atom from the allylic methyl group in the side chain is contemplated. Most oils contain 100-800 mg/100g sterols (See Table 4). Brassicasterol is specifically found in canola, rapeseed and mustard oils.

3.6. Carotenoids

Carotenoids are widespread in oilseeds, but are found in the highest amount in palm oil at 500-700 ppm levels. Both hydrocarbon-type carotenoids, namely α-and β-carotene, as well as xanthophylls are present. Carotenoids act as scavengers of singlet oxygen and hence are important in stability of oils exposed to light. While β-carotene, and αcarotene, are usually the dominant components, α-xanthophylls may be present in smaller amounts.

3.7. Chlorophylls

Chlorophylls are present in a variety of oils. In particular, extra virgin olive oil contains a large amount of chlorophylls and this is often associated with the high quality of this oil. However, in oils such as canola, immature seeds contain chlorophylls which end up in the oil and affect its stability. Meanwhile, grapeseed oil is generally green in color. Due to their photosensitizing effects, chlorophylls lead to oxidative deterioration of oils when exposed to light.

3.8. Phenolics

Phenolic compounds, other than tocols, may be present in edible oils. While tocols are always present in the free form phenolic acids, phenylpropanoids and flavonoids and related components occur in the free, esterified and glycosylated forms. These compounds reside mainly in the meal, but their presence in oils, such as olive oil, is well recognized. Olive oil contains a number of phenolics, including hydroxytyrosol. Furthermore, sesame oil contains sesamin, sesaminol and sesamol which render stability to the oil. Meanwhile, oat oil contains a number of ferulates that affect its stability.

3.9. Hydrocarbons

Hydrocarbons are another group of unsaponifiable matter that may occur in edible oils. This class of compound includes squalene which constitutes one third of the unsaponifiables in olive oil. The effect of squalene in stabilization of oils at high temperatures and in the body for protection of skin is recognized. However, in an oil system, under Schaal oven conditions, we did not find it to be antioxidative in nature.

Table 4. Removal of non-triacylglycerol components during edible oil refining.

Constituent	Removal (%)
Phospholipids	> 95
Free fatty acids	> 95
Carotenoids & Chlorophylls	> 95
Sterols	32 – 61
Tocols	35 - 47

Table 5. Tocophenols (T) and Ubiquinones (UQ) and sterols in edible oils (mg/100g)

Oil	α-T	β-T	γ-T	δ-T	UQ-9	UQ-10	Sterols
Canola	17-26	-	35-61.2	0.4-1.2	0.2	7.3	350-840
Soybean	10.1-10.2	0.27	47.3-59.3	26.4-35.2	0.8	9.2	150-420
Sunflower	48.7-78.3	0.25	1.9-5.1	0.7-1.0	2.1	0.4	ND
Palm oil*	180	-	320	7.0	ND	ND	30-260

*Palm oil also contains 124, 30, 280 at 70 mg/100g of α-, β-, γ- and δ-tocotrienols.

4.   EFFECT   OF   PROCESSING   ON   NON-TRIACYLYGLYCEROLS COMPONENTS AND STABILITY OF OILS

As explained earlier, during different stages of degumming, refining, bleaching and deodorization, 35-95% of non-triacylyglycerol components of edible oils may be removed (Reichert, 2002). The effects of processing, as determined by Ferrari *et al.* (1996), on the contents of tocopherols and sterols for canola, soybean and corn oils are summarized in Table 6. In addition, carotenoids in edible oils, especially in palm oil, might be depleted during processing. Bleaching of carotenoids might be carried out intentionally in order to remove the red color, however, carotenoids might be retained using cold pressing at temperatures of as low as 50°C.

In order to measure the effect of processing and contribution of minor components to the stability of edible oils, it is possible to strip the oil from its non-triacylyglycerols components. To achieve this, the oil is subjected to a multi-layered column separation. The procedure developed by Lampi *et al.* (1997) may be employed. We used a column packed with activated silicic acid (bottom layer, 40 g) followed by a mixture of Celite 545 / activated charcoal (20 g, 1:2 (w/w)), a mixture of Celite 545/powdered sugar (80 g, 1:2 (w/w)) and activated silicic acid (40g) as the top layer. Oil was diluted with an equal volumn of n-hexane and passed through the column that was attached to a water pump; the solvent was then removed. The characteristics of the oils before and after stripping indicated that while oxidative products and most of the minor components were removed, γ-tocopherol was somewhat retained in the oil.

Table 6. Changes in the content and composition of minor components of edible oils during processing (mg/100g).

Constituent	Crude	Refined	Bleached	Deodorized
Canola (Rapeseed)				
Tocopherols	13.6	128.7	117.8	87.3
Sterols	820.6	797.8	650.4	393.0
Soybean				
Tocopherols	222.3	267.7	284.0	195.2
Sterols	359.5	313.9	288.8	295.4
Corn				
Tocopherols	194.6	203.8	201.9	76.7
Tocotrienols	7.9	10.2	10.0	6.1
Sterols	1113.9	859.2	848.8	715.3

The oils (olive, borage and evening primrose) were subjected to accelerated oxidation under Schaal oven condition at 60°C or under fluorescent lighting. Results indicated that oils were more stable, as such, than their stripped counterparts when subjected to heating, but under light, the oxidative stability of oils stripped of their minor components was higher. Examination of the spectral characteristics of oils indicated that chlorophylls were present in the original oils and hence might have acted as photosensitizers leading to enhanced oxidation of unstripped oils (Khan and Shahidi,

1999, 2000). In case of red palm oil, however, removal of carotenoids by stripping resulted in a decrease of near 15 h in this induction period as measured by Rancimat at 100°C (Shahidi *et al.*, 2001). Therefore, minor components of edible oils and the nature of chemicals involved have a major influence on the stability of products during storage and food preparation.

5. REFERENCES

Ferrari, R. A., Schulte, E., Esteves, W., Brühl, L. and Mukherjee, K. D., 1996, Minor constituents of vegetable oils during industrial processing, *J. Am. Oil Chem. Soc.*, **73**:587-592.

Khan, M. A. and Shahidi, F., 1999, Rapid oxidation of extra virgin olive oil stored under fluorescent light, *J. Food Lipids*, **6**:331-339.

Khan, M.A. and Shahidi, F., 2000, Oxidative stability of stripped and non-stripped borage and evening primrose oils and their emulsions in water, *J. Am. Oil Chem. Soc.*, **77**:963-968.

King, M. F., Boyd, L. C., and Sheldon, B. W., 1992, Antioxidant properties of individual phospholipids in a salmon oil model system, *J. Am. Oil Chem. Soc.*, **69**:545-551.

Lampi, A. M., Hopia, A. and Piironen, V., 1997, Antioxidant activity of minor amounts of gamma-tocopherol on the oxidation of natural triacylglycerols, *J. Am. Oil Chem. Soc.,* **74**:549-555.

Reichert, R. D., 2002, Oilseed medicinals: In natural drugs, dietary supplements and in neuo functional foods. *Trends Food Sci. Technol.*, **13**:353-360.

Shahidi, F. and Finley, J. W., 2001, Omega-3 fatty acids: chemistry, nutrition and health effects. ACS Symposium Series 788. American Chemical Society, Washington, D.C.

Shahidi, F. and Shukla, V. K. S., 1996, Nontriacylglycerol constituents of fats/oils, *INFORM,* **8**:1227-1232.

Shahidi, F., Lee, C. L., Khan, M. A. and Barlow, P. J., 2001, Oxidative stability of edible oils as affected by their minor components. Presented at the Am. Oil Chem. Soc. Meeting & Expo, May 13-16, Minneapolis, MN.

FLAVOR OF VINEGARS

Especially on the volatile components

Shoji Furukawa and Rie Kuramitsu[*]

1. INTRODUCTION

Vinegar is an acidic seasoning with acetic acid as its major component. In the course of its brewing, acetic acid fermentation takes place following alcoholic fermentation, and subsequently it is inferred that fermented vinegars contain many flavor components. Although small in their amounts, besides acetic acid, different components constitute the characteristic flavor of vinegars. Numerous reports are available on the flavor components of vinegars.

By using capillary column and without concentration process, (Kahn *et al.*, 1996) identified 16 components including alcohols, esters, acids and acetoin from cider vinegar. Arima *et al*, (1967) and Yamaguchi *et al.* (1967) extensively studied the flavor components in the neutral and basic fractions in a vinegar made from sake cake. They extracted the neutral fraction with a mixture of ether and n-pentane (2:1, v/v) to isolate 51 compounds, of which 26 compounds were identified; these included 11 alcohols, 7 esters and 8 carbonyl compounds.

For the quality of fermented vinegars, standards are officially set for the acidity, non-salt soluble solid matter and others, and commercial vinegars are thought to meet these requirements. However, no standards are set for the flavor components. From the results cited in the literature, it is considered that the kinds and contents of flavor components in vinegars vary considerably depending on the starting material from which vinegars are manufactured, and also on breweries or manufacturing processes, even from identical starting materials.

To examine this problem, we identified and determined flavor components in 9 samples from 3 breweries each on 3 vinegars from brown rice, rice and grain using a chromatographic technique. Studies were also conducted on cider vinegar whose manufacturing has recently been increased due to its beneficial effect on health, and determined change of its flavor components during manufacturing.

[*] Shoji Furukawa and Rie Kuramitsu, Department of Chemistry, Faculty of General Educacion, Akashi College of Technology, Akashi, Hyogo 674-8501, Japan

Quality of Fresh and Processed Foods, edited by Shahidi et al.
Kluwer Academic/Plenum Publishers, 2004.

2. THE FLAVOR COMPONENTS IN COMMERCIAL VINEGARS

It was thought that the type and content of flavor components in vinegars differed from one another depending on the starting material and on breweries.   A total 9 samples of commercial vinegars of 3 types (brown rice, rice and grain) each from 3 breweries was used for analysis. Grain refers to cereals other than brown rice and rice, as shown in Table 1.   The gas chromatographic (GC) operating conditions were as follows.   Samples were analyzed using a Shimadzu GC -6A.   The column used was a PEG 20 M 10% on chromosorb W. (20 mm x 2.6 m).   A total of 4 µL of sample was injected; the injector temperature was 180°C and the carrier gas was nitrogen at 40 mL/min, hydrogen at 50 mL/min and air at 1.0 L/min.   The oven temperature was 60°C for 5 min, and then raised from 60 to 140°C at 3°C/min.

Table 1. Commercial vinegars studied

Sample No.	Type of vinegar	Brewery	Acetic acid, %
1	Brown Rice Vinegar	A	4.5
2		B	4.5
3		C	4.5
4	Rice Vinegar	A	4.5
5		B	4.5
6		C	4.5
7	Grain Vinegar	A	4.2
8		B	4.3
9		C	4.2

An example of the results of gas chromatography for brown rice vinegar is shown in Figure 1.   Before peak No.15 (acetic acid)14 peaks were detected, and of these 5 peaks, including ethyl acetate (peak 4) , were identified.   In addition, 3 analyses were repeated on each sample and the volatile components in the 9 vinegar samples determined.   The results are shown in Tables 2 to 4 and also in Figure 2.

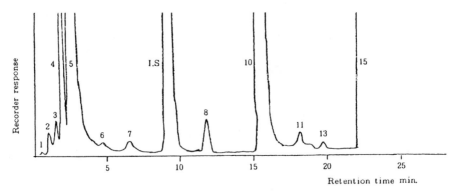

Figure 1. Chromatogram of a commercial brown rice vinegar

4 : ethyl acetate   5 : ethyl alcohol   6 : n-propyl alcohol   8 : isoamyl alcohol   10 : acetoin

The compound n-Propyl alcohol was present in small amounts in the samples, and as such was not detected in many samples. The concentrations of the other 4 identified components were distributed widely among samples, and analysis of variance on the kinds and breweries of vinegar showed that for ethyl acetate, ethanol and isoamyl alcohol differences were found in the concentration of components depending on breweries as well as the kind of vinegar within 1% (Table 2). For acetoin 5% significant difference was found only with kinds of vinegar. The concentration (mg per 100 mL) of ethyl acetate was highest in brown rice vinegar among other vinegars, showing 13-22 mg, and those of rice and grain vinegars were 3.8-12 mg and 2.1-11 mg, respectively. Among breweries, vinegars made by Brewery A showed the highest concentration of 11-22 mg and Breweries B and C showed 11-13 mg and 2.1-16 mg, respectively. The ethyl acetate concentrations in brown rice vinegar and in rice vinegar from Brewery B were similar. The concentration of ethyl alcohol was highest (61-206 mg) in brown rice vinegar followed by rice vinegar 41-133 mg and grain vinegar (22-130 mg). Among breweries the product by Brewery A showed the highest value of (130-206 mg) followed by that by Brewery B (61-96 mg) and that by Brewery C (22.1-187 mg).

The ethyl alcohol concentration in the brown rice vinegar by Brewery B was lower than that in rice vinegar. The vinegars with higher ethyl alcohol concentration were also higher in the ethyl acetate concentration.

The concentration of isoamyl alcohol was highest with 0.70-1.3 mg in brown rice vinegar followed by 0.01-0.92 mg in rice vinegar and 0.02-0.14 mg in grain vinegar (Tables 2 to 4). Among breweries, the products produced by Brewery C showed the highest concentrations, as exemplified for that of isoamyl alcohol. The concentration of acetoin was highest in brown rice vinegar (39-175 mg), and rice vinegar showed 1.9-74 mg and grain vinegar 2.1-21 mg.

Among breweries brown rice vinegar from Brewery C showed the highest concentration. With rice and grain vinegars Brewery B showed the highest concentration of acetoin. At 5% level, there was no significant difference among breweries in regard to their acetoin concentration. The concentrations of volatile components in vinegars were in the range of 2.1-22 mg for ethyl acetate, 22-206 mg for ethyl alcohol, below 0.3 mg for n-propyl alcohol, 0.01-1.3 mg for isoamyl alcohol and 1.9-175 mg for acetoin.

It has been reported that the ethyl acetate concentration is correlated with the residual concentration of ethanol. In the present studies the results shown in Figure 2 indicate an extremely close correlation between the concentrations of ethyl acetate and ethanol.

Many factors in the starting material and the manufacturing process influence the kind and content of volatile components in vinegar. Thus, there are set official standard amount of starting material for different vinegars. For example, for grain vinegar more than 400 g of grain is required per liter. However, the exact amount of starting material employed differs slightly among breweries.

The contents of ethyl alcohol, ethyl acetate and isoamyl alcohol appear to be influenced more by the difference in the amount of starting material employed than by that of the manufacturing process, as shown in this study.

By contrast, the major factor for the formation of acetoin is claimed to be the organic acids in the starting materials or those in the product of alcoholic fermentation. The content of acetoin, however, is thought to vary greatly depending on the method of manufacturing. It appears to be somewhat difficult for breweries to regulate the acetoin concentration to the level characteristic for each brewery on different types of vinegar. However, no difference was found in the acetoin content among breweries, perhaps due to the difficulty in manufacturing processes.

Table 2. Concentration of volatile components in commercial brown rice vinegars (mg/100mL)

Brewery	A				B				C			
	1	2	3	mean	1	2	3	mean	1	2	3	mean
ethyl acetate	21.95	24.26	19.84	22.02	15.49	12.75	11.49	13.24	16.37	12.92	17.61	15.63
ethyl alcohol	208.6	205.5	203.3	205.5	63.20	60.73	60.03	61.32	187.0	186.2	188.4	187.2
n-proryl alcohol	0.25	0.34	0.25	0.28	—	—	—	—	0.14	0.16	0.14	0.15
isoamyl alcohol	1.18	1.22	1.16	1.19	0.69	0.72	0.69	0.70	1.26	1.29	1.27	1.27
acetoin	38.95	39.02	39.22	39.06	60.23	59.58	60.88	60.23	176.0	176.5	172.2	174.9

Table 3. Concentration of volatile components in commercial rice vinegars (mg/100mL)

Brewery	A				B				C			
	1	2	3	mean	1	2	3	mean	1	2	3	mean
ethyl acetate	12.67	12.46	11.75	12.29	11.24	9.46	13.01	11.24	3.48	3.70	4.25	3.81
ethyl alcohol	133.5	132.1	133.6	133.1	95.34	93.52	95.30	94.72	41.49	42.48	39.47	41.15
n-proryl alcohol	—	—	0.04	0.01	—	—	—	—	0.05	0.17	0.09	0.10
isoamyl alcohol	0.01	trace	0.02	0.01	0.02	0.02	0.02	0.02	0.95	0.95	0.88	0.92
acetoin	1.70	1.70	2.30	1.91	74.29	73.43	75.39	74.37	22.44	22.50	21.87	22.27

Table 4. Concentration of volatile components in commercial grain vinegars (mg/100mL)

Brewery	A				B				C			
	1	2	3	mean	1	2	3	mean	1	2	3	mean
ethyl acetate	11.05	11.72	11.49	11.42	10.88	10.19	10.48	10.52	2.15	1.93	2.10	2.06
ethyl alcohol	130.2	129.6	129.0	129.6	96.70	94.80	95.25	95.58	21.90	23.01	21.26	22.06
n-proryl alcohol	—	—	—	—	0.04	—	—	0.01	trace	trace	—	trace
isoamyl alcohol	0.02	0.02	0.02	0.02	0.03	0.03	0.03	0.03	0.13	0.14	0.14	0.14
acetoin	2.21	2.01	2.21	2.14	20.30	21.51	20.04	20.62	4.11	4.16	4.02	4.10

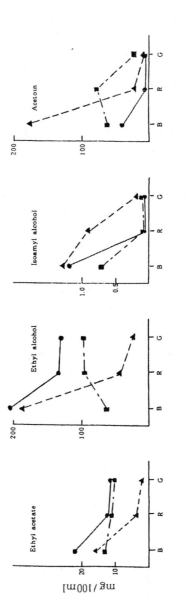

Figure 1-2. Concentration of volatile components in commercial vinegars

B : brown rice vinegar R : rice vinegar G : grain vinegar A Co. : ⬤—⬤ B Co. : ■—■ C Co. : ▲--▲

3. VOLATILE COMPONENTS OF COMMERCIAL CIDER VINEGAR

In a separate study, samples of commercial vinegars were analyzed to detect the relevant volatiles; and ethyl acetate, ethyl alcohol, n-propyl alcohol, isoamyl alcohol and acetoin were identified as volatile components. It was noted that the contents of these volatiles differed depending on the starting material and brewery. Seven samples of cider vinegar were examined and compared with 3 samples of rice vinegar were also used. The commercial cider vinegars studied are shown in Table 5. The samples of rice vinegars and cider vinegars were according to the Specification of the Japanese Department of Agriculture. The gas chromatograph used was the same as before and the oven temperature was modified to 80 °C for 10 min and programmed 80-140 °C at 3 °C/min. Other conditions were identical with those specified earlier. Samples were introduced into the gas chromatographic apparatus without any pretreatment.

An example of gas chromatographic analysis is shown in Figure 3. Of the 15 peaks, with exception of peak No.14 for acetic acid, components for peaks of No. 1, 8 and 9 were not identified. Components for the remaining 12 peaks were identified and these included acetaldehyde (peak No. 2) and acetone (peak No. 3) which were present in low amounts. Isobutyl acetate (peak No. 6) was detected only in 1 sample. Propionic acid (peak No. 15) and isobutyric acid (peak No. 16) were present in a very low amount. Of the 12 components that were identified, 7 components other than the above 5, namely, ethyl acetate, ethyl alcohol, n-propyl alcohol, isobutyl alcohol, isoamyl acetate, isoamyl alcohol, and acetoin were submitted to 3 repetitive analyses, respectively.

Table 6 shows the results (mean values) of analyses for the volatile components in 3 samples of commercial rice vinegar and 7 samples of commercial cider vinegar. The content of each component was distributed over a wide range in both rice and cider vinegars and it was different among samples from different breweries.

Table 5. Commercial vinegars studies

Brewery (Location)	Type of vinegar	materials	% as Acetic acid
A (Aichi)	Rice vinegar	Rice, Alcohol	4.5
B (Hyogo)	Rice vinegar	Rice, Alcohol	4.5
C (Osaka)	Rice vinegar	Rice, Corn	4.5
D (Aichi)	Cider vinegar	Apple juice	5.0
E (Kyoto)	Cider vinegar	Apple juice	4.5
F (Fukuoka)	Cider vinegar	Apple juice	4.2
G (Osaka)	Cider vinegar	Apple juice	4.8
H (Gifu)	Cider vinegar	Apple juice	4.5
I (Tokyo)	Cider vinegar	Apple juice, Alcohol	4.5
J (U. S. A)	Cider vinegar	Apple juice, water	5.0

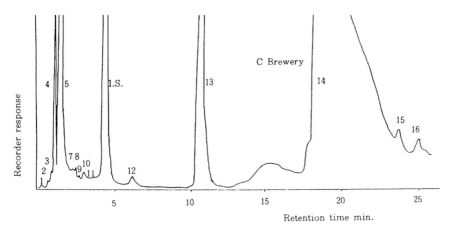

Figure 3. Chromatogram of a commercial rice vinegar
2 : Acetaldehyde   3 : Acetone   4 : Ethyl acetate   5 : Ethyl alcohol   6 : Isobutyl acetate   7 : Propyl alcohol
10 : Isobutyl alcohol   11 : Isoamyl acetate   12 : Isoamyl alcohol   13 : Acetoin   14 : Acetic acid
15 : Propionic acid   16 : Isobutyric acid

The contents of acetaldehyde in cider vinegars were 0.2-1.0 mg/100ml and were similar to those in rice vinegars.   The contents of ethyl acetate were 3.9-17.7 mg in cider vinegars and 4.5-13.4 in rice vinegars; the mean content in cider vinegar was 9.0 mg and was nearly identical with that in rice vinegar.   The contents of ethyl alcohol were 50-279 mg in cider vinegar and 54-162 mg in rice vinegars; the mean values in both were similar.   A sample of cider vinegar produced by a foreign company (Brewery J) contained small amounts of ethyl acetate and ethyl alcohol, and these contents in domestic cider vinegars were slightly higher than those in rice vinegars.   As expected, the ethyl acetate contents in both cider and rice vinegars correlated were with the residual ethyl alcohol contents. Isobutyl acetate was detected only in the sample of cider vinegar from Brewery J (foreign company).   The content was about 1.0 mg and was fairly high for a volatile component.   n-propyl alcohol was contained in 0.14 and 0.32 mg in cider vinegars from Breweries D and J, respectively, and the contents were lower than 0.1 mg in other cider vinegar and rice vinegar samples.   The contents of isoamyl acetate in the cider vinegars from Breweries D and J were 0.15 mg and 0.18 mg, respectively, and those in other cider vinegars and rice vinegars were less than 0.1 mg.

The contents of isobutyl alcohol in cider vinegars by Breweries D and J were 0.62 mg and 0.77 mg, respectively, and that in rice vinegar from Brewery C 0.22 mg; the contents in other cider and rice vinegars were all less than 0.1 mg. The contents of isoamyl alcohol in cider vinegars from Breweries F and I were less than 0.1 mg but were 0.16-2.35 mg in those from other breweries.   In rice vinegar the sample from Brewery C showed a content of 0.46 mg and samples from other breweries showed contents of lower than 0.1 mg.   A tendency was found that the content of isoamyl alcohol was generally higher in cider vinegar than in rice vinegar. The average content of propionic acid in cider vinegars was 0.1-0.2 mg, and appeared to be similar to that in rice vinegars.   The content of isobutyric acid was nearly the same as that of proionic acid.

Table 6. Concentration of volatile components in commercial vinegars (mg/100mL)

Type	Rice vinegar				Cider vinegar							
Brewery	A	B	C	Mean	D	E	F	G	H	I	Mean	J
Acetaldehyde	+	+	+	+	++	+	-	+	+	-	+	+
Acetone	+	+	+	+	++	±	±	++	±	-	+	++
Ethyl acetate	13.35	8.17	4.51	8.68	17.70	12.94	3.86	7.82	7.20	8.21	9.62	4.95
Ethyl alcohol	161.7	97.45	53.48	104.2	279.2	152.4	49.77	73.08	68.81	129.5	125.5	72.07
Isobutyl acetate	-	-	-	-	-	-	-	-	-	-	-	++
Propyl alcohol	-	-	0.03	0.01	0.14	-	-	0.05	-	-	0.03	0.32
Isobutyl alcohol	0.01	0.04	0.22	0.09	0.62	0.03	-	0.02	0.06	0.01	0.12	0.77
Isoamyl acetate	-	-	0.03	0.01	0.15	-	-	0.04	0.01	-	0.03	0.18
Isoamyl alcohol	-	0.03	0.46	0.16	2.35	0.11	-	0.16	0.30	0.01	0.49	2.34
Acetoin	1.17	2.49	24.43	9.36	13.37	2.28	5.78	2.56	6.18	3.36	5.59	96.40
Propionic acid	-	+	++	+	++	+	-	+	+	-	+	++
Isobutyric acid	±	+	++	+	++	±	+	+	+	±	+	++

Concentration: ++ = > 1.0 + = 1.0~2.0 ± = < 0.2 mg/100 ml - = not detected

The acetoin contents in cider vinegars from Breweries D and J were 13.4 and 96.4 mg, respectively, and that in rice vinegar from Brewery C was 24.4 mg. Other cider vinegar and rice vinegar samples showed less than 6 mg of acetoin content. In comparison with domestic cider vinegars, foreign product (Brewery J) showed an extremely higher content of acetoin. In domestic cider vinegars the acetoin content was slightly higher than that in rice vinegars (except the product by Brewery c).

In comparing the contents of volatile components, the amounts of ethyl acetate and ethyl alcohol in domestic cider vinegars were slightly higher than those in rice vinegars. Isobutyl acetate, propyl alcohol, isobutyl alcohol and isoamyl alcohol were present in more than 0.1 mg only in the cider vinegars by Breweries D and J and in the rice vinegar by Brewery C; therefore, it was difficult to make a general comparison, but it appeared that cider vinegars contained slightly more of these components than rice vinegars. Isoamyl alcohol was present in higher amounts in cider vinegars than in rice vinegars. Except the products by Breweries C and D (with particularly higher contents than the products by other breweries), the contents of acetoin in cider vinegars were slightly higher than those in rice vinegars. Foreign made cider vinegars contained more acetoin and higher alcohols than domestic cider vinegars. As in cider vinegars by Breweries D and J and in rice vinegar by Brewery C, the products with higher contents of such higher alcohols as n-propyl alcohol isobutyl alcohol and isoamyl alcohol contained more such acids as proionic acid and isobutyric acid that corresponded to the higher alcohols and isoamyl acetal of an ester of a higher alcohol, was detected.

The results suggest that greater parts of the higher alcohols, are formed in alcoholic fermentation and that they are partially converted into the corresponding acids or acetate esters in the acetic acid fermentation that follows. Accordingly, it is considered that the contents of higher alcohols in vinegars are closely related with the contents of corresponding acids or acetate esters.

From the results of the present studies, it may be concluded that the kind and content of volatile components in vinegars differ greatly depending on the types (cider and rice) of vinegar as well as on breweries (Breweries A, B, C and so on)by which they are produced.

4. CHANGES OF FLAVOR COMPONENTS DURING ACETIC ACID FERMENTATION

Higher alcohols, aldehydes and other flavor components which are formed in alcoholic fermentation are considered to be changed further during acetic acid fermentation to generate flavor components characteristic to vinegars.

The formation of flavor components, particularly volatile components, in vinegars were greatly influenced by the kinds (cider and rice) of vinegars as well as by breweries (Breweries A, B and C). The results suggested that the difference in manufacturing process in different breweries influence the quality of vinegars. Here, the formation of volatile components, change of volatile components, particularly higher alcohols, and that of the oxidizing activity of acetic acid bacteria on various alcohols during acetic acid fermentation were studied by using apple juice after alcoholic fermentation.

The bacteria was Strain A-4, belonging to Acetobacter aceti. To an apple juice after alcoholic fermentation were added acetic acid and water to adjust the solution to 3% of alcohol content and 2% of acidity. The adjusted solution (500 ml) was placed in a 2 L-shaking flask and cultured at 30°C (with shaking of 20 cm amplitude at 60 rpm).   Volatiles were analyzed by conventional methods using a Hitachi K-53 gas chromatograph.   For volatile components 500 ml of the adjusted solution, after fermentation, was neutralized (to pH 7.3) and extracted with diethyl ether-pentane (2:1, v/v) and the extract was dehydrated with sodium sulfate and evaporated to remove the solvent.   The concentrated residue was submitted to gas chromatography.   The conditions of gas chromatography were: column, PEG1500, celite 545 (20%) (3mmx2m); temperature 40°C for 5 min and then repeat to 130°C at 3°C/min; injection post temperature 180°C; Flame ionization detector at 180°C with carrier gas, hydrogen at 20 mL/min and nitrogen at 200 mL/min. For measurement of oxidizing activity, a suspension of isolated bacteria was used to estimate the activity by using a manometer.

Changes of general components during fermentation of apple juice are shown in Figure 4. Bacterial mass increased rapidly after the start of fermentation and reached the steady phase after 15 h.   The concentration of alcohol decreased while acidity increased, and after 20 h the acidity reached a maximum.

The oxidizing activity was estimated on 12 alcohols from the results of gas chromatographic analyses.   Table 7 shows the changes of the oxidizing activity during acetic acid fermentation of the apple juice after alcoholic fermentation.   After 12 h, concentration of methyl alcohol was very low while that of 4 other normal alcohols was slightly higher.   Against the concentration of ethyl alcohol those of propyl, butyl and amyl alcohols were 40, 60 and 90%, respectively.

Table 7. Changes in oxygen uptake rate during acetic acid fermentation of cider (µL 0°C/mg cell/h)

Alcohol	Time (h)	
	12	22
Methyl	6.3	5.5
Ethyl	3309.3	1350.3
n-Propyl	1459.3	1024.9
n-Butyl	1929.5	1171.2
n-Amyl	3008.9	1375.0
iso-Propyl	22.8	10.4
iso-Butyl	51.8	40.9
iso-Amyl	79.9	57.6
sec-Butyl	31.7	11.3
Active amyl	46.9	39.3
β-Phenyl ethyl	19.7	12.5
2,3-Butylene glycol	1.1	1.9

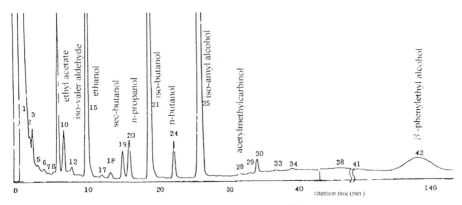

Figure 4. Gas chromatogram of cider.

Table 8. Changes in the contents of alcohol, esters and carbonyl compounds during acetic acid fermentation of cider (µL/100mL)

	Time (h)			
	0	12	22	49
Ethyl acetate	14.21	63.20	3.14	0.28
iso-Valer aldehyde	5.32	5.37	4.59	4.37
iso-Butyl acetate	0.25	0.26	1.04	0.75
Ethyl n-butyrate	0.49	0.48	0.35	0.29
Sec-Butanol	2.43	2.35	2.07	1.69
n-Propanol	3.40	2.08	0.23	Trace
iso-Butanol	35.16	32.21	23.43	22.97
n-Butanol	2.79	1.84	0.28	0.19
iso-Amyl alcohol	58.32	58.14	36.19	34.59
Acetyl methyl carbinol	0.37	3.83	3.67	3.47
n-Hexyl alcohol	1.15	0.45	Trace	Trace
β-Phenyl ethyl alcohol	13.80	13.47	12.90	12.90
Ethanol (vol%)	3.11	1.38	0.00	0.00

Alcohols of other types showed lower values. After 22 h the oxidizing activity of bacteria on all alcohols decreased. Figure 4 shows changes of flavor components on an example of gas chromatographic result. Table 7 shows changes of the content of 12 flavor components which were detected during acetic acid fermentation of the apple juice after alcoholic fermentation. Of higher alcohols, the contents of isoamyl,isobutyl, n-propyl and phenyl alcohols were slightly higher. These concentrations of alcohols decreased along with the progress of acetic acid fermentation; their contents decreased in parallel to ethyl alcohol to half of the starting values after 12 h similar to ethyl alcohol, and after 22 h by which time ethyl alcohol had disappeared their contents were trace. The iso-type alcohols did not decrease for 12 h and decreased considerably from 12 to 22 h after the start of fermentation. The changes in the content of these alcohols corresponded well to the values of the oxidizing activity. The content of ethyl acetate rapidly increased during initial 12 h and reached 4 times that of the original content followed by a gradual decrease to trace amount after 49 h from the start (Table 8). In the apple juice the acetoin content increased for 12 h and remained unchanged thereafter.

Takeuchi *et al.* (1968) studied the non-volatile organic acids in vinegars and reported that organic acids are important flavor components in vinegars. During the early half of acetic acid fermentation their contents did not change much, but gradually decreased in the latter half period when alcohol had disappeared and oxidation to acetic acid had taken place. In submerged culture non-volatile organic acids showed a rapid decrease when the alcohol concentration became lower than 0.3%. The flavor components, particularly volatiles (alcohols and their acetate esters), markedly decreased when alcohol disappeared similar to non-volatile organic acids. For manufacturing of high quality vinegars it is advisable to keep changes of volatile components to a minimum. Therefore, it is recommended to discontinue the fermentation when the ethyl alcohol concentration reached 0.3% and to subject the product to paging thereafter.

5. CONCLUSIONS

Volatile components for nine samples of 3 kinds of commercial vinegars (brown rice, rice and grain) which were produced by 3 different breweries included ethyl acetate, ethyl alcohol, n-propyl alcohol, isoamyl alcohol and acetoin. The type and contents of these components were different depending on the kind of vinegar (brown rice, rice and grain) and on the breweries (Breweries A, B and C) in which they were produced. Brown rice vinegars contained these components in the highest amounts followed by rice and grain vinegars.

Seven samples of cider vinegar and 3 samples of rice vinegar were also analyzed; twelve volatile components were identified, including ethyl acetate, ethyl alcohol, isobutyl acetate, n-propyl alcohol, isobutyl alcohol, isoamyl alcohol and acetoin. It was found that the volatile components were greatly different depending on the kinds of vinegar (cider and rice) as well as the breweries (Breweries A, B and C) in which they were produced.

To understand the mechanism of formation of volatile components and changes in their content, particularly higher alcohols and their acetate esters, during brewing for cider vinegar were examined. The volatile components decreased along with the progress of fermentation, and it was observed that after the acidity reached maximum the normal-type alcohols decreased markedly while the iso-type alcohols decreased rather slightly. Thus, it is recommended that changes in the volatile components be minimized by discontinuing fermentation when the concentration of ethyl alcohol reaches 0.3% and then submitting the solution to aging.

6.   REFERENCES

Arima, K., Yamaguchi, G., and Tamura, G., 1967, Studies on Flavor Components of Rice Vinegar Part I.   Neutral Fraction, *Agric. Biol. Chem. Jpn. (Engl. Transl.)*, **41**: 660-666.

Kahn, J. H., Nickol, G.B., and Conner, H..A., 1966, Analysis of Vinegar by Gas-Liquid Chromatography, *J. Agric. Food Chem.*, **14**:460-465.

Takeuchi, T., Furukawa, S., and Ueda, R., 1968,   Studies on Non-volatile Organic Acids in Vinegar (III).   Changes in Amount of Non-volatile Organic Acids during Fermentation and Storage of Cider Vinegar, *J. Ferment. Technol. Jpn. (Engl. Transl.)*, **46**: 288-292.

Yamaguchi, G., Tamura, G., and Arima, K., 1967,   Studies on Flavor Components of Rice Vinegar Part II. Basic Fraction, *Agric. Biol. Chem. Jpn. (Engl. Transl.)*, **41**: 667-670.

TEXTURAL QUALITY ASSESSMENT FOR FRESH FRUITS AND VEGETABLES

Judith A. Abbott[*]

1. INTRODUCTION

Texture is critical to the acceptability of fruits and vegetables, both fresh and cooked. This article focuses primarily on texture measurement of fresh (raw, uncooked) fruits and vegetables. It is important to remember that fruits and vegetables continue to metabolize, synthesize, and catabolize after harvest. In order to study the chemical and physiological mechanisms of textural changes during the development and senescence of fruits and vegetables, it is critical to be able to measure the texture. Although the term is widely used, *texture* is not a single, well-defined attribute. It is a collective term that encompasses the structural and mechanical properties of a food and their sensory perception in the hand or in the mouth. A comprehensive and useful definition of texture based on one by Bourne (1982) might be:

> *Texture is the group of physical characteristics that result from the structural elements of the food; are sensed by the sensation of touch in the hand or in the mouth; are related to the deformation, disintegration, and flow of food under force; and are measured by functions of force, distance, and time.*

Some of the many terms used to describe sensory texture of various fruits or vegetables are hard, firm, soft, crunchy, crisp, limp, mealy, tough, leathery, melting, gritty, stringy, sticky, dry, and juicy. However, there are not accepted complementary methods for instrumental measurement of each sensory textural attribute. In fact, there is some disagreement among sensory, horticultural, and engineering uses of certain textural terms, particularly *firmness* which is discussed later.

Textural attributes of raw fruits and vegetables are determined by structural, physiological, and biochemical characteristics of living cells and their changes over time through the stages of development, maturation, and senescence. Specifically, texture is determined by cell wall strength and elasticity, cellular arrangement and density, cell-to-cell bonding, and turgor. Thus, development of measurement methods requires an understanding of the anatomy of tissues within the fruit or vegetable, structure of their cells, and biological changes that occur following harvest, as well as some understanding of sensory texture perception.

---

[*] Produce Quality and safety Laboratory, USDA, Agricultural Research Service, Beltsville, Maryland, 20705-2350, USA.

Quality of Fresh and Processed Foods, edited by Shahidi et al.
Kluwer Academic/Plenum Publishers, 2004.

It is important to identify the main elements of tissue strength and to determine which mechanical properties are responsible for the textural attributes of interest. The number of terms used to describe texture of fruits and vegetables illustrate just how variable texture can be. Psychometrics, that as the study of the measurement of human response, is a fairly young science and has not been extensively applied to fresh fruits and vegetables, so the contributions of mechanical properties to sensory texture are not well understood.

2. FRUIT AND VEGETABLE TISSUES

Fruits and vegetables encompass organs or tissues from all plant parts: flowers, fruits, seeds, roots, tubers, rhizomes, stems, buds, and leaves. The common factor is that all are relatively soft when eaten, largely due to the presence of parenchyma cells. Parenchyma cells are unlignified and their primary walls are separated by a morphologically distinct region known as the middle lamella which is rich in pectic substances. The size and arrangement of parenchyma cells within the tissue influences the mechanical strength of produce. For example, apple cells are large (~ 300 µm diam.), elongated along the fruit radius, and organized in columns interspersed with up to 25% intercellular air space (Khan and Vincent, 1993). The elastic modulus of apple tissue is higher and strain at failure is lower when tissue specimens are compressed in a direction radial to the fruit axis compared to a parallel or tangential orientation (Abbott and Lu, 1996).

From a chemical viewpoint, the primary cell wall of parenchyma cells is composed of a matrix of cellulose, hemicellulose, and pectin. Changes that occur in the cell wall during ripening, storage, and cooking are critical to the texture of the final product. During maturation of some vegetative parts, especially stems and petioles, cell walls become lignified (Okimoto, 1948; Price and Floros, 1993), resulting in toughening, such as woodiness in asparagus, broccoli, pineapple, and rutabaga. During fruit ripening, cell wall changes include solublization and degradation of pectins and a net loss of galactose and arabinose, and there may be a decrease in the molecular weight distribution of hemicelluloses (Harker et al., 1997a). No single enzyme has been identified as the major cause of fruit softening, suggesting that cell wall breakdown may result from reduced synthesis of some components and from coordinated action of several enzymes. Numerous enzymes have been suggested as being critical to the degradation of the cell wall, including polygalacturonases, glycosidases, β-galactosidase, xyloglucan, endotransglycosylase, and cellulases (Dey and del Campillo, 1984; Huber, 1992; Seymour and Gross, 1996; Harker et al., 1997a).

Postharvest treatments involving dipping or infiltrating with calcium are known to maintain firmness during storage of a wide range of fruit (Conway et al., 1994), likely due to stabilizing pectin bonds. Examination of fracture surfaces following tensile testing of apple cortex indicated that tissue failure from calcium-treated fruit was due to cell rupture; whereas failure in control apples was due to cell debonding (Glenn and Poovaiah, 1990).

Turgor pressure prestresses the cell wall by applying a continuous force against the structures. Tissue with low turgor is perceived as soft, spongy, flabby, or withered. High turgor tends to make the cells brittle and failure occurs at lower force and smaller strain in compression (Lin and Pitt, 1986). However, cells with high turgor require higher force to fail in tension (De Belie et al., 1999). Turgor is unlikely to influence tissue strength if the mechanism of failure is cell-to-cell debonding rather than fracturing across individual cells (Glenn and Poovaiah, 1990; Harker and Hallett, 1992), unless the increase in turgor causes the cell to swell and reduces cell-to-cell contact area. The strength of the cell wall relative to the adhesion between neighboring cells will determine whether tissue fails through cell rupture or cell-to-cell debonding.

Many fruits and vegetables contain a number of tissue zones. These tissues can differ in strength and biological properties and may need to be considered individually when measuring texture. For example, the core of pineapple may become woody while the radial tissues soften; the locular gel in tomato liquefies as the pericarp softens. The skin plays a key role in holding the flesh together in many types of produce, particularly in soft fruit and citrus, and it usually has a distinctly different texture from the rest of the product. The presence of the cuticle of epidermal cells and thickened cell walls of hypodermal cells contribute to the strength of simple skins. In harder inedible skins, collenchyma, sclerenchyma, tannin-impregnated cells, and cork may be present.

The presence of tough strands of vascular tissue strengthen the flesh, but often results in a fibrous texture. For example, toughness of asparagus spears is principally due to fiber content and fiber lignification (Lipton, 1990). In most commercial fruits, with the exception of pineapple (Okimoto, 1948) and mango, fibrousness of the flesh is not a major problem. However, in some fruits when the flesh becomes soft, the vascular fibers become more noticeable, and objectionable, because of a contrast effect, such as in peach and muskmelons (Diehl and Hamann, 1979). Some fruits such as pear and guava contain sclerenchymous stone cells that cause a gritty texture (Harker *et al.*, 1997a). Such cells become more noticeable when the surrounding parenchyma becomes relatively soft.

3. SENSORY EVALUATION OF TEXTURE

People sense texture in numerous ways: the *look* of the product, the feel in the hand, the way it feels as they cut it, the sounds as they bite and chew, and, most important of all, the feel in their mouth as they eat it. It is generally accepted that texture relates primarily to mechanical properties of the tissue. Harker et al. (1997a) reviewed texture of fruits and included an extensive discussion of oral sensation of textural attributes.

Texture may be the limiting factor in acceptability if one or more textural attributes is outside the individual's range of acceptability for that particular combination of attribute and commodity. People have quite different expectations and impose different limits for different commodities. The relationship of instrumental measurements to specific sensory attributes and the relationship of those sensory attributes to consumer acceptability are still being investigated. Instrumental tests may be imitative, that is they imitate human testing methods, or they may measure fundamental mechanical properties or they may indirectly estimate texture by measuring some related optical or chemical property. Only people can *judge* quality, but instruments that *measure* quality-related attributes are vital for research and for inspection (Abbott *et al.*, 1997).

4. INSTRUMENTAL MEASUREMENT OF TEXTURE

The ability to quantify textural attributes is critical to study the chemical and physiological mechanisms of texture. The diversity of tissues involved, the variety of attributes required to fully describe textural properties, and the changes in these attributes as the product ripens and senesces contribute to the complexity of texture measurement. Instrumental measurements are preferred over sensory evaluations for both research and commercial applications because instruments reduce variations due to differences in human perception; are more precise; and can provide a common language among researchers, companies, and customers. There have been numerous reviews of methods for instrumental measurement of fruit and vegetable texture (Abbott *et al.*, 1997; Bourne, 1980; Harker *et al.*, 1997a).

Most instrumental measurements of texture have been developed empirically. While they may provide satisfactory assessments of the quality of produce, they often do not fulfill engineering requirements for fundamental measurements (Bourne, 1982). On the other hand, fundamental material properties measurements were developed to study the *strength* of materials for construction or manufacture. Once the failure point of a construction material is exceeded, there is little interest in the subsequent behavior of the material. Scientists that deal with food are interested in initial failure, but they are also interested in the continuous breakdown of the food in the mouth in preparation for swallowing. Bourne (1982) pointed out that "food texture measurement might be considered more as a study of the *weakness* of materials rather than strength of materials." In fact, both strength and breakdown characteristics are important components in the texture of fruits and vegetables.

4.1. Mechanical Properties of Fruits and Vegetables

Fruits and vegetables exhibit viscoelastic behavior under mechanical loading, which means that the force, distance, and time involved in loading determine the value of any measurement. The role of time is in the rate of application and/or duration of the load. For example, the impact of an apple falling from a tree represents very rapid, short term loading, whereas the weight of a meter-high pile of potatoes on an individual potato at the bottom of a bin or the force of a carton wall against tightly packed oranges are long-term loads. The product will respond quite differently to the two forms of loading. In purely elastic materials, loading rate and duration are not critical. Because of the viscoelastic character of fruit and vegetable tissues, every effort should be made to hold the speed of the test constant in manual texture measurements, such as the Magness-Taylor puncture test (Blanpied *et al.*, 1978); and the rate of loading should be specified and controlled in mechanized measurements.

There are many types of mechanical loading: puncture, compression, shearing, torsion (twisting), extrusion, crushing, tension, bending, vibration, and impact. There are four basic values that can be obtained from mechanical properties tests: force (load), deformation (distance, displacement, penetration), slope (ratio of force to deformation), and area under the force/deformation curve (energy). The engineering terms for these are stress, strain, modulus, and energy, respectively. Stress is force per unit area, either area of contact or sample cross-section, depending on the test. Strain is deformation as a percentage of initial height or length of the sample. There are several forms of modulus (tangent, secant, chord, or initial tangent, Young's), but all refer to a stress/strain ratio. Force and deformation values are more commonly used in applications dealing with foods than stress and strain values and are sufficient, provided that the contact area and the distance the probe travels are constant and sample dimensions are similar from sample to sample (Sample here means the portion of tissue tested, not necessarily the size of the fruit or vegetable.). In most horticultural texture tests, the deformation is fixed and the force value is reported; examples are widely used puncture tests, the Kramer shear test (Floros *et al.*, 1992; Harker *et al.*, 1997b; Yamaguchi *et al.*, 1977), and the instrumental Texture Profile (Friedman *et al.*, 1963; Bourne, 1968) measurement. In a few horticultural tests, a known force is applied to the product and the deformation after a specified time is reported; an example is the tomato creep test (Hamson, 1952; Ahrens and Huber, 1990).

An idealized force/deformation (F/D) curve for a cylindrical piece of tissue compressed at constant speed is shown in Fig. 1. F/D curves for puncture tests look similar and are interpreted similarly. The portion of the initial slope up point A in the figure represents nondestructive elastic deformation; point A is called the elastic limit. Beyond the elastic limit, permanent tissue damage begins.   Nondestructive tests should be made within the elastic limit.   F/D characteristics beyond the elastic limit may be more important than those before it, because they simulate the destruction that occurs in bruising or eating (Bourne, 1968).   There *may* be a bioyield point (point B) where cells start to rupture or to move with respect to their neighbors, causing a noticeable decrease in slope.   Point C in Fig. 1 marks rupture, where major tissue failure causes the force to decrease substantially.   Beyond rupture, the force may again increase, level off, or decrease as deformation increases.   At the maximum deformation point specified by the user, the probe is withdrawn and the force diminishes until contact is lost.   Products containing a mixture of parenchyma and fibers or stone cells may have quite jagged F/D curves, with several local maxima and ruptures as the probe encounters resistant clusters of stone or fiber cells.

Firmness of horticultural products can be measured at different force or deformation levels in all three regions of Fig. 1, depending on the purpose of the measurement and the definitions of the textural attributes.   It is important to recognize and understand the fundamental properties measured by both destructive tests and nondestructive methods, the differences between them, and the factors that can affect the tests.

4.2. Destructive Texture Tests for Fruits and Vegetables

Numerous mechanical instruments have been developed over the past century for measuring textural attributes of horticultural products.   Some are reviewed below.

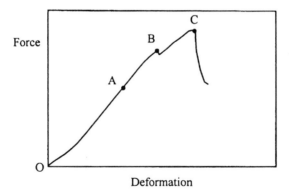

Figure 1. Idealized force/deformation curve demonstrating A elastic limit, B bioyield, and C rupture or massive tissue failure.

4.2.1. Puncture Tests

Puncture testers based on the original Magness-Taylor pressure tester (Magness and Taylor, 1925; Haller, 1941), more correctly called the *Magness-Taylor fruit firmness tester*, are used commercially to measure firmness of numerous fruits and vegetables to estimate harvest maturity or for postharvest inspection.   There are several adaptations of the Magness-Taylor (MT) tester that differ in instrument size and shape, manual or mechanical use, and analog or digital readout, Table 1 (Bourne, 1974; Breene *et al.*, 1974; Abbott *et al.*, 1976; Harker *et al.*,

1996; Lehman-Salada, 1996). Some of the instruments are no longer manufactured but are listed for identification purposes. In addition, a MT probe can be mounted in a universal force/deformation testing instrument such as those listed in Table 2. The term *Magness-Taylor firmness* is used generically for the measurements made with the several variants of the MT. All use the same rounded-tip probes (Fig. 2) — note that these are *not* complete hemispheres — and measure the maximum force required to press the probe 7.94 mm (5/16 inch) into the flesh (Haller, 1941). An 11.11-mm (7/16 inch) diameter probe with a radius of curvature of 8.73 mm is used for apples. A 7.94-mm (5/16 inch) diameter probe with a radius of curvature of 5.16 mm is used for cucumbers, kiwifruits, mangoes, papayas, peaches, pears, plums, and sometimes avocados. A thin slice of skin (about 2 mm thick and slightly larger diameter than the probe) should be removed from the area to be tested except for cucumbers, which are tested with the skin intact. It is recommended that steady force should be applied to insert the probe to the inscribed depth mark in 2 s (Blanpied *et al.*, 1978). The probe can be mounted in a universal force/deformation testing machine and inserted to a depth of 8 mm; speeds between 50 and 250 mm/minute are recommended (Smith, 1985), although 25 mm/minute has often been used in research. Because of the different curvatures of the MT probes and the fact that shear and compression contribute to the total force in variable proportions, it is not possible to convert measurements made with one size MT probe to the other MT size, or to accurately convert to or from values for probes of other geometries (Bourne, 1982). A random sample of 20 to 30 fruit of similar size and temperature should be tested on two opposite sides, depending on uniformity of the lot. Peaches are often more variable around the circumference than other fruit so the larger number is recommended (Blanpied *et al.*, 1978). Numerous puncture tests with smaller curved probes, flat-faced cylindrical probes, conical probes (Sakurai and Nevins, 1992; Wann, 1996), and other shapes (Holt, 1970) are found in the literature. None has achieved the acceptance of the Magness-Taylor fruit firmness test.

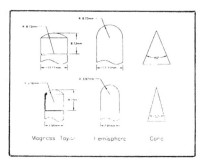

Figure 2. Popular probes used for puncture testing of fruits and vegetables. The term Magness-Taylor firmness is used generically for the measurements made with the several variants of the MT, all using the same rounded-tip probes (left column) — note that these are not hemispherical. Hemispherical (center column), flat-faced cylindrical, and conical probes (right column) and blades of various dimensions are also used for testing produce.

Table 1. Examples of fruit firmness testers (puncture testing devices).

Device	Use, data output	Available*	Manufacturer
Effe-gi	manual, digital	yes	Alfonsina, Ravenna, Italy
EPT or Electronic Pressure Tester	manual, digital	yes	Lake City Technical Products, Kelowna, British Columbia, Canada
Magness-Taylor (may be USDA Tester in very early references)	manual, digital	no	D. Ballauff Mfr. Co., Laurel, Maryland, USA
McCormick	manual, analog	no	McCormick Fruit Technologies, Yakima, Washington, USA
FT	manual, analog	yes	Wagner Instruments (Greenwich, Connecticut, USA)
UC or University of California tester	manual, analog	no	Western Industrial Supply Co., San Francisco, California, USA

* Commercially available at present.

Table 2. Examples of universal testing instruments in which Magness-Taylor or other probes and testing cells can be mounted.

Manufacturers
Ametek Test and Calibration Instruments, Paoli, Pennsylvania, USA
Chatillon, Largo, Florida, USA
Food Technology Corporation, General Kinetics Inc.,
Instron Corp., Canton, Massachusetts, USA
Lloyd Instruments Material Testing Products, Fareham, Hampshire, England
Stable Micro Systems Ltd., Godalming, Surrey, England

4.2.2. Compression Tests

Although compression tests are not commonly used by the fruit and vegetable industry, they are widely used in horticultural research. They can be made on intact products or on tissue specimens. Whole fruit compression involves pressing an intact fruit or vegetable between parallel flat plates or between a pair of probes until a specified force or deformation is achieved or until rupture occurs. Force/deformation curves for parallel plate compression of excised tissue of several fruits and vegetables are shown in Fig. 3. Modulus of elasticity, modulus of deformability, and stress index for fruit (ASAE, 2000) can be calculated from such measurements, although often only maximum force or distance is reported. Whole fruit compression is affected by fruit morphology, size, and shape, as well as by cellular structure, strength, and turgor. Rupture force of intact watermelons (Sundstrom and Carter, 1983), whole tomato compression (Jackman *et al.*, 1990; Ahrens and Huber, 1990), and apple tissue compression (Abbott *et al.*, 1984; Abbott and Lu, 1996) are examples of compression measurements.

Fruit and vegetable tissues are usually tested as elastic materials but are, in fact, viscoelastic. Viscoelastic behavior means that force, distance, and *time*, in the form of rate, extent, and duration of load, determine the value of measurements. If the viscous element is a significant contributor to the texture, as in intact tomato and citrus, measurement of continuing deformation under a constant force (creep) (Hamson, 1952; Kojima *et al.*, 1991; El Assi *et al.*, 1997) or decrease in force under a fixed deformation (relaxation)(Wu and Abbott, 2002; Sakurai and Nevins, 1992; Errington *et al.*, 1997) is appropriate.

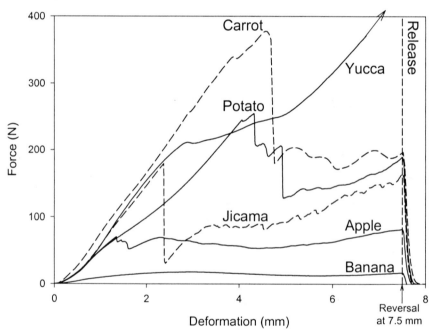

Figure 3. Force/deformation curves for uniaxial compression of tissue cylinders (15 mm diam. × 10 mm high) of several fruits and vegetables at a constant loading rate of 1 mm s^{-1} to 75% compression, then reversal and release of load at 1 mm s^{-1}. Note that yucca, a woody root, greatly exceeded the 250-N capacity of the load sensor. Note also that jicama is nearly linear until a massive rupture (failure) occurs; whereas potato is distinctly nonlinear.

Wu and Abbott (2002) measured force-relaxation on slices of tomatoes of different initial ripeness which had been stored for 0 to 19 days as intact fruit or as fresh-cut slices. Fig. 4 illustrates force-relaxation curves on the day of harvest. They developed a three-parameter empirical equation to describe the shape of the force-relaxation curves. The three relaxation parameters were more sensitive to textural differences between the storage durations than was the simple elastic property force at maximum loading.

Creep measurement is an alternative technique to force relaxation. In creep, the specimen is loaded to a fixed force and the deformation required to maintain that force over time is measured. Creep measurement is somewhat less precise than force relaxation measurement with many universal testing instruments because the load is maintained by incremental movements of the probe, thus is not truly constant.

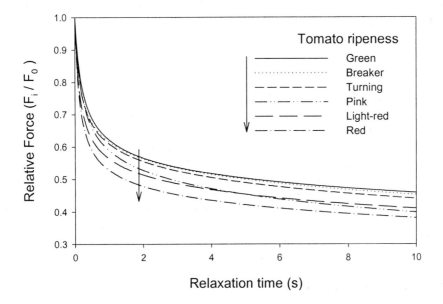

Figure 4. Force relaxation curves for outer pericarp of tomato slices (7 mm thick) loaded with a 4-mm diameter cylinder to 3 mm deformation at a rate of 1 mm s^{-1}, then the deformation was held for >10 s. Force relaxation rates and amounts changed with ripening for this tomato cultivar. Curves were normalized by dividing each point on the curve (F_i) by the initial loading force for that curve (F_0 = force at time 0), i.e., each point represents percentage of initial force at 3 mm deformation. More rapid application of the deformation is customarily used to minimize relaxation during the deformation period. (Wu and Abbott, 2002)

Despite its name, the twist test recently developed by Studman and Yuwana (1992) actually tests the rupture force of the tissues within a sample. A blade on a spindle is pushed into the fruit and rotated; firmness is estimated by measuring the rotational angle or torsional force required to crush the flesh. It has been tested on apple and kiwifruit (Harker et al., 1996;).

4.2.3. Kramer Shear or Compression-shear Test

The Kramer shear cell actually applies compression, shear, and extrusion to the sample. The functional part of the test equipment is a sample cell with multiple dull blades that pass entirely through the sample volume. The original Kramer cell is approximately $67 \times 67 \times 63$ mm with ten 2.9-mm thick blades (Kramer, 1951); several adaptations with differing volumes and differing blade numbers and sizes have been used. The cells can be mounted in any universal testing instrument (Table 2) of suitable force capacity. Usually only the maximum force to pass the blades entirely through the sample is reported, although the entire force/deformation curve can be obtained. Forces are dependent on quantity and organization of the sample within the sample cell (Szczesniak et al., 1970; Voisey and Kloek, 1981). A sample size of 100 g is recommended for the standard cell size. Because of the combination of destructive forces applied to the food, it sometimes relates fairly well to sensory assessments of texture. Kramer shear is rather more widely used in the food processing industry than in horticulture (Floros et al., 1992; Harker et al., 1997b; Makus and Morris, 1993; Nakhasi et al., 1991; Rosen and Kader, 1989; and Yamaguchi et al., 1977).

4.2.4. Tension Tests

Tensile tests measure the force required to pull a sample apart. Failure can be through cell rupture, cell separation, or a combination of both. Tensile measurement has not been widely applied to fruits and vegetables because it is not intuitively related to bruising or chewing and because of difficulties in making the measurements, it requires gripping the ends of the sample, gluing moist tissues to probes, or shaping the sample for gripping with claw-like holders without damaging the tender tissues. Researchers often examine the broken ends where they have been pulled apart to determine the mode of fracture (Schoorl and Holt, 1983; Stow, 1989; Harker and Hallett, 1992). Microscopic analyses of the broken ends (Lapsley et al., 1992; Harker and Sutherland 1993; Harker and Hallett 1994; Harker et al., 1996) reveal that tissue from unripe fruit fractures due to individual cells breaking, cells from ripened fruits which tend to be crisp (apple and watermelon) usually break or rupture, and cells from ripened soft fruits (banana, nectarine and kiwifruit) tend to separate at the middle lamellae.

4.3. Nondestructive Measurements

Most force/deformation measurements, such as the Magness-Taylor fruit firmness test and the Kramer shear test, are destructive or are slow, like the Cornell creep tester. Recently a great deal of research has gone into developing nondestructive methods to estimate the textural quality of fruits and vegetables. Commercial packing houses need rapid, nondestructive texture sensing for real-time sorting to ensure the quality of the produce they sell. None of these nondestructive methods has attained wide commercial acceptance for on-line sorting to date, although some are in advanced testing stages, but they are quite useful for research on postharvest physiology because fruit can be measured at the beginning of experiments to determine the maturity or condition and because the same fruit or vegetable can be measured repeatedly, minimizing or eliminating the variation among individuals.

4.3.1. Laser-air Puff

A nondestructive, noncontact firmness detector was recently patented (Prussia *et al.*, 1994) that uses a laser to measure the deflection caused by a short puff of high-pressure air, similar to some devices used by ophthalmologists to detect glaucoma. This is essentially a compression tester. Under fixed air pressure, firmer products deflect less than softer ones. Laser-puff readings correlated fairly well with destructive Magness-Taylor firmness values for apple, cantaloupe, kiwifruit, nectarine, orange, pear, peach, plum, and strawberry (Fan *et al.,* 1994; Hung *et al.*, 1998).

4.3.2. Impact or Bounce Test

When one object collides with another object, its response is related to its mechanical properties, mass, and the contact geometry. Numerous studies have been conducted on the impact responses of horticultural products and several impact parameters have been proposed as firmness values, including peak force, coefficient of restitution (ratio of velocities just before and after impact), contact time, and the impact frequency spectrum. Selection of parameters seems to depend on commodity, impact method, and the firmness reference used by the investigators. Most impact tests involve dropping the product onto a sensor (Rohrbach, 1981; Delwiche *et al.*, 1987; Zapp et al., 1990; McGlone and Schaare, 1993; Patel *et al.*, 1993) or striking the product with the sensor (Bajema and Hyde, 1998; Brusewitz *et al.*, 1991; Delwiche *et al.*, 1989). Delwiche et al. (1989, 1991) developed a single-lane firmness sorting system for pear and peach. Impact measurements often do not correlate highly with the Magness-Taylor puncture measurement (Hopkirk *et al.*, 1996).

4.3.3. Sonic or Acoustic Tests

Sonic (or acoustic) vibrations provide a means of measuring fruit and vegetable firmness. The traditional watermelon ripeness test is based on the acoustic principle, where one thumps the melon and listens to the pitch of the response. A number of sonic instruments and laboratory prototype sorting machines have been developed and tested (Abbott *et al.*, 1968, 1992; Affeldt and Abbott, 1992; Armstrong *et al.*, 1990; Peleg, 1989; Peleg *et al.*, 1990; Zhang *et al.*, 1994; Shmulevich *et al.*, 1995; Stone *et al.*, 1998; Schotte *et al.*, 1999; De Belie *et al.*, 2000; Muramatsu *et al.*, 2000). When an object is caused to vibrate, amplitude varies with the frequency of the vibration and will be at a maximum at specific frequencies determined by a combination of the shape, size, and density of the object and, in the case of fruits and vegetables, by their turgor pressure (Fig. 5A). The points of maximum amplitudes are referred to as *resonances*. For objects of similar size and shape, the firmer the material the higher the resonance frequencies. Thus, resonant frequencies decrease as fruit ripens (Fig. 5A). Resonance measurement can be achieved by applying an impulse or thump that contains a range of frequencies. Modulus of elasticity determined from resonant frequencies has correlated well with that measured by conventional compression tests (Fig. 5B), but often were correlated poorly with MT puncture forces. Despite considerable research, sonic vibration has not been used for on-line sorting of other horticultural products.

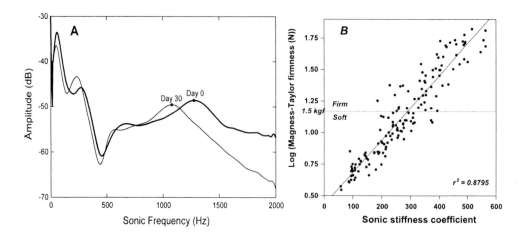

Figure 5. A. Sonic resonance curves for kiwifruit held for 30 d at 0 °C, showing the shift to lower frequency resonant frequencies with ripening. B. Relationship of sonic stiffness coefficient ($f^2 m^{2/3} \times 10^{-5}$) to Magness-Taylor maximum force.

5. SUMMARY

Academically, quantitative measurement of texture is essential for the study of the chemical and physiological mechanisms of texture. Commercially, quantitative measurement of texture is essential to ensure the quality of produce at packout. The diversity of tissues involved, the variety of attributes required to fully describe textural properties, and the changes in these attributes as the product ripens and senesces contribute to the complexity of texture measurement. Texture is a human assessment of the structural elements of a food. It is generally accepted that texture relates primarily to mechanical properties, so instrumental measurements relate mainly to mechanical properties. Fruits and vegetables exhibit viscoelastic behavior under mechanical loading, which means that the force, distance, and time involved in loading determine the value of any measurement. Because of their viscoelastic character, every effort should be made to hold the speed of the test constant in manual texture measurements and the rate of loading should be specified and controlled in mechanized measurements. There are many types of mechanical loading: puncture, compression, shearing, torsion (twisting), extrusion, crushing, tension, bending, vibration, and impact. The most widely used texture measurement for fruits and vegetables, after manual squeezing of course, is the Magness-Taylor fruit firmness test, which measures the maximum force to puncture the product in a specified way. The Kramer shear or shear-compression test is widely used in the processed foods industry, but is less commonly used by horticulturists. Nondestructive methods are highly desired both for sorting and for postharvest research. Compression tests of excised tissue pieces are frequently used in research. Nondestructive testing using impact, vibrational behavior, light scattering, and optical methods are being investigated but none has been widely accepted to date. Multiple instrumental measurements may be necessary to adequately the diversity of textural attributes sensed by the human consumer.

6. REFERENCES

Abbott, J. A., Affeldt, H. A. and Liljedahl, L. A., 1992, Firmness measurement of stored 'Delicious' apples by sensory methods, Magness-Taylor, and sonic transmission. *J. Am. Soc. Hort. Sci.*, **117**:590-595.

Abbott, J . A., Bachman, G. S., Childers, N. F., Fitzgerald, J. V., and Matusik, F. J., 1968, Sonic techniques for measuring texture of fruits and vegetables, *Food Technol.*, **22(5)**:101-112.

Abbott, J. A. and Lu, R., 1996, Anisotropic mechanical properties of apples. *Trans. Am. Soc. Agr. Eng.*, **39**:1451-1459.

Abbott, J. A., Lu, R., Upchurch, B. L., and Stroshine, R. L., 1997. Technologies for nondestructive quality evaluation of fruits and vegetables. *Hort. Rev.*, **20**:1-120.

Abbott, J. A., Watada, A. E., and Massie, D. R., 1976, Effe-gi, Magness-Taylor, and Instron fruit pressure testing devices for apples, peaches, and nectarines, *J. Am. Soc. Hort. Sci.*, **101**: 698-700.

Abbott, J. A., Watada, A. E., and Massie, D. R., 1984, Sensory and instrument measurement of apple texture. *J. Am. Soc. Hort. Sci.*, **109**:221-228.

Affeldt, H. A., Jr. and Abbott, J. A., 1992, Agricultural commodity condition measurement, *U.S. Patent* 5,152,401.

Ahrens, M. J., and Huber, D. J., 1990, Physiology and firmness determination of ripening tomato fruit, *Physiol. Plant.*, **78**:8-14.

Armstrong, P., Zapp, H. R., and Brown, G. K., 1990, Impulsive excitation of acoustic vibrations in apples for firmness determination, *Trans. Am. Soc. Agric. Eng.*, **33**:1353-1359.

ASAE, 2000, Compression test of food materials of convex shape, *Am. Soc. Agric. Eng., Standard 368.4. Am. Soc. Agric. Eng.*, St. Joseph, MI.

Bajema, R. W., and Hyde, G. M., 1998, Instrumented pendulum for impact characterization of whole fruit and vegetable specimens, *Trans. Am. Soc. Agric. Eng.*, **41**:1399-1405.

Blanpied, G. D., Bramlage, W. J., Dewey, D. H., LaBelle, R. L., Massey ,L. M., Mattus, Jr, G. E., Stiles, W. C. and Watada, A. E., 1978, A standardized method for collecting apple pressure test data, *N. Y. Food and Life Sciences Bul.*, 74.

Bourne, M. C., 1968, Texture profile of ripening pears, *J. Food Sci.* **33**:223-226.

Bourne, M.C., 1974, Comparison of results from the use of the Magness-Taylor pressure tip in hand- and machine-operation, *J. Texture Studies*, **5**:105-108.

Bourne, M. C., 1980, Texture evaluation of horticultural crops, *HortScience*, **15**:51-57.

Bourne, M. C., 1982, Food Texture and Viscosity; Concept and Measurement. *Academic Press*, New York.

Breene, W. M., Jeon, I. J., and Bernard, S. N., 1974, Observations on texture measurement of raw cucumbers with the fruit pressure tester, *J. Texture Studies*, **5**:317-327.

Brusewitz, G. H., McCollum, T. G., Zhang, X., 1991, Impact bruise resistance of peaches, *Trans. Am. Soc. Agric. Eng.* **34**:962-965.

Conway, W. S., Sams, C. E., Wang, C. Y., and Abbott, J. A., 1994, Additive effects of postharvest calcium and heat treatment on reducing decay and maintaining quality in apples, *J. Am. Soc. Hort. Sci.*, **119**:49-53.

de Belie, N., Tu, K., Jancsok, P., and de Baerdemaeker, .J, 1999, Preliminary study on the influence of turgor pressure on body reflectance of red laser light as a ripeness indicator for apples, *Postharv. Biol. Technol.*, **16**:279-284.

de Belie, N., Schotte, S., Coucke, P., and de Baerdemaeker, J., 2000, Development of an automated monitoring device to quantify changes in firmness of apples during storage, *Postharv. Biol. Technol.*, **18**:1-8.

Delwiche, M. J., McDonald, T., and Bowers, S. V., 1987, Determination of peach firmness by analysis of impact force, *Trans. Agric. Soc. Agr. Eng.*, **30**:249-254.

Delwiche, M. J., Singh, N., Arevalo, H., and Mehlschau, J., 1991, A second generation fruit firmness sorter, *Am. Soc. Agr. Eng.*, Paper 91-6042.

Delwiche, M. J., Tang, S., and Mehlschau, J. J., 1989, An impact force response fruit firmness sorter, *Trans. Am. Soc. Agric. Eng.*, **32**:321-326.

Dey, P. M., and Campillo, E. del, 1984, Biochemistry of the multiple forms of glycosidases in plants, *Advances in Enzymology*, Vol **56**:141-249.

Diehl, K. C., and Hamann, D. D., 1979, Relationships between sensory profile parameters and fundamental mechanical parameters for raw potatoes, melons and apples, *J. Texture Studies*, **10**:401-420.

El Assi, N. E., Huber, D. J., and Brecht, J. K., 1997, Irradiation-induced changes in tomato fruit and pericarp firmness, electrolyte efflux, and cell wall enzyme activity as influened by ripening stage, *J. Am. Soc. Hort. Sci.*, **122**:100-106.

Errington, N., Mitchell, J. R., and Tucker, G. A., 1997, Changes in the force relaxation and compression responses of tomatoes during ripening: the effect of continual testing and polygalacturonase activity, *Postharv. Biol. Technol.*, **11**:141-147.

Fan, S., Prussia, S. E, and Hung, Y. C., 1994, Evaluating the UGA laser-puff food firmness detector, *Trans. Am. Soc. Agric. Eng.*, Paper 94-6540.

Floros, J. D., Ekanayake, A., Abide, G. P., Nelson, P. E., 1992, Optimization of a diced tomato calcification process, *J. Food Sci.*, **57**:1144-1148.

Friedman, H. H., Whitney, J. E., and Szczesniak, A. S., 1963, The texturometer—a new instrument for objective texture measurement, *J. Food Sci.*, **28**:390-396.

Glenn, G. M., and Poovaiah, B. W., 1990, Calcium-mediated postharvest changes in texture and cell wall structure and composition in 'Golden Delicious' apples, *J. Am. Soc. Hort. Sci.*, **115**: 962-968.

Haller, M. H., 1941, Fruit pressure testers and their practical applications. *USDA Circular* 627, Washington, D.C.

Hamson, A.R. , 1952, Measuring firmness of tomatoes in a breeding program, *Proc. Am. Soc. Hort. Sci.*, **60**:425-433.

Harker, F. R., and Hallett, I. C.,1992, Physiological changes associated with development of mealiness of apple fruit during cool storage, *HortScience* 27 (12) p. 1291-1294.

Harker, F. R. and Hallett, I. C., 1994, Physiological and mechanical properties of kiwifruit tissue associated with texture change during cold storage, *J. Am. Soc. Hort. Sci.*, **119**: 987-993.

Harker, F. R., Maindonald, J. H., and Jackson, P. J., 1996, Penetrometer measurement of apple and kiwifruit firmness: operator and instrument differences. *J. Am. Soc. Hort. Sci.*, **121**: 927-936.

Harker, F. R., Redgwell, R. J., Hallett, I. C., and Murray, S .H., 1997a, Texture of fresh fruit, *Hort. Rev.*, **20**:121-224.

Harker, F. R., Stec, M. G. H., Hallett, I. C., and Bennett, C. L., 1997b, Texture of parenchymatous plant tissue: a comparison between tensile and other instrumental and sensory measurements of tissue strength and juiciness, *Postharv. Biol. Technol.*, **11**:63-72.

Harker, F. R. and Sutherland, P. W., 1993, Physiological changes associated with fruit ripening and the development of mealy texture during storage of nectarines, *Postharv. Biol. Technol.*, **2**:269-277.

Holt, C. B., 1970, Measurement of tomato firmness with a Universal testing machine, *J. Texture Studies* **1**:491-501.

Hopkirk, G., Maindonald, J. H., and White, A., 1996, Comparison of four new devices for measuring kiwifruit firmness, *N. Z. J. Crop Hort. Sci.*, **24**:273-286.

Huber, D. J., 1992, The inactivation of pectin depolymerase associated with isolated tomato fruit cell wall: implications for the analysis of pectin solubility and molecular weight, *Physiol. Plant.*, **86**:25-32.

Hung, Y. C., McWatters, K. H., and Prussia, S. E., 1998, Sorting performance of a nondestructive laser air-puff firmness detector, *Appl. Eng. Agr.*, **14(5)**:513-516.

Jackman, R. L., Marangoni, A. G., and Stanley, D. W., 1990, Measurement of tomato fruit firmness, *HortScience* **25(7)**:781-783.

Khan, A. A., and Vincent, J. F. V., 1993, Compressive stiffness and fracture properties of apple and potato parenchyma, *J. Texture Studies.*, **24**:423-435.

Kojima, K., Sakurai, N., Kuraishi, S., Yamamoto, R., and Nevins, D. J., 1991, Novel technique for measuring tissue firmness within tomato (Lycopersicon esculentum Mill.) fruit, *Plant Physiol.* **96**:545-550.

Kramer, A., Burkhardt, G. J., and Rogers, H. P., 1951, The shear-press, a device for measuring food quality, *Canner,* **112**:34-36,40.

Lapsley, K. G., Escher, F. E., and Hoehn, E., 1992, The cellular structure of selected apple varieties, *Food Struct*, **11(4)**:339-349.

Lehman-Salada, L., 1996, Instrument and operator effects on apple firmness readings, *HortScience* **31**:994-997.

Lin, T. T., and Pitt, R. E., 1986, Rheology of apple and potato tissue as affected by cell turgor pressure, *J. Texture Stud.*, **17**:291-313.

Lipton, W. J., 1990, Postharvest biology of fresh asparagus, *Hort. Rev.*, **12**:69-155.

Magness, J. R., and Taylor, G. F., 1925, An improved type of pressure tester for the determination of fruit maturity, *USDA Cir.* 350.

Makus, D. J. and Morris, J. R., 1993, A comparison of fruit of highbush and rabbiteye blueberry cultivars, *J. Food Quality*, **16**:417-428. firmness (as shear)

McGlone, V. A., Schaare, P. N., 1993, The application of impact response analysis in the New Zealand fruit industry, *Trans. Am. Soc. Agr. Eng.*, 93-6537.

Muramatsu, N., Sakurai, N., Wada, N., Yamamoto, R., Tanaka, K., Asakura, T. ,Ishikawa-Takano, Y., and Nevins, D. J., 2000, Remote sensing of fruit textural changes with a laser doppler vibrometer, *J. Am. Soc. Hort. Sci.*, **125**:120-127.

Nakhasi, S., Schlimme, D., and Solomos. T., 1991, Storage potential of tomatoes harvested at the breaker stage using modified atmosphere packaging, *J. Food Sci.*, **56**:55-59.

Okimoto, M. C., 1948, Anatomy and histology of the pineapple inflorescence and fruit. *Bot. Gaz.*, **110**:217-231.

Patel, N., McGlone, V. A., Schaare, P. N., and Hall, H., 1993, "BerryBounce": a technique for the rapid and nondestructive measurement of firmness in small fruit, *Acta Hort.*, **352**:189-198.

Peleg, K., 1989, Method and apparatus for automatically inspecting and classifying different objects, *U.S. Patent* 4,884,696.

Peleg, K., Ben-Hanan, U., and Hinga, S., 1990, Classification of avocado by firmness and maturity, *J. Texture Studies,* **21**:123-139.

Price, J. L., and Floros, J. D., 1993, Quality decline in minimally processed fruits and vegetables, *Dev. Food Sci.*, **32**:405-427.

Prussia, S. E., Astleford, J. J., Hung, Y-C., and Hewlett, R., 1994, Non-destructive firmness measuring device, *U.S. Patent* 5,372,030.

Rohrbach, R. P., 1981, Sorting blueberries to improve fresh market shelf life, *Trans. Am. Soc. Agric. Eng.* Paper 81-1501.

Rosen, J. C., and Kader, A. A., 1989, Postharvest physiology and quality maintenance of sliced pear and strawberry fruits, *J. Food Sci.*, **54**:656-659.

Sakurai, N., and Nevins, D. J., 1992, Evaluation of stress-relaxation in fruit tissue, *HortTechnology,* **2**:398-402.

Schoorl, D., and Holt, D. J., 1983, A practical method for tensile testing of apple tissue, *J. Texture Studies,* **14**:155-164.

Schotte, S., de Belie, N., and de Baerdemaeker, J., 1999, Acoustic impulse-response technique for evaulation and modelling of firmness of tomato fruit, *Postharv. Biol. Technol.*, **17**:105-115.

Seymour, G. B., and Gross, K. C., 1996, Cell wall disassembly and fruit softening. *Postharv. News Info.*, **7**:45N-52N.

Shmulevich, I., Galili, N., and Benicho.,N., 1995, Development of a nondestructive method for measuring the shelf-life of mango fruit, p.275-287. In: Proc. *Food Processing Automation IV Conf.*, Chicago, IL, Nov. 3-5.

Smith, S. M., 1985, Measurement of the quality of apples: Recommendations of an EEC working group, Commission of the European Communities, Brussels.

Stone, M. L., Armstrong, P. R., Chen, D. D., and Brusewitz, G. H., 1998, Peach firmness prediction by multiple location impulse testing, *Trans. Am. Soc. Agric. Eng.*, **41**:115-119.

Stow, J., 1989, The involvement of calcium ions in maintenance of apple fruit tissue structure, *J. Exp. Bot.*, **40**:1053-1057.

Studman, C. J, and Yuwana, 1992, Twist test for measuring fruit firmness, *J. Texture Studies.*, 23(2):215-227.

Sundstrom, F. J., and Carter, S. J, 1983, Influence of K and Ca on quality and yield of watermelon, *J. Am. Soc. Hort. Sci.*, **108**:879-881.

Szczesniak, A. S., 1963, Classification of textural characteristics, *J. Food Sci.*, **28**:385-389.

Szczesniak, A. S., 1963, Objective measurement of food texture, *J. Food Sci.*, **28**:410-420.

Szczesniak, A. S., Humbaugh, P. R., and Block, H. W., 1970, Behavior of different foods in the standard compression cell of the shear press and the effect of sample weight on peak area and maximum force, *J. Texture Studies,* **1**:356-378.

Voisey, P. W., and Kloek, M., 1981, Effect of cell size on the performance of the shear-compression texture test cell, *J. Texture Studies,* **12**:133-139.

Wann, E. V., 1996, Physical characteristics of mature green and ripe tomato fruit of normal and firm genotypes, *J. Am. Soc. Hort. Sci.*, **121**:380-383.

Wu, T., and Abbott, J. A., 2002, Firmness and force relaxation characteristics of tomatoes stored intact or as slices, *Postharv. Biol. Technol.*, **24**:59-68.

Yamaguchi, M., Hughes, D. L., Yabumoto, K., and Jennings, W. G., 1977, Quality of cantaloupe muskmelons: variability and attributes, *Sci. Hortic.*, **6**:59-70.

Zapp, H. R., Ehlert, S. H., Brown, G. K., Armstrong, P. R. and Sober, S. S., 1990. Advanced instrumented sphere (IS) for impact measurements, *Trans. Am. Soc. Agric. Eng.*, **33**: 955-960

Zhang, X., Stone, M. L., Chen, D., Maness, N. O., and Brusewitz, G. H., 1994, Peach firmness determination by puncture resistance, drop impact, and sonic impulse, *Trans. Am. Soc. Agric. Eng.*, **37**:495-500.

IRRADIATION OF APPLE CIDER: IMPACT ON FLAVOR QUALITY

Terri D. Boylston, Hui Wang, Cheryll A. Reitmeier, and Bonita A. Glatz[*]

1. INTRODUCTION

Serious outbreaks of illness in 1996 and 1999 linked to contamination of apple cider by *E. coli* O157:H7 have contributed to intense public concern regarding the safety of fresh apple cider. In response to this concern, the Food and Drug Administration (FDA) has proposed more stringent regulations on the production of apple cider including a requirement to reduce microbial contaminants by 5 logs (Terpstra, 1997). Currently, FDA requires a warning statement on all fruit and vegetable juice products that have not been pasteurized (FDA, 1998). The safety concerns associated with fresh apple cider have contributed to a need to evaluate new approaches to cider processing to provide a safe product without sacrificing quality (Terpstra, 1997).

Pasteurization and irradiation are two processing treatments that are effective in reducing the microbial load of foods. Pasteurization of apple cider will reduce the pathogenic microorganisms by the desired 5 logs, but has an adverse effect on the color, flavor, and viscosity of the cider (Fisher and Golden, 1998). Irradiation of apple cider is an effective means to reduce pathogenic microorganisms (Buchanan *et al.*, 1998), including E. coli (Wang, 1999). However, further research is needed to determine the effective dose and the effect of irradiation on the sensory quality of apple cider.

The volatile flavor compounds in apples and apple products are dependent on variety, maturity, apple quality, processing, and storage conditions (Williams *et al.*, 1980; Watada *et al.*, 1981; Poll and Flink, 1983; Poll, 1985; Boylston and McCracken, 1994). Esters, alcohols, and aldehydes are among over 200 volatile flavor compounds that contribute to characteristic apple flavor (Young *et al.*, 1996). Specific compounds of particular importance to apple flavor include butyl acetate, 2-methylbutyl acetate, hexyl acetate, ethyl 2-methyl butyrate, ethyl butyrate, butanol, hexanol, hexanal, and trans-2-hexenal (Poll, 1983, 1985; Young *et al.*, 1996; Lopez *et al.*, 1998). The low aroma threshold of the esters and aldehydes makes these compounds important contributors to the flavor of apple juice and other apple products (Poll, 1988). The distribution of

---

[*] Department of Food Science & Human Nutrition, Iowa State University, Ames, IA 50011-1061.

Quality of Fresh and Processed Foods, edited by Shahidi et al.
Kluwer Academic/Plenum Publishers, 2004.

281

volatile flavor compounds and relative percentages of certain flavor compounds have an impact on the perceptions of fruit aromas and off-aromas in apple products by sensory panelists. Increases in the relative amounts of alcohols were correlated with the increased detection of off-aroma in the apple juice samples (Poll and Flink, 1984). For apple aroma condensates, increasing concentrations of butyl acetate, 3-methyl butyl acetate, hexyl acetate, hexanal and 2-hexenal contributed to desirable apple aroma, while increasing concentrations of ethanol, hexanol, and ethyl acetate decreased desirable apple aroma (Petró-Turza *et al.*, 1986). Therefore, the overall perception of aroma is dependent on the threshold and aroma characteristics of the individual compounds and the distribution of volatile flavor compounds present.

During processing and storage of apple juice, degradation of esters and aldehydes and formation of Maillard reaction products contribute to changes in sensory quality. Pasteurization of apple juice resulted in a decreased fruit-aroma score accompanied by a decreased content of esters and an increased cooked aroma, attributed to the formation of furfural and hydroxymethylfurfural during heating (Poll, 1983). In addition, 5-methyl-2-furfural, benzaldehyde, and 2,4-decadienal, compounds formed during heating, have been identified in cooked apple slices (Nursten and Woolfe, 1972). Irradiation of apple juice resulted in slight differences in odor and flavor quality at 0.5 to 1.5 kGy and detectable differences at 2.0 kGy. Trained panelists noted that the juice irradiated at 2.0 kGy had a dried apple flavor with an uncharacteristic after-taste (Zegota, 1991). The 'irradiated' off-flavor in apple juice detected by panelists has been attributed to the formation of tetrahydrofuran (Herrmann *et al.*, 1977).

Aroma and taste are important quality factors for apple juice and cider (Poll and Flink, 1983). Successful processing treatments to improve the safety of apple cider must not have an adverse effect on the flavor quality of the cider. The hypothesis of this research is that electron beam irradiation is an effective processing treatment to provide an apple cider product that is microbiologically safe and has the flavor and sensory characteristics typical of fresh, apple cider. Apple cider irradiated at 2.0 and 4.0 kGy will be compared to untreated and pasteurized apple cider. The effects of irradiation and pasteurization on the flavor and sensory characteristics of cider will be addressed in this chapter.

2. MATERIAL & METHODS

2.1. Apple Cider Processing Treatments

Apple cider, containing 0.1% potassium sorbate as a preservative, was obtained from an Iowa cider processor (Deal's Orchard, Jefferson, IA) during Fall 1999. Untreated, irradiated, and pasteurized cider was evaluated for flavor and sensory quality.

The cider was irradiated at the Iowa State University Linear Accelerator Facility (Ames, IA). The cider (150 mL) was packaged in transparent polyethylene bags (Nasco, ID No. B736, 532 mL capacity) and subjected to electron beam irradiation at 2.0 and 4.0 kGy at an energy level of 10MeV. Following treatment, the cider was stored at 4°C and analyzed within 3 days.

2.2. Sensory Evaluation

A ten-member trained sensory evaluation panel evaluated quality attributes of the apple cider. Panelists were selected from staff and students at Iowa State University and participated in three 1-hour training sessions prior to evaluation of the samples. Attributes evaluated included sweetness, acidity, apple flavor, cooked flavor, and off-flavor. An unstructured 15-cm line with anchor points of 'low' at 1-cm and 'high' 14 cm for each attribute was used for evaluation of the samples. Sensory evaluation data were collected using Compusense 5 data collection and analysis software (Compusense Inc., Guelph, Ontario, Canada).

2.3. Soluble Solids Content and Titratable Acidity

Soluble solids content and titratable acidity of the apple cider were determined as instrumental measures of sweetness and acidity. Soluble solids content was measured using a 0-32° Brix refractometer. Titratable acidity was determined by titrating the cider (20 mL) with 0.1 N NaOH to an endpoint of pH 7.0. Titratable acidity was expressed as grams of malic acid per 100 mL cider.

2.4. Volatile Flavor Analysis

Solid-phase microextraction (SPME, Supelco, Inc., Bellefonte, PA) techniques were used for the isolation of volatile flavor compounds. Apple cider (25 mL) was transferred to a 100-mL headspace bottle and sealed with a teflon septum. Following a 15-min equilibration of the sample, the volatiles were adsorbed onto the solid phase microextractor (SPME) fiber (100 μ polydimethylsiloxane) for 30 min. Samples were held in a 40°C water bath during equilibration and adsorption. A gas chromatograph equipped with a splitless injection port and flame ionization detector was used for the analysis of volatile flavor compounds (Model 6890, Hewlett-Packard, Inc., Wilmington, DE). The volatiles were thermally desorbed (225°C) for 1 min via the GC injection port onto a fused-silica capillary column (SPB-5, 30 m x 0.25 mm x 0.25 μm film thickness, Supelco, Inc.). The column pressure was set at 18.0 psi with a helium flow rate of 1.9 mL/min. The oven was initially held at 30° C for 3 min, increased at a rate of 5° C/min to a final temperature of 200°C. The detector temperature was 220°C. Flow rates of detector gases were air, 400 mL/min; hydrogen, 30 mL/min; and nitrogen make-up gas, 23 mL/min. Volatile flavor compounds were identified using authentic standards (Sigma-Aldrich, Milwaukee, WI; AccuStandard, Inc., New Haven, CT) and confirmed with GC/MS analyses.

A gas chromatograph-mass spectrometer (Trio 1000, Fisons Instruments, Danvers, MA) with a quadrupole mass analyzer was used for the confirmation of the identity of the volatile compounds. GC conditions were as for the chromatographic analysis. The conditions for the mass spectrometer were set as follow: source electron energy, 70 eV; source electron current, 150 μA; ion source temperature, 220°C; interface temperature, 220°C; source ion repeller, 3.4 V; electron multiplier voltage, 600 V; and scan range, 41-250 m/z. Mass spectra of the volatile flavor compounds were compared to the NBS Library and a flavor and fragrance database (FlavorWORKS, 1999) for identification.

2.5. Statistical Analysis

Irradiation and pasteurization treatments of the apple cider were replicated three times.   Analysis of variance and Fisher's least square difference tests ($p<0.05$) were conducted to determine the effects of the processing treatments on the content of volatile flavor compounds, intensity of sensory attributes, and soluble solids content and titratable acidity (SYSTAT, 1999).

3.   RESULTS AND DISCUSSION

Irradiation and pasteurization treatments resulted in small changes in the sweetness and acidity of the apple cider, as determined by sensory (Table 1) and instrumental (Table 2) analyses. Panelists indicated a significantly lower intensity of sweetness and a slightly higher, but not significant, intensity of acidity in the cider irradiated at 4.0 kGy in comparison to the untreated cider.   Differences in sweetness and acidity were not detected by the panelists for the untreated, pasteurized, and 2.0 kGy irradiated ciders. Instrumental analyses of sweetness and acidity showed small but significant differences in these attributes.   Untreated cider had a lower soluble solids content than the pasteurized and irradiated ciders. Titratable acidity of the irradiated (2.0 and 4.0 kGy) ciders was higher than either the untreated or pasteurized ciders. Perception of sweetness and acidity by sensory panelists is influenced by the balance of soluble solids and titratable acidity (Poll, 1981), and thus small differences in instrumental measurements of sweetness and acidity may not have been detected by the panelists.

Table 1. Sensory evaluation of sensory quality of apple cider

Attribute	Untreated	Pasteurized	Irradiated 2.0 kGy	Irradiated 4.0 kGy
Sweetness[1]	10.9[a]	9.9[ab]	9.5[ab]	8.3[b]
Acidity[1]	5.1[a]	6.0[a]	5.9[a]	6.2[a]
Apple Flavor[1]	10.8[a]	8.1[b]	7.9[b]	6.3[c]
Cooked Flavor[1]	2.3[c]	7.2[a]	4.9[b]	6.7[ab]
Off-Flavor[1]	1.7[b]	3.2[b]	2.6[b]	5.3[a]

[1] 0=low, 15=high.
[a-c] Means within the same row having the same superscript are not significantly different (P>0.05).

Pasteurization and irradiation treatments had the greatest impact on the panelists' perception of the flavor attributes of the apple cider (Table 1). The untreated cider had significantly ($p<0.05$) higher intensity of apple flavor and lower intensity of cooked flavor and off-flavor. Pasteurization and irradiation treatments decreased the intensity of apple flavor and increased the intensity of cooked flavor and off-flavors. Cooked flavor and off-flavors were most prevalent in the pasteurized and 4.0 kGy irradiated ciders, respectively.

Table 2. Soluble solids content and titratable acidity of apple cider

Component/ Attribute	Untreated	Pasteurized	Irradiated 2.0 kGy	Irradiated 4.0 kGy
Soluble solids[1]	12.62[b]	12.95[a]	12.87[a]	12.97[a]
Titratable acidity[2]	0.354[b]	0.353[b]	0.367[a]	0.369[a]
SS/TA ratio	3.56	3.67	3.51	3.51

[1] % Soluble solids
[2] g Malic acid/100 mL juice
[a,b] Means within the same row having the same superscript are not significantly different ($p > 0.05$).

The changes in the flavor attributes of the irradiated and pasteurized apple ciders (Table 3) detected by the sensory evaluation panel can be attributed to changes in the content of the volatile flavor compounds that resulted during processing. Esters contribute to the desirable fruity, apple-like flavor notes in apple cider (Poll, 1985; Mangas et al., 1996). This class of compounds not only accounts for a majority of the volatile compounds present, but also is an important contributor to flavor because of the low threshold of esters. Four esters (butyl acetate, 2-methyl butyl acetate, hexyl acetate, and ethyl hexanoate) decreased significantly following pasteurization. Contents of these esters decreased slightly, but were not significantly different following irradiation at 2.0 or 4.0 kGy. Decreases in the intensity of apple flavor by the sensory evaluation panel may be attributed to the decrease in the ester contents.

The aliphatic alcohols and aldehydes identified in the cider have green, grassy, herbaceous, and woody aroma characteristics (Poll, 1985). These compounds contribute to the characteristic flavor of apples and apple products and are also lipid oxidation products. As products of lipid oxidation, these compounds are often considered to be undesirable. The aromatic aldehydes, including benzaldehyde and phenylacetaldehyde, have aromatic and fragrant flavor characteristics and contribute to the desirable aroma characteristics of apples and apple products. The contents of these alcohols and aldehydes did not change significantly as a result of the pasteurization and irradiation treatments (Table 3).

Heterocyclic compounds, including 2-furfural and 5-hydroxymethylfurfural, increased during pasteurization and irradiation treatments and contribute to the increased cooked flavors detected by the sensory evaluation panel (Table 3). These compounds are products of the Maillard reaction and have been identified in pasteurized apple juice stored at elevated temperatures (Poll, 1985). Tetrahydrofuran, which has been detected in irradiated apple juice (Herrmann et al., 1977), was not detected in the irradiated apple ciders in this study.

For many of the volatile flavor compounds, there were no significant differences between the untreated and irradiated apple ciders. However, (RI 1570) was significantly higher in both irradiated cider treatments and may be responsible for the off-flavor detected in the irradiated ciders by the sensory evaluation panels.

Table 3. Volatile flavor compounds of apple cider

Compound	Untreated	Pasteurized	Irradiated 2.0 kGy	Irradiated 4.0 kGy
Esters				
Butyl acetate	100.0[a]	56.6[b]	102.6[a]	97.8[a]
2-Methylbutyl acetate	151.5[a]	59.9[b]	125.2[a]	134.0[a]
Hexyl acetate	178.2[a]	47.7[b]	166.4[a]	144.0[a]
3-Methylbutyl propionate	3.9[a]	0.0[c]	1.3[bc]	3.0[ab]
Methyl butanoate	6.7[a]	1.6[a]	0.0[a]	1.6[a]
Pentyl 4-methyl pentanoate	1.8[a]	7.9[a]	2.2[a]	4.1[a]
Ethyl hexanoate	12.0[a]	3.3[c]	7.0[ab]	9.2[a]
Pentyl hexanoate	7.8[a]	14.7[a]	11.1[a]	11.4[a]
Hexyl hexanoate	17.3[a]	21.8[a]	18.8[a]	19.0[a]
Aldehydes				
Hexanal	2.6[a]	7.4[a]	5.6[a]	7.1[a]
t-2-Octenal	3.0[a]	8.1[a]	2.2[a]	3.8[a]
2,4-Decadienal	3.8[a]	10.0[a]	1.7[a]	3.0[a]
Benzaldehyde	3.7[a]	0.0[b]	3.2[a]	3.2[a]
Phenylacetaldehyde	8.4[a]	3.8[a]	3.5[a]	4.4[a]
Alcohols				
c-3-Hexen-1-ol	2.7[a]	9.9[a]	7.8[a]	6.0[a]
Hexanol	93.0[a]	83.3[a]	91.3[a]	75.2[a]
Octanol	4.5[a]	8.1[a]	2.5[a]	6.2[a]
Furans				
2-Furfural	0.0[b]	18.2[a]	0.0[b]	0.0[b]
Furfuryl alcohol	4.3[a]	2.4[a]	1.8[a]	5.8[a]
5-Methylfurfural	8.6[a]	2.8[a]	5.1[a]	13.2[a]
5-Hydroxymethylfurfural	1.5[a]	16.6[a]	4.0[a]	6.6[a]
Terpenes				
α-Farnesene	31.1[a]	45.4[a]	54.9[a]	155.1[a]
Unidentified Compounds				
Kovats RI 1205	0.0[a]	4.1[a]	1.8[a]	2.9[a]
Kovats RI 1293	60.1[b]	103.2[a]	67.0[b]	57.8[b]
Kovats RI 1570	3.0[b]	11.3[b]	86.0[a]	70.1[a]
Total	778.3[a]	622.1[a]	846.2[a]	918.2[a]

[a,b] Means within the same row having the same superscript are not significantly different (p>0.05).

4. CONCLUSIONS AND FUTURE RESEARCH

This research demonstrated that apple cider irradiated at 2.0 and 4.0 kGy has sensory attributes that are more similar to untreated apple cider than pasteurized apple cider. However, the chemical reactions responsible for the off-flavors detected by the panelists in the irradiated cider are not completely understood. Future research should address the complex relationship between sensory perception and volatile flavor compounds and identify the impact of irradiation on the flavor quality of apple cider.

5. ACKNOWLEDGMENTS

The research was funded in part by a grant from USDA/CSREES National Research Initiative Competitive Grants Program, Project No. 98-35502-6604. Journal Paper No. J-19309 of the Iowa Agriculture and Home Economics Experiment Station, Ames, Iowa, Project No. 3546, and supported by Hatch Act and State of Iowa Funds.

6. REFERENCES

Boylston, T. D. and McCracken, V. A., 1994, Consumer acceptance and quality evaluation of new apple varieties, *Inst. Food Technol. Annual Meeting*, Atlanta, GA, Abstr. No. 12A-4.

Buchanan, R. L., Edelson, S. G., Snipes, K., and Boyd, G., 1998, Inactivation of *Escherichia coli* O157:H7 in apple juice by irradiation, *Appl Environ. Microbiol.*, **64**:4533-4535.

Fisher, T. L. and Golden, D. A., 1998, Fate of Escherichia coli O157:H7 in ground apples used in dicer production, *J. Food Prot.*, **61**:1372-1374.

FlavorWORKS. 1999. Ver. 2.0, Flavometrics, Inc, Anaheim Hills, CA.

Food and Drug Administration, 1998, Food labeling: warning and notice statements; labeling of juice products. *Fed. Reg.*, **63**:20486-20493.

Herrmann, J., Grigorova, S., and Grigorova, L., 1977, New methods for evaluation and analysis of organoleptic qualities of foodstuffs, and for the prediction of their changes. IX. Dependence of the irradiation flavour in apple juice and concentrate on the gamma irradiation dose, *Nahrung*, **20**:619-627.

Lopez, M. L., Lavilla, M. T., Ribva, M., and Vendrell, M., 1998, Comparison of volatile compounds in two seasons in apples: Golden Delicious and Granny Smith, *J. Food Qual.*, **21**:155-166.

Mangas, J. J., Gonzáliz, M. P., Rodríguez, R., and Blanco, D., 1996, Solid-phase extraction and determination of trace aroma flavour components in cider by GC-MS, *Chromatographia*, **42**:101-105.

Nursten, H. E. and Woolfe, M. L., 1972, An examination of the volatile compounds present in cooked Bramley's seedling apples and the changes they undergo on processing, *J. Sci. Food Agric.*, **23**:803-822.

Petró-Turza, M., Szárföldi-Szalma, I., Madarassy-Mersich, E., Teleky-Vámossy, G., and Füzesi-Kardos, K, 1986, Correlation between chemical composition and sensory quality of natural apple aroma condensates, *Die Nahrung*, **30**:765-774.

Poll, L., 1981, Evaluation of 18 apple varieties for the suitability for juice production, *J. Sci. Food Agric.*, **32**:1081-1090.

Poll, L., 1983, Influence of storage temperature on sensory evaluation and composition of volatiles of McIntosh apple juice, *Lebensm –Wiss. u.-Technol.*, **16**:220-223.

Poll, L., 1985, The influence of apple ripeness and juice storage temperature on the sensory evaluation and composition (volatile and non-volatile components) of apple juice, *Lebensm.-Wiss. u. Technol.*, **18**:205-211.

Poll, L., 1988, The effect of pulp holding time on the volatile components in apple juice (with and without pectolytic enzyme treatment), *Lebensm.-Wiss. u.-Technol.*, **21**:87-91.

Poll, L. and Flink, J. M., 1983, Influence of storage time and temperature on the sensory quality of single-variety apple juice, *Lebensm.-Wiss. u.-Technol.*, **16**:215-219.

Poll, L. and Flink, J. M., 1984, Aroma analysis of apple juice: Influence of salt addition on headspace volatile composition as measured by gas chromatography and corresponding sensory evaluations, *Food Chem.*, **13**:193-207.

SYSTAT, 1999, Ver. 9.01, SPSS, Inc, Chicago, IL.

Terpstra, E., 1997, Fresh cider safety: The past and what the future holds, *Great Lakes Fruit Grower News,* March 1997, p. 18.

Wang, H., 1999. Personal communication.

Watada, A. E., Abbot, J. A., Harenburg, R. E., and Lusby, W., 1981, Relationships of apple sensory attributes to headspace volatiles, soluble solids, and titratable acids, *J. Amer. Soc. Hort. Sci.,* **106**:130-132.

Williams, A. A., Lewis, M. J., and Tucknott, O. G., 1980, The neutral volatile components of cider apple juices, *Food Chem.,* **6**:139-151.

Young, H., Gilbert, J. M., Murray, S. H., and Ball, R. D., 1996, Causal effects of aroma compounds on Royal Gala apple flavours. *J. Sci. Food Agric.,* **71**:329-336.

Zegota, H., 1991, Evaluation of the sensory properties of irradiated apple juice concentrate, *Z. Lebensm. Untersuch. Forsch.* **192**:7-10.

FRUIT AND VEGETABLE EDIBLE WRAPS: APPLICATION TO PARTIALLY DEHYDRATED APPLE PIECES

T.H. McHugh, C.W. Olsen and E. Senesi[*]

1. INTRODUCTION

With 96 million Americans now over the age of 45 and 113 million projected by 2010, health foods may be one of the largest, most lucrative markets of all time (Sloan, 2001). During the past five years Americans have moved towards positive eating, not just avoiding negative macro ingredients such as fat, but ensuring the inclusion of positive ingredients such as fruits and vegetables. Nearly seven in ten consumers say they are trying to eat more fruits and vegetables (FMI, 2000).

More consumers are looking for convenience in the form of pre-cut minimally processed fruits and vegetables. The market demand for minimally processed fruits and vegetables has rapidly expanded due to busy lifestyles, increasing purchase power, and health conscious consumers (Baldwin *et al.,* 1995). It is estimated that by the year 2000 the market share of fresh cut product will reach 25% of all produce sales in the U.S. retail market (Ahvenainen, 1996). Preservation of minimally processed produce presents unique challenges to the food industry.

One method of extending the shelf life of minimally processed produce is through the use of edible films and coatings (Guilbert *et al.*, 1996). Edible films can regulate transfer of moisture, oxygen, carbon dioxide, lipid, aroma and flavor compounds in food systems, thus increasing food product shelf-life and improve food quality. Generally, an edible coating is defined as a thin layer of edible material formed on a food; whereas, an edible film is a preformed thin layer of edible material placed on or between food components (Krochta, 1997).

[*] Authors McHugh and Olsen are affiliated with the USDA, Agricultural Research Service, Western Regional Research Center, 800 Buchanan Street, Albany, CA 94710
Author Senesi is affiliated with the Institute for Agricultural Products Technology (IVTPA), Via Venezian 26, 20133 Milano, Italy

DISCLAIMER: Names are necessary to report factually on available data; however, the USDA neither guarantees nor warrants the standard of the product, and the use of the name by USDA implies no approval of the product to the exclusion of others that may also be suitable.

Quality of Fresh and Processed Foods, edited by Shahidi et al.
Kluwer Academic/Plenum Publishers, 2004.

Numerous studies have investigated the potential of edible coatings to extend the shelf life of minimally processed fruits (Wong *et al.*, 1994; Baldwin *et al.*, 1995; and Ahvenainen, 1996). Ponting *et al.* (1972) investigated the preservative effects of ascorbic acid, calcium and sulfites to protect the color of refrigerated apple slices. More recently the use of sulfites for the prevention of browning on fruits and vegetables was banned in the United States, stimulating efforts to identify alternative browning inhibitors. Baldwin *et al.* (1996) demonstrated that a cellulose-based edible coating, Nature Seal 1020, could prolong the shelf life of fresh cut apple when stored in overwrapped trays by about 1 week at 4°C. Luo and Barbosa-Canovas (1996 and 1997) discovered that browning inhibition using a mixture of ascorbic acid and 4-hexylresorcinol was comparable to that achieved by sulfites. Ascorbic acid plus calcium chloride coatings were shown to inhibit color deterioration and loss of firmness in apple cubes by Rocha *et al.* (1998).

McHugh *et al.* (1996) developed the first edible films made from fruit purees and characterized their water vapor and oxygen permeability properties. Apple-based edible films were excellent oxygen barriers, but not very good moisture barriers.

More recently McHugh and Senesi (2000) developed and characterized the water vapor permeability properties of apple based edible films containing lipids. The water vapor barrier properties of apple based films were significantly improved through the addition of fatty acids, fatty alcohols, beeswax and vegetable oil. Fresh cut apple pieces were coated with solutions or wrapped in preformed films and changes in moisture and color during storage were measured. Wraps were significantly more effective than coatings in reducing moisture loss and browning. Color was preserved for 12 days at 5°C.

Apples and other fruits and vegetables are frequently processed through dehydration. Apples can be preserved by drying. The peeled and cored apples are prepared as rings, segments, chops or cubes and are commonly treated with a weak solution of citric acid and bisulfite to inhibit enzymatic browning (Kaushal and Sharma, 1995). Recently the effects of osmotic dehydration on the mechanical properties of apple pieces were studied (Alvarez *et al.*, 2000).

This study is the first to explore the potential of a combined process involving partial dehydration followed by wrapping in edible films to extend the shelf life of fresh cut apple pieces. These hurdle technologies were hypothesized to have beneficial effects on deterioration of apple piece color, moisture and texture during storage.

2.   MATERIALS & METHODS

2.1 Film materials

Cling peach (32°Brix), golden apple (38°Brix) and strawberry (8°Brix) purees from Sabroso Company (Medford, OR) were utilized as the primary ingredient in edible fruit based edible films. Brocolli (6°Brix) and carrot (9°Brix) purees from DMH Ingredients (Buffalo Grove, IL) were used to form edible vegetable films. Ascorbic acid (BASF Corporation, Mount Olive, NJ) and citric acid (Archer Daniels Midland, Decatur, IL) were utilized as browning inhibitors. Soy vegetable oil (Hunt Wesson, Inc., Fullerton, CA) was incorporated into apple films to improve their water barrier properties. Granny Smith apples were purchased from a local supermarket (Andronico's, Berkeley, CA).

2.2 Fruit and vegetable films

Fruit and vegetable purees were diluted with distilled water to facilitate even spreading over the casting surface. Aqueous solutions of the following concentrations were prepared: 13% (w/w) peach, 26% (w/w) apple, 40% strawberry (w/w), 40% (w/w) broccoli and 33% (w/w) carrot. A vacuum was applied to each solution to remove dissolved air prior to film formation. Films were cast on 15.5 cm (internal diameter) rimmed, smooth, Teflon plates. Fifty grams of solution was pipetted onto each plate. The solutions were spread evenly with a glass rod and allowed to dry overnight at 24±1°C and 40±2% relative humidity (RH). Dried films were peeled intact from the casting surface.

2.3 Apple wraps

Larger films were formed to wrap apple pieces. Apple puree solutions were prepared at 26% (w/w) concentrations as described above. Apple films to be used as wraps also contained ascorbic and citric acids at 0.5% (w/w) level. Vegetable oil was added at a concentration of 10% (w/w) and the mixture was homogenized 2 min at 12,000 rpm using a Polytron 3000 (Brinkman Instruments, Wesbury, NY). Homogenized apple mixtures (160 g) were then cast on 29 cm x 29 cm square teflon plates and allowed to dry for 24 h at 24±1°C and 40±2% RH. Dried films were peeled intact from the casting surface.

2.4 Apple piece dehydration

Large Granny Smith apples were peeled, cored and sectioned into sixteenths. Apple sections were cut into rectangular pieces (approximately 25 mm by 20mm by 15 mm). Apple pieces were then dipped in antibrowning solution containing 10 g/L ascorbic acid and 5 g/L citric acid for 30 min. Following dipping, apple pieces were removed from the solution and dried at 70°C in a food dehydrator (Excalibur, Sacramento, CA) until a weight loss of 30% (w/w) or 50% (w/w) was achieved. Drying time was approximately 30 min for 30% (w/w) weight loss and 60 min for 50% (w/w) weight loss.

2.5 Apple piece wrapping

Dehydrated apple pieces were manually wrapped in precut apple based films (7 cm x 10 cm). Apple films adhered well to the cut apples pieces. Both nonwrapped (control) and wrapped apple pieces were tested.

2.6 Apple piece storage

Three replicates of each sample were placed on each labeled petri dish bottom. Three petri dishes of control and wrapped samples were used for each analysis. Petri dishes were stored at either 5±1°C and 30±5% relative humidity or 30±1°C and 50±5% RH for up to 12 days. A SmartReader 2 data logger (ACR Systems Inc., Surrey, BC, Canada) was used to monitor temperature and relative humidity conditions during storage.

2.7 Film thickness measurements

Film thicknesses were measured with a micrometer (L.S. Starrett Co., Athol, MA) to the nearest 0.001 mm at 5 random positions around the film and average values were used in all calculations.

2.8 Water activity measurement

Water activity was measured using an AquaLab Model CX-2 water activity meter (Decagon Devices, Inc., Pullman, WA) at 24±1°C and 40±2% RH. Nine apple pieces were tested for each sample.

2.9 Water vapor permeability determination

Water vapor permeability (WVP) was determined using the WVP Correction Method (McHugh *et al.*, 1993). A cabinet containing a variable speed fan was used to test film WVP. Cabinet temperature was maintained at 22±1°C. Fan speed was set to achieve an air velocity of 152 m/min to ensure uniform relative humidity throughout the cabinets. Each cabinet contained a hygrometer to monitor RH conditions. Cabinets were pre-equilibrated to 0% RH using calcium sulfate desiccant.

Circular test cups were made from polymethylmethacrylate (Plexiglas). A film was sealed to the cup base with a ring containing a 19.6 cm^2 opening using 4 screws symmetrically located around the cup circumference. Both sides of the cup contacting the film were coated with silicon sealant. Deionized water was placed in the bottoms of the test cups to expose the film to a high percentage RH inside the test cups. Each cup contained 6 mL deionized water. Average stagnant air gap heights between the water surface and the film were measured. Test cups holding films were then inserted into the pre-equilibrated 0% RH desiccator cabinets. Within 2 h, steady state was achieved. Each cup was weighed 5 times at > 2 h intervals. Four replicates of each film were tested.

RHs at the film undersides and WVPs were calculated using the WVP Correction Method (McHugh *et al.,* 1993). This method was necessary for testing films with low water vapor resistance since it accounted for the water vapor partial pressure gradient within the stagnant air layer of the test cup.

2.10 Moisture loss determination

Moisture loss was measured daily by weighing the petri dishes containing either the control or wrapped apple pieces. Triplicates for each treatment were tested.

2.11 Colorimetric measurements

A Minolta 508d spectrophotometer (Ramsey, NJ) was used to measure the CIELAB color parameters, L^*, a^*, and b^* (Francis, 1995). A standard white calibration plate was employed to calibrate the spectrophotometer. Three measurements of apple pieces were taken at three locations for each sample, making a total of nine measurements per sample. Duplicate samples for each treatment were measured. Wraps were removed prior to color measurements. Color was measured daily during the storage period.

2.12 Texture profile analyses

An Instron 4502 (Canton, MA) was utilized to analyze the texture profile of the apple pieces. Wraps were removed prior to texture profile analysis. Apples were compressed in two cycles at 12 mm/min to 80% of their original height. Force, time, and distance measurements were recorded every 6 s. The data was analyzed as described by Bourne (1978) for fracturability, hardness, adhesiveness, springiness, gumminess and chewiness and measurements were expressed as texture profile values. Triplicate measurements were made for each sample type.

2.13 Statistical analyses

ANOVA and Fisher LSD analyses were performed using SuperANOVA at $P < 0.05$ (Abacus Concepts, Berkeley, CA).

3. RESULTS AND DISCUSSION

3.1 Permeability properties of fruit and vegetable films

The water vapor permeability (WVP) properties of different types of fruit and vegetable films were compared in Table 1. Each of these films was composed of 100% fruit or vegetable puree. The vegetable films (carrot and broccoli) had significantly lower WVP values than the fruit films (peach and strawberry). Broccoli films were the best water barriers, followed by carrot, peach and strawberry.

Table 1. Water Vapor Permeability Properties of 100% Fruit and Vegetable Edible Films at 22°C.

Film Type	Thickness[a] (mm)	RH inside cup[ab]	Water Vapor Permeability[ab] (g-mm/kPa-d-m^2)
Peach	0.103	74%	67.9 ± 3.7^c
Strawberry	0.117	75%	71.5 ± 4.0^c
Carrot	0.110	78%	57.0 ± 4.5^d
Broccoli	0.104	83%	40.8 ± 5.6^e

[a]Thickness and RH data are mean values. WVP data are mean values ± standard deviation.

[b]Relative humidity at the inner surface and WVP values were corrected for stagnant air effects using the WVP Correction Method (McHugh et al., 1993).

[c,d,e]WVP means with different letters are significantly different at p<0.05 using Fisher LSD multiple comparison test.

Previous studies have compared the WVP values for a variety of different fruit types at a variety of relative humidity conditions (McHugh *et al.*, 1996). Peach and apricot films exhibited significantly better water barrier properties than apple and pear films. WVP properties of peach films were very low and below 50% relative humidity (RH). Above 50% RH, WVP values increase exponentially under increasing relative humidity conditions. The exponential effect of relative humidity on water vapor and oxygen permeability values have been observed in other hydrophilic film forming materials, such as polyamide (De Leiris, 1986), nylon (Myers *et al.*, 1962), cellulose (Karel *et al.*, 1959), collagen (Lieberman *et al.*, 1972) and whey protein (McHugh *et al.*, 1993).

Peach puree was a good oxygen barrier, particularly at low to moderate RHs (McHugh *et al.*, 1996). Oxygen permeability values (O2P) of peach films were lower than those for methylcellulose, whey protein and most lipid based barriers. The O2Ps of peach films were comparable to those of many synthetic films such as high density polyethylene, cellophane and ethylene vinyl alcohol. These previous results implicate the potential of other types of fruit and vegetable based films to act as effective oxygen barriers in food systems.

3.2 Moisture loss from partially dehydrated and wrapped apple pieces

Dehydration of apple pieces was performed as a hurdle technology prior to wrapping apple pieces in apple films. By dehydrating prior to wrapping we hypothesized that the apple shelf life would be significantly extended versus fresh cut apple slices. Partial dehydration to 30% and 50% weight loss was performed on apple pieces; the water activity of these apples were 0.983 and 0.973, respectively. These values serve as initial values for all of our experiments.

Storage temperature, dehydration level and wrapping significantly affected moisture loss from apple pieces during 12 days of storage (Figures 1 and 2). Temperature had a significant effect on moisture loss from apple pieces which had been dehydrated to 30% and 50% moisture loss. As storage temperature increased from 5 to 30°C, moisture loss increased significantly. The effect of temperature on 50% dehydrated apple pieces was more apparent than that on 30% dehydrated apple pieces; although at both dehydration levels the effect was significant. Fifty percent dehydrated apple pieces lost significantly less moisture during storage than 30% dehydrated samples. At both storage temperatures and both levels of dehydration, wrapping significantly reduced moisture loss from partially dried apple pieces.

These results support those of a previous study which demonstrated that apple wraps significantly reduced moisture loss from fresh cut apple pieces during storage at 5°C (McHugh and Senesi, 2000). Dehydration prior to wrapping did not significantly improve the shelf life of the final apple pieces over that of fresh cut apple pieces. After 12 days of storage at 5°C fresh cut apple pieces lost 34% moisture; whereas, 30% dehydrated apple pieces lost 37% moisture and 50% dehydrated pieces lost 44% moisture. Wrapped dehydrated apple pieces lost a greater amount of moisture during storage than wrapped fresh cut apple pieces. Differences in the effectiveness of the wraps on fresh cut versus dehydrated apple pieces may have resulted from the varying degree of adhesion of the wraps on the apple pieces. Apple wraps adhered readily to fresh cut apple pieces as a result of the natural moisture present in the pieces; whereas, they adhered less readily on partially dehydrated apple pieces due to the surface dryness of these pieces. This may have resulted in less effective barrier properties of films wrapped around partially dehydrated apple pieces.

3.3 Colorimetric changes in partially dehydrated and wrapped apple pieces

The apple wraps resembled the color of apple sauce. The wraps themselves did not significantly brown during storage at either 5 or 30°C. Wrapped apple pieces were unwrapped prior to colorimetric measurements.

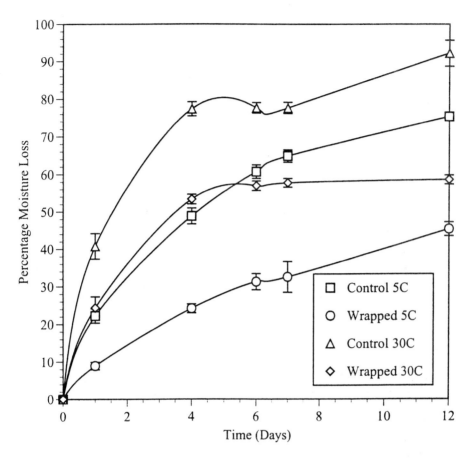

Figure 1. Moisture loss from 30% dehydrated apple pieces during storage: Control, 5°C (□); Wrapped, 5°C (○); Control, 30°C (△); Wrapped, 30°C (◇).

Wrapping and level of dehydration did not significantly affect browning of apple pieces during storage. Only storage temperature showed a significant effect. Apple piece L^* values (brightness) decreased significantly and a^* values (redness) increased significantly as 30% dehydrated apple pieces browned during storage at 30°C for twelve days (Table 2). At lower storage temperatures lesser degrees of browning occured. Application of apple wraps to 30% dehydrated apple pieces did not significantly reduce browning during twelve days of storage at 30°C (Table 2). Similar results were observed during storage at 5°C and for 50% dehydrated apple pieces (data not show).

These results contrasted with those of a previous study where wraps were very effective in reducing browning in fresh cut apple pieces (McHugh and Senesi, 2000). In this previous study, wraps completely eliminated browning for up to ten days at 5°C. Two different hypotheses were developed to explain the differences observed in the effectiveness of the wraps on fresh cut versus dehydrated apple pieces. The first hypothesis suggests that the varying degree of adhesiveness of the wraps on the fresh cut versus the dehydrated apple pieces resulted in the differences. Apple wraps adhered

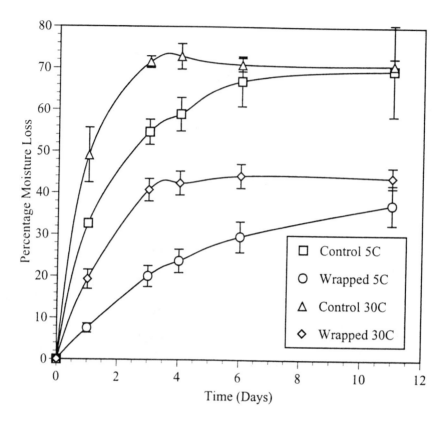

Figure 2. Moisture loss from 50% dehydrated apple pieces during storage: Control, 5°C (□); Wrapped, 5°C (○); Control, 30°C (Δ); Wrapped, 30°C (◇).

readily to fresh cut apple pieces as a result of the natural moisture present in the pieces. Apple wraps adhered less readily to partially dehydrated apple pieces due to the surface dryness of these products. This may have reduced their effectiveness as barriers on the dehydrated pieces. Secondly, the partially dehydrated apple pieces were dipping in ascorbic acid/citric acid solutions prior to wrapping; whereas the fresh cut pieces were not predipped. This dip presumably reduced the overall degree of browning in both the control and the wrapped partially dehydrated apple pieces and may have resulted in the differences observed between the two studies.

3.4 Textural profile analyses of partially dehydrated and wrapped apple pieces

Instrumental texture profile analysis (TPA) is one of the most highly-valued instrumental techniques for simultaneous measurement of numerous textural parameters (Fizman *et al.*, 1998). Sensory results have frequently correlated well with TPA results (Ocon *et al.*, 1995 and Kim *et al.*, 1996).

Table 2. Browning in Control and Wrapped 30% Dehydrated Apple Pieces After 12 Days of Storage at 30°C.

Sample	Color Value	Time Zero	Twelve Days
Control	L*	79.59 ± 1.18^b	67.53 ± 4.84^c
Wrapped	L*	79.59 ± 1.18^b	69.28 ± 4.99^c
Control	a*	-3.13 ± 0.27^e	6.14 ± 2.18^d
Wrapped	a*	-3.13 ± 0.27^e	6.89 ± 1.49^d

[a] Wraps contained 70% apple puree/27% vegetable oil/1.5% ascorbic acid/1.5% citric acid weight/weight on dry solids basis. Wraps were removed prior to colorimetric measurements.
[b,c,d,e] Moisture loss means ± standard deviations with different letters are significantly different at p<0.05 using Fisher LSD multiple comparison test.

Fracturability is a critical parameter reflecting the sensorial acceptability of apple pieces. Fracture resistance was demonstrated by Alvarez *et al.* (2000) to be a useful mechanical property of apple tissue, providing a good measure of the deterioration of Granny Smith tissue during storage. Fracturability values were significantly affected by wrapping, storage temperature and storage time. During storage, fracturability values decreased significantly for all apple pieces. At 5°C storage, the wrapped apple pieces remained fracturable throughout the twelve days of storage. After five days at 5°C storage fracturability was significantly higher in wrapped 30% dehydrated apple pieces than in control apple pieces (Table 3). Storage at 30°C resulted in a loss of fracturability after one day storage for both wrapped and control, 30% dehydrated apple pieces. Similar results were observed for 50% dehydrated apple pieces (data not shown).

Hardness values were not significantly affected by wrapping or storage temperature. Over time, hardness values decreased significantly as apple pieces lost moisture during storage.

Three way analysis of variance revealed that adhesiveness and springiness values were not significantly affected by storage temperature or storage time. Both were; however, affected by wrapping. Wrapped dehydrated apple pieces were significantly more adhesive and more springy than control apple pieces. No significant interactions were observed.

Gumminess and chewiness values, on the other hand, were significantly affected by both temperature and time, but not by wrapping. Significant interactions between time and temperature were also apparent from the ANOVA results of gumminess and chewiness data. Gumminess and chewiness values increased during storage and were significantly higher after storage under higher temperature conditions. Wrapping did not significantly affect apple piece gumminess or chewiness at either level of dehydration.

Table 3. Fracturability Values for Control and Wrapped 30% Dehydrated Apple Pieces During Storage at 5°C.

Sample	Storage Time (days)	Fracturability[a] (N)
Control	Zero	63.50 ± 7.87^b
Wrapped	Zero	63.50 ± 7.87^b
Control	Five	40.27 ± 2.26^d
Wrapped	Five	54.39 ± 8.71^c
Control	Twelve	0^f
Wrapped	Twelve	18.37 ± 1.49^e

a Wraps contained 70% apple puree/27% vegetable oil/1.5% ascorbic acid/1.5% citric acid weight/weight on dry solids basis. Wraps were removed prior to mechanical property measurements.

b,c,d,e,f Moisture loss means ± standard deviations with different letters are significantly different at $p<0.05$ using Fisher LSD multiple comparison test.

4. CONCLUSIONS

Fruit and vegetable wraps can be formed from many fruits and vegetables. Vegetable films are generally better water barriers than fruit films.

Application of apple wraps to partially dehydrated apple pieces significantly reduced moisture loss during storage; however, it did not have a significant effect on apple piece browning. Improvements in fracturability, springiness and adhesiveness were observed in wrapped apple pieces. Wrapped apple pieces also maintained their fresh flavor and aroma better than unwrapped controls.

Fruit and vegetable wraps can be used to extend the shelf-life of cut fruits and vegetables, as well as enhance their nutritional value and increase their consumer appeal. They also can be used on other food systems such as nuts, baked goods and confectionery products.

5. REFERENCES

Ahvenainen, R., 1996, New approaches in improving the shelf life of minimally processed fruit and vegetables, *Trends Food Sci. Technol.*, **7**:179-187.

Alvarez, M. D., Saunders, D. E. J., and Vincent, J. F. V., 2000, Fracture properties of stored fresh and osmotically manipulated apple tissue, *Eur. Food Res. Technol.*, **211**:284-290.

Baldwin, E. A., Nisperos, M. O., Chen, X., and Hagenmaier, R. D., 1996, Improving the storage life of cut apple and potato with edible coating, *Postharvest Bio. Technol.*, **9**:151-163.

Baldwin, E. A., Nisperos-Carriedo, M. O., and Baker, R. A., 1995, Use of edible coatings to preserve quality of lightly (and slightly) processed products, *Crit. Rev. Food Sci. Nutr.*, **35**:509-524.

Bourne, M. C., 1978, Texture profile analysis, *Food Technol.*, **32(7)**:62-78.

De Leiris, J. P., 1986, Water activity and permeability, Ch. 12, in: *Food Packaging and Preservation*, M. Mathlouthi, ed., Elsevier Science Publishing Co., Inc., New York, pp. 371-394.

Fizman, S. M., Pons, M., and Damasio, M. H., 1998, New parameters for instrumental texture profile analysis: instantaneous and retarded recoverable springiness, *J. Texture Studies.*, **29**:499-508.

FMI., 2000, Trends in the United States: Consumer attitudes and the supermarket. Food Mktg. Inst., Washington, D.C.

Guilbert, S., Gontard, N., and Gorris, L. G. M., 1996, Prolongation of the shelf-life of perishable food products using biodegradable films and coatings, *Lebensmit Wissenschaft und-Technol.*, **29**:10-17.

Kaushal, B. B., and Sharma, P. C., 1995, Apple, in: *Fruit Science and Technology,* D. K. Salunkhe, and S. S. Kadam, eds., Marcel Dekker, Inc., New York, NY., pp. 91-122.

Karel, M., Proctor, B. E., and Wiseman, G., 1959, Factors affecting water vapor transfer through food packaging films, *Food Technol.,* **13(1)**: 69-74.

Kim, Y. S., Wiesenborn, D. P., Lorenzen, J. H., and Berglund, P., 1996, Suitability of edible bean and potato starches for starch noodles, *Cereal Chem.,* **73**:302-308.

Krochta, J. M., and De Mulder-Johnson, C., 1997, Edible and biodegradable polymer films: challenges and opportunities, *Food Techol.,* **51**:61-74.

Lieberman, E. R., Gilbert, S. G., and Srinivasa, V., 1972, The use of gas permeability as a molecular probe for the study of cross linked collagen structures, *N.Y. Acad. Sci. II.,* **34**:694-708.

Luo, Y., and Barbosa-Canovas, G. V., 1997, Enzymatic browning and its inhibition in new apple cultivars slices using 4-hexylresorcinol in combination with ascorbic acid, *Food Sci. Technol. Intern.,* **3**:195-201.

Luo, Y., and Barbosa-Canovas, G. V., 1996, Preservation of apple slices using ascorbic acid and 4-hexylresorcinol, *Food Sci. Technol. Intern.,* **2**:315-321.

McHugh, T. H., and Senesi, E., 2000, Apple wraps: A novel method to improve the quality and extend the shelf life of fresh cut apples, *J. Food Sci.,* **65**:480-485.

McHugh, T. H., Huxsoll, C. C., and Krochta, J. M., 1996, Permeability properties of fruit puree edible films, *J. Food Sci.,* **61**:88-91.

McHugh, T. H., Avena-Bustillos, R., and Krochta, J. M., 1993, Hydrophilic edible films: Modified procedure for water vapor permeability and explanation of thickness effects, *J. Food Sci.,* **58**:899-903.

Myers, A. W., Meyer, J. A., Rogers, C. E., Stannett, V., and Szwarc, M., 1962, The permeation of water vapor. Ch. VI., in: *Permeability of Plastic Films and Coated Paper to Gases and Vapors*, M. Kouris, ed., Technical Association of Pulp and Paper Industry, New York, pp. 62-77.

Ocon, A., Anzaldua-Morales, Gastelum, G., and Quintero, A., 1995, The texture of pecan nuts measured by instrumental and sensory means, *J. Food Sci.,* **60**:1333-1336.

Ponting, J. D., Jackson, R., and Watters, G., 1972, Refrigerated apple slices: preservative effects of ascorbic acid, calcium and sulfites, *J. Food Sci.,* **37**:434-436.

Rocha, A. M. C. N., Brochado, C. M., and Morais, A. M. M. B., 1998, Influence of chemical treatment on quality of cut apple (cv. Johnagored), *J. Food Qual.,* **21**:13-28.

Sloan, A. A., 2001, Top 10 trends to watch and work on, *Food Techol.,* **55**:38-58.

Wong, D. W. S., Camirand, W. M., and Pavlath, A. E., 1994, Development of Edible Coatings for Minimally Processed Fruits and Vegetables, in: *Edible Coatings and Films to Improve Food Quality.*, J. M. Krochta, E. A. Baldwin, and M. O. Nisperos-Carriedo, eds., Technomic Publishing Co., Lancaster, PA., pp. 65-87.

QUALITY OF FRESH CITRUS FRUIT

Robert D. Hagenmaier and Robert A. Baker[1]

1. INTRODUCTION

Fresh citrus fruit, comprising oranges, grapefruit, mandarins, limes, lemons and minor varieties, are among the most popular fruits. With increasing year round competition from other fruits, maintaining this market position will require that fresh fruit quality be optimal. Determinants of fruit quality can be divided into those affecting external quality and those defining internal quality. Both of these are critical, since external quality influences initial purchasing decision, while internal quality determines consumption and repeat sale.

External quality of citrus fruit is affected by peel color, firmness, peel defects and shine. Peel color develops at maturity as the carotenoids characteristic of the species or variety are synthesized, and chlorophyll is lost. However, in warmer climates such as Florida, mature fruit peel color may continue to be obscured by chlorophyll if night temperatures do not drop below 13°C (Stewart, 1980). Gassing with ethylene is often necessary to eliminate the chlorophyll. Valencia oranges, which mature in the spring following the year of flowering, may develop full peel color during colder months even though the fruit interior is immature. Conversely, mature fruit may regreen with the advent of warmer weather, requiring steps to remove the chlorophyll. Since citrus is nonclimacteric, fruits do not undergo the rapid maturation, senescence, and softening which are characteristics of climacteric fruits. Loss of firmness in harvested fruit is often due primarily to loss of water from the peel. Although citrus fruits do have surface cuticular wax, which provides some barrier to water loss, most of this is lost during washing. Washed fruit loses moisture and firmness much faster than unwashed fruit (Hagenmaier and Baker, 1993). To compensate for this, and to enhance appearance, fruit is always coated with a commercial "wax" before shipment. This not only reduces weight loss, but provides surface shine.

[1]Robert D. Hagenmaier and Robert A. Baker, Citrus and Subtropical Products Laboratory, Winter Haven, FL 33881. South Atlantic Area, Agricultural Research Service, U.S. Department of Agriculture. Mention of a trademark or proprietary product is for identification only and does not imply a guarantee or warranty of the product by the U.S. Department of Agriculture. The U.S. Department of Agriculture prohibits discrimination in all its programs and activities on the basis of race, color, national origin, gender, religion, age, disability, political beliefs, sexual orientation, and marital or family status.

Quality of Fresh and Processed Foods, edited by Shahidi et al.
Kluwer Academic/Plenum Publishers, 2004.

Internal or eating quality of fruit is influenced by the ease of peeling and segment separation, texture, seediness, color and flavor. Peel removal and segmenting can be a significant deterrent to consumption. A recent survey found that citrus was perceived second only to pineapple in difficulty of consumption (M. Ismail, personal communication). Perhaps indicative of this, shipments of fresh Florida grapefruit have declined 43% in the last 10 years (Anon., 2000). One potential solution to this problem is marketing of citrus as a fresh-cut or minimally processed product. Production of high quality fresh-cut orange and grapefruit is possible with the use of an enzyme infusion technology which utilizes commercial pectinase enzymes to loosen the peel (Bruemmer, 1981; Baker and Bruemmer, 1989; Baker and Wicker, 1996). Peeled whole or segmented fruit may help fresh grapefruit remain a viable market commodity by expanding institutional usage.

Internal color of citrus is not usually reflected in the peel color, especially for Florida fruit, yet still affects repeat purchases. Whether for juice or consumption as whole fruit, consumers associate brighter, redder fleshed oranges and red fleshed grapefruit with higher quality (Stewart, 1980). The Rohde Red Valencia with higher internal levels of the orange-red pigment cryptoxanthin (Stewart *et al.*, 1975) is competing with and partially replacing traditional Valencia plantings. Newer varieties of red grapefruit such as Star Ruby have levels of the red pigment lycopene 11-20 times that of older red varieties such as Ruby Red, and double the level of beta-carotene (Rouseff *et al.*, 1992). Loss of lycopene during maturation is also lower in these newer varieties (Stein *et al.*, 1986). Thus, in addition to having higher color quality, these varieties could serve as a significant dietary source of these important phytonutrients.

Citrus flavor is dependent on the total soluble solids (sugars and acids), the ratio of these constituents, and volatile compounds characteristic of the particular species or variety, such as citral in lemon. In addition, flavor may be influenced by undesirable volatiles and bitter or astringent nonvolatile components. The characteristic flavor volatiles of each citrus variety are complex mixtures of terpene hydrocarbons, alcohols, aldehydes and esters, among others. Over 200 compounds have been identified and quantified in volatiles of grapefruit juice (Maarse and Visscher, 1985), and over 300 compounds in volatiles of lemon juice (Mussinan, 1981). No citrus has been studied more than orange, in attempts to characterize its unique and complex flavor. Critical to the quality of fresh fruit flavor is the balance of flavor components. It has been stated that "...the quality of citrus fruits is determined, to a considerable extent, by the presence of too much or too little of certain constituents" (Maier and Brewster, 1977). Nowhere is this more true than in orange flavor. Although over 45 compounds have been identified and quantified from orange juice (Moshonas and Shaw, 1994), no single compound or selected group of compounds can be identified as being responsible for its full flavor. However, the above authors did conclude that 9 oil-soluble and 7 water-soluble compounds were important to orange juice flavor.

Citrus fruit quality is influenced by a number of preharvest factors, including cultivar, rootstock, tree age, geographic location of production, tree nutrition, climate and fruit maturity at harvest. It is to be expected that some cultivars will have better flavor, color, or surface appearance than others. Less obvious is the influence of rootstock on fruit quality. Fast growing rootstocks such as rough lemon, Palestine sweet lime, citron and *Citrus macrophylla* may provide larger more vigorous trees, but may affect fruit quality. Fruit from cultivars grown on these rootstocks tends to be larger, but has a thicker and rougher peel, and has lower total soluble solids and acids. Conversely, fruit from scions grown on slow growing rootstocks such as trifoliate orange, citranges or citrumelos are smaller, with smoother peel, and have higher total soluble solids and acids. While fruit from younger producing trees tends to be larger than that of older trees, quality is often lower due to lower soluble solids. Desert grown fruits have much thicker peels than fruits gown in more humid environments, but often have higher flavedo color. Mineral nutrition of trees can also markedly influence fruit quality, with nitrogen affecting fruit color, and potassium and phosphorus affecting total soluble solids and acidity (Nagy and Shaw, 1990).

Postharvest handling procedures, such as ethylene degreening treatments or waxing, may have significant impacts on citrus fruit quality. Although degreening with ethylene may be necessary to break down peel chlorophyll, the process accelerates senescence, and may increase incidence of spoilage.

Fresh fruit for shipping is almost always coated with "wax" to reduce shrinkage and impart shine. The most common coatings or 'waxes' applied to the fruit in U.S. citrus packinghouses have as their major ingredients a mixture of shellac and wood rosin. These coatings were designed to impart high gloss, which the buyer often evaluates shortly after application of the coating.

When coatings are applied to fruit, they form an additional barrier through which gases must pass. Because coatings differ in gas permeance and ability to block openings in the peel, they have different effects on gas exchange (Hagenmaier and Baker, 1993). Citrus fruit with shellac-rosin coatings generally has lower internal O_2, higher internal CO_2 and higher ethanol content than fruit with wax coatings (Hagenmaier and Baker, 1994). High ethanol content, in turn, is an indication of off-flavor (Ke and Kader, 1990; Ahmad and Khan, 1987; Cohen et al., 1990). The low gas permeance of shellac-rosin coatings might promote pitting of fresh citrus fruits (Petracek et al., 1995, 1998). The current study evaluated experimental wax coatings as alternates to shellac and wood rosin-based coatings for grapefruit.

2. MATERIALS AND METHODS

Three experimental wax microemulsions were selected for application (Hagenmaier, 1998). The first was a polyethylene-candelilla wax (PE-candelilla) microemulsion composed of 12.2% polyethylene (grade A-C316 from Allied Signal Inc, Morristown, NJ), 8.9% candelilla wax (grade S.P. 75 from Strahl & Pitsch Inc., W. Babylon, NY), 3.8% oleic acid (Emersol 6321, from Henkel Corp., Cincinnati, OH), 1.1% myristic acid (Hystrene 9014 from Witco Corp., Memphis, TN or Emery 655 from Henkel Corp., Cincinnati, OH) and 1.1% NH_3, balance water. The second, a candelilla wax microemulsion contained 18.7% candelilla wax, 1.8% oleic acid, 1.0% myristic acid and 0.8% NH_3, balance water. The third was a carnauba wax microemulsion with 19.8% carnauba wax (Yellow No. 3, Strahl & Pitsch Inc.), 2.5% oleic acid, 1.7% myristic acid, 0.7% NH_3 and 0.5% polyethylene glycol (Carbowax 600, Union Carbide, Danbury, CT), balance water. These wax-based coatings were compared with non-coated fruits and also with a commercial citrus fruit coating that contained shellac and wood rosin (in proportion 3:1), fatty acid soaps, propylene glycol and other minor ingredients, but no wax.

Florida 'Marsh' grapefruit harvested at four times from February through May, three times with fruit from central Florida and twice from the 'Indian River' district were used. None of the fruit were treated with ethylene for degreening. All were cleaned with rotating polyethylene brushes, using fruit-wash detergent (type 241 or 178, Decco Div., Elf Atochem, Monrovia, CA) and dipped in 1000 ppm imazalil (Fungaflor 500EC, Janssen Pharmaceutica, Titusville, NJ). Coatings were applied to fruit by hand (latex gloves); the mean amount applied was 0.45 g per grapefruit. Fruit shine, in gloss units (G.U.) was measured with a reflectance meter (micro-TRI-gloss, BYK Gardner Inc., Silver Spring, MD), fitted with a shield having a 19 mm diameter hole (Hagenmaier and Baker, 1994). Initial reflectance measurements were made the day after application of the coatings (10 numbered fruit per treatment per experiment). Coating 'fracture' was determined subjectively after hitting together two fruit 3×, rubbing one on the other in circles 10×, twisting together 3×, wiping the contact surfaces with a black cloth, and rating the amount of coating found on the cloth (1.0 = none; 2.0 = minimal; 3.0 = significant but acceptable; 4.0 = heavy and unacceptable; and 5.0 = virtually all coating removed). Three measurements per treatment were taken.

Samples for internal O_2 and CO_2 (10 fruit per treatment, each experiment) were withdrawn with a syringe (previously flushed with N_2 to remove traces of oxygen) from fruits that were submerged in water. The O_2 and CO_2 concentrations were measured with a Hewlett Packard 5890 gas chromatograph fitted with a CTR-1 column (6 ft long, 1/4" and 1/8" diameter, outer and inner columns, respectively, Alltech, Deerfield, IL). Samples were applied with a loop injector. Column flow rate was 140 ml/min. Temperatures were 40°C and 120°C, respectively, for the column and thermal conductivity detector. Peak areas obtained from standard gas mixtures were determined before and after analysis of the samples. Oxygen concentration was calculated from the O_2-Ar peak area after correction for 0.93% Ar in the atmosphere.

Firmness was taken as the force in newtons (N) required to compress the fruit by 10 mm between two flat surfaces closing together at a rate of 20 mm/min (Model 1011, Instron Corp., Canton, MA). There were ten fruits per treatment, with firmness of each measured only once (subsequent measurements on the same fruit gave progressively lower results).

Ethanol samples consisted of two juice samples per treatment, each pooled from 5 pieces of fruit. The juice was spiked with 1000 ppm n-propanol as internal standard, centrifuged and injected into the gas chromatograph using a FFAP column (Hewlett Packard) and flame ionization detector. Column flow was 4 ml/min. Column temperature was 55°C for injection, then increased 3°C/min to 70°C and held for 1 min.

Flavor was evaluated after sniffing and tasting the juice by 14 trained panelists, who marked their rating on a 15 cm line, the low end of which was labeled 'fermented or over ripe' and the high end 'very fresh or very good.' For training purposes, juice from freshly harvested fruit and from stored fruit with elevated ethanol content was also presented. Weight loss samples consisted of three trays each with 3 fruit for each treatment, weighed one and 8 days after application of the coating.

Statistix 4.1 (Analytical Software, Tallahassee, FL) was used for computation of statistical parameters. Standard errors (S.E.) are shown as error bars on the graphs except when covered by the symbols. The changes in weight and gloss that occurred during storage were computed as changes on numbered fruit, and treated as paired comparisons. The reported probabilities (p) are for the null hypotheses.

3. RESULTS AND DISCUSSION

The parameters related to internal quality, namely flavor, ethanol and internal gas concentrations, were most different for non-coated fruit and the shellac-rosin coating. The mean values for the three experimental wax coatings (carnauba wax, candelilla wax and polyethylene-candelilla wax) were similar to one another and intermediate between non-coated fruit and the shellac-rosin coating. The carnauba coating tended to fracture more than the other two wax coatings, and the candelilla-coated fruit had lowest weight loss. Candelilla gloss decreased less than any other treatment during storage (Table 1). Linear regression on harvest date showed that ethanol, flavor, internal gases, firmness and other properties were not significantly affected by time of harvest.

Table 1. Mean properties of 'Marsh' grapefruit after storage at 21°C [a].

Treatment	Flavor score	EtOH[b] (ppm)	Firm-ness (N)[b]	Fracture score	Internal gases		Gloss		Wt. loss (%)/day
					CO_2 (%)	O_2 (%)	Initial (GU)	Change (GU)	
Shellac-Rosin	6.2a	1061a	50a	4.5a	10.6a	0.9a	12.4a	-2.9a	0.38b
Candelilla	8.3b	676ab	69b	1.1b	7.5ab	9.3b	10.7b	-0.8d	0.08d
Carnauba	7.8ab	439bc	67b	4.3a	5.5bc	11.7bc	11.1ab	-1.0c	0.18c
PE-Cand	9.1b	416bc	65b	1.3b	4.8bc	10.9bc	11.7ab	-1.6b	0.17c
No coating	9.0b	265c	48a	1.0b	3.5c	14.7c	7.5c	-1.0c	0.57a

[a] Storage times for flavor, ethanol, and fracture: 8 days ; for firmness and internal gases: 13 days; for gloss: 1 and 14 days. Weight measured days 1, 8 and 14. Values in a column are not different if they share a common letter (p>0.05, Tukey).
[b] The mean values of ethanol and firmness on the day coatings were applied were 108 ppm and 71.5 N, respectively.

Grapefruit coated with the shellac-rosin coating had many disadvantages: lowest flavor score, highest internal CO_2, lowest internal O_2 and highest ethanol. In addition, compared to the other coated fruit it had least firmness, highest decrease in gloss during storage, and more fracture than the candelilla-containing wax coatings (Table 1, Fig. 1-2). Gloss measured one day after application of the coating was highest for the shellac-rosin coating, but this advantage was lost during storage (Table 1). The gloss of grapefruit with any of the three experimental wax coatings after 14 days storage at 21°C was not significantly different from that of fruit with the shellac-rosin coating.

The non-coated grapefruit did not have the coating present as an extra barrier to diffusion of gases to and from the fruit and for that reason had internal gas concentrations more like the surrounding atmosphere than any of the coated fruit, and for the same reason these fruit also had the fastest weight loss (Table 1). Candelilla wax, well-known to be a good moisture barrier, gave the lowest weight loss (Table 1).

Considering the data for all coatings, the ethanol content after 8 days of storage increased nearly linearly with internal CO_2 (Table 1, Fig. 1). The correlation coefficient was 0.81 (p < 10^{-5}) with a slope of 108 ± 17 ppm/% CO_2. Statistical separation of means was about the same for ethanol content and mean internal CO_2, which suggests that the two measurements have similar information value (Table 1).

Ethanol content was also related to internal O_2 concentration. At values of internal O_2 <1%, the mean ethanol content was 1220 ppm. At O_2 ≥1%, the mean ethanol content was only 470 ppm and the correlation between ethanol and internal oxygen was not significantly different than zero (data not shown).

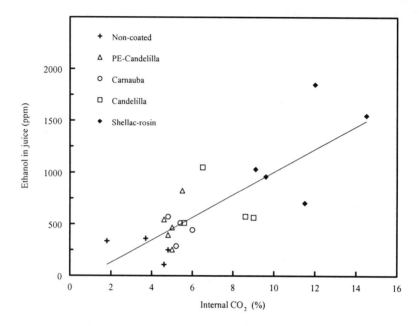

Figure 1. Mean ethanol content of the juice from 'Marsh' grapefruit stored 8 days at 21°C. The mean ethanol content before storage was 108 ppm.

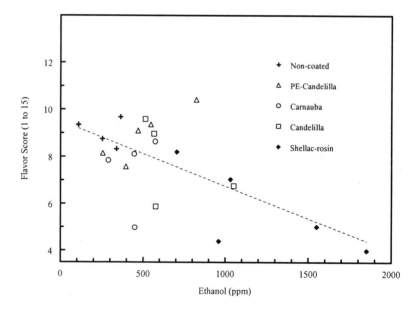

Figure 2. Mean flavor of the juice from 'Marsh' grapefruit stored 8 days at 21°C .

Grapefruit flavor decreased 2.9±0.9 units per 1000 ppm increase in ethanol, (r = −0.61, p<0.005; Fig. 2). Measurement of ethanol content was much less time-consuming than determining taste panel scores, and also seemed a more objective measurement. Statistical separation of means was slightly different for these two measurements, but both showed no difference between wax-based coatings for flavor or ethanol content.

The polyethylene-candelilla combination is intended to balance the relative advantages and disadvantages of these two waxes. Polyethylene wax has relatively high oxygen permeance (Hagenmaier and Shaw, 1991), and the high-melting, brittle types tend to have high gloss. Candelilla wax, with its relatively low melting point of about 68°C tends to be somewhat soft and therefore less brittle, and it is a good barrier to water vapor (Table 1). Mixtures of these two waxes had some of the advantages of both (Table 1) and were also relatively easy to emulsify. The experimental carnauba wax coating was somewhat too fragile (Table 1), although that problem has been overcome with commercial carnauba waxes by including some shellac in the formulation.

4. CONCLUSIONS

Citrus fruit quality is affected by a number of pre- and post-harvest influences. Some of these, such as cultivar, rootstock, and geographical location, are determined at time of planting; others, such as the weather, are beyond control. However, postharvest treatment of fruit, particularly waxing, can significantly influence stored fruit quality. Current citrus coatings containing shellac and rosin may adversely affect fruit flavor. 'Marsh' grapefruit with a polyethylene-candelilla wax coating had relatively good gloss and flavor after storage at 21°C. With a candelilla wax coating, the flavor was good and weight loss was very low. A typical, commercial shellac-rosin resulted in good short-term gloss but relatively poor flavor at this temperature.

5. REFERENCES

Ahmad, M., and Khan, I., 1987, Effect of waxing and cellophane lining on chemical quality indices of citrus fruit, *Plant Foods Human Nutr.,* **37**:47-57.

Anon., 2000, Fresh grapefruit- tomorrow and today. *Citrus Industry*, **81**(11):6-10.

Baker, R. A., and Bruemmer, J. H., 1989, Quality and stability of enzymically peeled and sectioned citrus fruit, in: *Quality Factors of Fruits and Vegetables--Chemistry and Technology*, J. J. Jen, ed., ACS Symposium Series 405, American Chemical Society, Washington, DC, pp.140-148.

Baker, R. A., and Wicker, L., 1996, Current and potential applications of enzyme infusion in the food industry, *Trends in Food Science and Technology*, **7**:279-284.

Bruemmer, J. H., 1981, Method of preparing citrus fruit sections with fresh fruit flavor and appearance, U.S. Patent No. 4,284,651.

Cohen, E., Shalom, Y., and Rosenberger, I., 1990, Postharvest ethanol buildup and off-flavor in 'Murcott' tangerine fruits, *J. Am. Soc. Hort. Sci.,* 115:775-778.

Hagenmaier, R. D., and Shaw, P. E., 1991, Permeability of coatings made with emulsified polyethylene wax, *J. Agric. Food Chem.,* **39**:1705-1708.

Hagenmaier, R. D., and Baker, R. A., 1993, Reduction in gas exchange of citrus fruit by wax coatings, *J. Agric. Food Chem.,* **41**:283-287.

Hagenmaier, R. D., and Baker, R. A., 1993, Cleaning method affects shrinkage rate of citrus fruit, *HortScience,* **28**:824-825.

Hagenmaier, R. D., and Baker, R. A., 1994, Internal gases, ethanol content and gloss of citrus fruit coated with polyethylene wax, carnauba wax, shellac or rosin at different application levels, *Proc. Fla. State Hort. Soc.,* **107**:261-265.

Hagenmaier, R. D., 1998, Wax microemulsion formulations used as fruit coatings, *Proc. Fla. State Hort. Soc.,* **111**:251-255.

Ke, D., and Kader, A. A., 1990, Tolerance of 'Valencia' oranges to controlled atmospheres as determined by physiological responses and quality attributes, *J. Amer. Soc. Hort. Sci.,* **115**:779-783.

Maarse, H., and Visscher, C. A., 1985, *Volatile Compounds in Food- Quantitative Data, Vol. 4.* TNO-CIVO Food Analysis Institute, Zeist, The Netherlands.

Maier, V. P., and Brewster, L. C., 1977, The biochemical basis of fruit composition and quality, *Proc. Int. Soc. Citriculture,* **3**:709-711.

Moshonas, M. G., and Shaw. P. E., 1994, Quantitative determination of 46 volatile constituents in fresh, unpasteurized orange juices using dynamic headspace gas chromatography, *J. Agric. Food Chem.,* **42**:1525-1528.

Mussinan, C. J., Mookherjee, B. D., and Malcolm, G. I., 1981, Isolation and identification of the volatile constituents of fresh lemon juice, in: *Essential Oils,* B. D Mookherjee and C. J. Mussinan, eds., Allured, Wheaton, IL, pp. 198-228.

Nagy, S., and Shaw, P. E., 1990, Factors affecting the flavour of citrus fruit, in: *Food Flavours. Part C: The Flavour of Fruits,* I. D. Morton and A. J. MacLeod, eds., Elsevier, pp. 93-124.

Petracek, P. D., Wardowski, W. F., and Brown, G. E., 1995, Pitting of grapefruit that resembles chilling injury, *HortScience,* **30**:1422-1426.

Petracek, P. D., Dou, H., and Pao, S., 1998, Influence of applied waxes on postharvest physiological behavior and pitting of white grapefruit, *Postharvest Biol. Technol.,* **14**:99-106.

Rouseff, R. L., Sadler, G. D., Putnam, T. J., and Davis, J. E., 1992, Determination of beta-carotene and other hydrocarbon carotenoids in red grapefruit cultivars, *J. Agric. Food Chem.,* **40**:47-51.

Stein, E.R., Brown, H.E., and Cruse, R.R., 1986, Seasonal and storage effects on color of red-fleshed grapefruit juice, *J. Food Sci.,* **51**:574-576.

Stewart, I., 1980, Color as related to quality in citrus, in: *Citrus Nutrition and Quality,* S. Nagy and J. A. Attaway, eds., ACS Symposium Series No. 143, American Chemical Society, Washington, DC, pp. 129-149.

Stewart, I., Bridges, G. D., Pieringer, A. P., and Wheaton, T. A., 1975, 'Rohde Red Valencia', an orange selection with improved juice color, *Proc. Fla. State Hort. Soc.,* **88**:17-19.

EVALUATION OF WATER WASHES FOR THE REMOVAL OF ORGANOPHOSPHORUS PESTICIDES FROM MAINE WILD BLUEBERRIES

Russell A. Hazen, L. Brian Perkins, Rodney J. Bushway, and Alfred A. Bushway[*]

1. INTRODUCTION

Organophosphate pesticides (OPs) are derivatives of phosphoric acid with a basic structure shown in Figure 1. The majority of OPs are insecticides and have been used extensively in agriculture, commercial and residential applications since the late 1970's, when organochlorine pesticides began to fall out of favor (Organophosphate Fact Sheet, 1996). Although the use of carbamate insecticides has grown dramatically throughout the past decade, the OPs continue to be insect control staples, with worldwide sales topping $2.9 billion in 1992 (Agrow, 1994). The organophosphates tend to be much more toxic to vertebrate populations, but are also as effective and much less expensive than the carbamate class of insecticides.

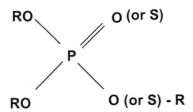

R = methyl or ethyl (vast majority)   OR = leaving group
                                        alkyl, alkoxy, alkylthio, aryl or heterocylclic
                                        aryloxy, arylthio or a hetercyclic analog

Figure 1. Basic structure of an organophosphate pesticide

[*] Russell A. Hazen and L. Brian Perkins are Ph.D. Candidates in Food and Nutrition Sciences, Rodney J. Bushway and Alfred A. Bushway are Professors of Food Science, University of Maine, Orono ME 04469-5736

Quality of Fresh and Processed Foods, edited by Shahidi et al.
Kluwer Academic/Plenum Publishers, 2004.

Until recently, little concern was given to the cumulative or chronic effects of the OP insecticides on human populations. During the past decade the general public has shown renewed concern about exposure to all pesticides, with particular apprehension to residues in the food supply. The government publication, *Pesticides in the Diets of Infants and Children* (National Academy of Sciences, 1993), has certainly exacerbated these fears. Additionally, in 1997 the Environmental Protection Agency (EPA) prioritized its reassessment process for all pesticides by listing most of the commonly used OPs as "Priority-Group 1" (Raw and Processed Food Schedule for Pesticide Tolerance). The EPA's recent decision to severely restrict the use of chlorpyrifos (Environmental Protection Agency, 2000), may be a good indication of the regulations that other related insecticides face in the near future.

The organophosphate insecticides registered for use in Maine's wild blueberry industry include phosmet, azinphos-methyl, malathion, and diazinon. Growers rely especially heavily on both phosmet and azinphos-methyl to control blueberry maggot. Because these two insecticides are used close to harvest, it is common to find residues ranging to near-tolerance levels on a large percentage of unprocessed fruit (Bushway and Perkins, 1999). The bulk of the berries are individually quick frozen (IQF) after a chlorinated wash that is designed to remove debris and to lower the microbial load. Although it is assumed that such processing also removes the bulk of pesticide residues, no studies have been performed to confirm this.

This chapter details work which examines the effectiveness of distilled water and water containing 100 and 200 ppm active chlorine washes on the removal of phosmet, azinphos-methyl, malathion and diazinon from blueberry fruit.

2. MATERIALS AND METHODS

2.1 Fortification

Twenty-seven kilogram of fresh, pesticide-free blueberries were purchased form a certified-organic grower in Deblois, Maine. Analytical grade phosmet, azinphos-methyl, malathion, and diazinon were obtained from the EPA repository, Fort Mead, MD. Stock solutions were prepared by weighing 50 mg of each insecticide into separate 50 ml volumetric flasks. Each Stock solution was brought to volume with ACS grade methanol. To create a low fortification solution, 1.0 ml phosmet, 0.5 ml azinphos-methyl, 0.8 ml malathion, and 0.05 ml diazinon were removed and mixed with 50 ml of methanol. This process was repeated with 20 ml phosmet, 10 ml azinphos-methyl, 16 ml malathion, and 1.0 ml diazinon to make the high fortification solution.

The device used to spray the blueberries with the pesticide solutions was created by attaching the squeeze trigger sprayer from an ordinary (unused) household spray bottle, to a 50 ml screw-top test tube. Teflon tubing was attached to the sprayer and extended to the bottom of the test tube.

Two 2 kg lots of fruit were spread on 30 x 60 stainless steel trays. One tray was inoculated at 0.5 phosmet, 0.25 ppm azinphos-methyl, 0.4 ppm malathion, and 0.025 ppm diazinon, by spaying the berries with the entire low spiking solution. The other tray of berries was treated in an identical manner with the high spiking solution, resulting in theoretical residual levels equal to the EPA tolerances; 10.0 ppm phosmet, 5.0 ppm azinphos-methyl, 8.0 ppm malathion, and 0.5 ppm diazinon. After fortification, the samples were allowed to sit overnight before processing, to allow as much residual methanol as possible to evaporate.

2.2 Processing

The processing and sampling scheme for this study is depicted in Figure 2. Three-350 g sub-samples were removed from the low-spiked berries and spread on 3-30 x 30 cm wire screens. After removing a control for immediate pesticide analysis from each tray, the berries were sprayed with 500 ml of either distilled water, 100 ppm chlorine, or 200 ppm chlorine. Chlorine solutions were prepared using household bleach as specified by Camire (1994). After one minute of contact with the washing solutions, a sample was removed for pesticide analysis. After blast freezing for 30 min, fruit from the three wash treatments were once again removed for residue analysis. The entire wash and analysis procedure was repeated for the blueberries fortified at a high level with the insecticides.

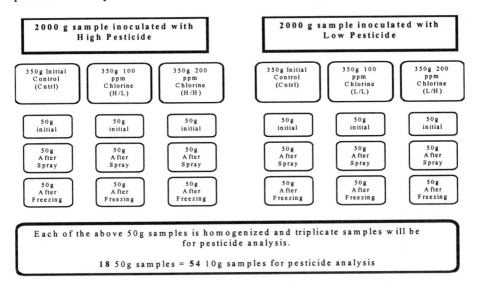

Figure 2. Sampling Scheme for Fortified, Processed Blueberries

2.3 Analysis

Analysis of the blueberry samples was accomplished by using a (GC-AED) method developed for the determination of organophosphate insecticides in a variety of fruits and vegetables (Bushway and Fan, 1998). The GC-AED was a Hewlett Packard (HP) model 5890 Series II chromatograph interfaced with a HP 5921 atomic emission detector and a HP 7673 autosampler. A HP-5 capillary column (cross-linked 5% methyl phenyl silicone; 25 m by 0.32 mm by 0.017 um film thickness) was used for analyte separation. Data were processed with a HP AED 5890 Chemstation. GC-AED conditions were as follows: carrier gas, He at a flow rate of 1.0 ml/min; makeup gas, He at a flow rate of 75 ml/min; reagent gases, O_2, H_2. The system was operated in the splitless mode with the injector temperature set at 250°C. Two microliters of sample were injected into the system, with the following temperature profile: initial temperature, $100°$ for 0.5 min, followed by an increase of $30°$ per min to a final temperature of $250°$. The final temperature was held for 4 min. The detector temperature was $250°$.

A simple extraction and solid phase extraction (SPE) were used to prepare the samples for analysis. Fifty grams of fruit were homogenized in a blender. A 10 g sub sample was weighed into a 50 ml disposable polypropylene centrifuge tube followed by the addition of 20 ml of ACS grade methanol. The combined substances were then extracted for 3 min with a Polytron homogenizer (Kinematica, CH-6010 Kriens, Lucerne, Switzerland) and then centrifuged for 10 min at 5,000 x g. A 10 ml aliquot was removed and added to 90 ml of distilled water. The entire 100 ml mixture was passed through an activated tC18 Sep-Pak SPE cartridge (Waters Assoc., Milford, MA) after an activation sequence of 5 ml methanol and 5 ml of water. After rinsing the loaded Sep-Pak with 5 ml of water, the cartridge was dried for 20 min under vacuum before eluting with ethyl acetate. The first 1.0 ml of eluate was collected for GC-AED analysis.

2.4 Statistical Analysis

The mean and standard deviation of each of the pesticide treatments were determined using Microsoft©Excel 97. Statistical analysis was conducted using ANOVA and Tukey's HSD Multiple comparisons in the SYSTAT© 8.0 (SPSS Inc., 1998) software package.

3. RESULTS

Means and standard deviations for each insecticide before and after wash treatments are given in Table 1. All of the washing processes resulted in a significant ($P<0.05$) reduction in residual levels for all four insecticides tested. Distilled water rinses averaged a reduction of 29.66%. Washes were more effective on samples receiving near tolerance levels of the four insecticides examined. Figure 3 uses Phosmet to illustrate this finding, showing that distilled water rinses resulted in greater reductions at the high spiking level (27.9-45.0%) than at the low spiking level (21.6-29.0%). This phenomenon may be partially explained by the lack of surfactants and stickers in our spiking solutions that would be present in the formulations used in actual agricultural situations. Although they did not publish the actual reductions, a recent study by Krohl *et al.* (2000) showed similar results for malathion and diazinon for several fruits and vegetables collected from the field and washed with tap water.

The addition of chlorine to the wash water at 100 and 200 ppm concentrations each resulted in an additional mean reduction of 15% for all of the insecticides tested. Although these findings were not statistically significant, they showed a trend that should be further examined. Sample size and sample replication should be increased to allow for sample variability. The goal of this work was an initial investigation designed to mimic the processing plant conditions which blueberries were subjected to, where chlorine concentrations were very low (100-150 ppm), and contact times were minimal (approximately 1 min). There are few published studies which examine the effects of chlorine washes on OP reduction, but, once such paper showed 50-100% reductions of Azinphos-methyl on apples washed with 500 ppm chlorinated water (Ong, et al., 1996). Most of chlorine/OP degradation literature describes research with drinking water and

Table 1. Means and standard deviations for each insecticide before and after wash treatments given in parts per billion (ppb)

		DI Water - 5% Tolerance		100 PPM - 5% Tolerance		200 PPM - 5% Tolerance	
		Initial	After Spray	Initial	After Spray	Initial	After Spray
Diazinon	Mean[a]	53.9	65.1	72.1	66.6	75.1	49.8
	SD	27.5	1.6	11.1	10	10.4	6.6
Malithion	Mean	356.9	358.6	429.5	364.6	447.8	315
	SD	151	8.3	32.4	24	24.2	17.9
Phosmet	Mean	620	485.9	491.4	411.6	517.3	367.3
	SD	227.3	18.2	3.9	16.5	61.4	22.8
Azinphosmethyl	Mean	57.1	193.7	225	198.6	261.3	179.6
	SD	12	39.9	12.8	26.7	34.1	20.9
		DI Water - Near Tolerance		100 PPM - Near Tolerance		200 PPM - Near Tolerance	
		Initial	After Spray	Initial	After Spray	Initial	After Spray
Diazinon	Mean	199.7	141.8	292.6	158.3	227.9	121.4
	SD	19.3	7.8	12.1	17.2	13	4.9
Malithion	Mean	1568.9	1103.7	2250.8	1330.4	1859.7	1043.1
	SD	79.9	45.1	90.7	82.9	87.4	36.4
Phosmet	Mean	2413.2	1740.4	3462.2	2005.9	2808.4	1545.8
	SD	161.1	82	152.9	132.4	159.1	58
Azinphosmethyl	Mean	772	523.8	1162.6	605.9	917.8	465.1
	SD	95.2	14.9	107.9	43.8	50.9	26.9

[a]Values are the mean of three samples

sewage treatment, where chlorine concentrations tend to be much higher (4000 ppm) and contact times (15-45 min) are much longer (Frederick, 1997; Tehobanoglous and Burton, 1991).

Food processors currently employ chlorine washing in an effort to reduce the microbial load common to fresh produce. However, recent work has shown that this process may only be minimally effective, producing reductions equivalent to deionized water (Beuchat, 1998). In fact, recently completed studies by the author have also found chlorine sprays to be only marginally effective at reducing microbial load on blueberries (Hazen et al., 2001) resulting in no statistically significant differences in either total aerobic mesophilic bacteria or yeast (Hazen, 2000) after exposure to 100 and 200 ppm chlorine sprays. This ineffectiveness could be due to the quantity of soluble organic matter commonly associated with delicate fruits and vegetables. Freezing of samples generally resulted in a further small reduction in residues for all insecticides studied; however these reductions were not significant ($p > 0.05$).

It should also be noted that interactions between chlorine and organophosphorous pesticides might often result in the formation of highly toxic oxygen analogs. In fact, a study which followed the oxidation of diazinon in aqueous systems showed that the activation energy required for oxidation to diazoxon was much lower than that necessary for the hydrolysis of diazinon (Zhang and Pehkonen, 1999).

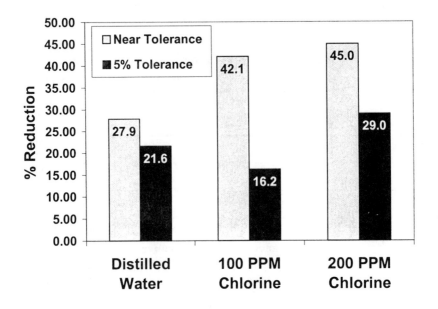

Figure 3. %Reduction for Phosmet After Exposure to Water Washes

4. CONCLUSION

Washing wild blueberries with ordinary water can clearly reduce organophosphate residues. The effectiveness of active chlorine, hydrogen peroxide and commercially available wash products designed for home consumer use must be evaluated. Future studies should concentrate on phosmet and its oxygen analog, since this insecticide is one of the most prevalent among wild blueberry growers.

5. REFERENCES

Organophosphate Fact Sheet, *Pesticide News*, 1999, 34.
    http://www.gn.ape.org/pesticidestrust/artacts/organoph.htm.
Agrow No. 199, January 7, 1994.
National Academy of Sciences, *Pesticides in the Diets of Infants and Children* (National Academy Press, 1993).
Raw and Processed Food Schedule for Pesticide Tolerance, *Federal Register* 6(149).
Environmental   Protection   Agency,   Office   of   Pesticide   Programs.   (June   8,   2000)
    http://www.epa.gov/pesticides/op/chlorphrifos/consumerqs.htm.
Bushway, R., and Perkins, B., 1999, Analysis of Blueberries for Pesticide Residues, University of Maine,
    Department of Food Science, Unpublished data.
Camire, A., *Food Safety News,* 1994, Facts on Using Chlorine in Food Manufacturing Sanitation.
Bushway, R. J., and Fan, Z., 1998, Complementation of GC-AED and ELISA for the determination of diazinon
    and chlorpyrifos in fruits and vegetables, *J. Food Protec.,* **61**(6):708-711.
Krohl, W. J., Arsenault, T. L., Pylpiw, H. M., Jr., and Incorvia-Mattina, M. J., 2000, Reduction of pesticide
    residues on produce by rinsing, *J. Agric. Food Chem.,* **48**:4666-4670.
Ong, K. C., Cash, J. N., Zabik, M. J., Siddiq, M., and Jones, A. L., 1996, Chlorine and ozone washes for
    pesticide removal from apples and processed applesauce, *Food Chem.* **55**(2):153-160.

Frederick, P. W., 1997, Drinking water disinfection with chlorine, *Health and Envir. Digest.*

Tehobanoglous, G. and Burton, F. F., *Water Engineering* 3rd ed. (McGraw-Hill, New York, 1991) pp. 501.

Beuchat, L. R., 1998, Surface decontamination of fruits and vegetables eaten raw: a review. Food Safety Unit, World Health Organization, Food safety issues. "WHO/FSF/FOS/98.2." (1998)

Hazen, R. A., Bushway, A. A., and Davis-Dentici, K., 2001, Evaluation of the Microbiological Quality of IQF Processed Maine Wild Blueberries: an in plant study, *Small Fruits Review*, 1:47-59.

Hazen, R. A., 2000, Effects of Chlorine Spray and Freezing on the Microbiological Quality of Maine Wild Blueberries, Thesis, Department of Food Science and Human Nutrition, University of Maine, Orono, ME.

Zhang, Q., and Pehkonen, S. O., 1999, Oxidation of diazinon by aqueous chlorine: kinetics, mechanics and product studies, *J. Agric. Food Chem.*, 47:1760-1766.

SUGAR QUALITY IN SOFT DRINK MANUFACTURE: THE ACID BEVERAGE FLOC PROBLEM

Les A. Edye[*]

1. INTRODUCTION

A floc or floccule is a small portion of matter resembling a tuft of wool or a wispy cloud. In soft drink or acidic beverage manufacture the term is used to describe a visible defect in the product. This visible defect may be particulate and sedimentary or tuft-like and suspended in the beverage, and may be attributed to microbial contamination or to water and sugar ingredients that are of unsuitable quality for beverage manufacture. Microbial contamination of soft drinks is considered to be outside the scope of this review. Simple tests of microbial contamination and a review of the microbiology of soft drinks are described by Ashurst (1998).

The emphasis in this contribution is focused on floc in soft drinks that is not a result of microbial contamination. This type of floc is referred to by the sugar and beverage manufacturing industries as acid beverage floc (ABF). While ABF is harmless, it is nevertheless a visible defect and consumers reject the soft drink product for aesthetic reasons. Since in many parts of the world soft drink manufacturers are large consumers of sugar (sucrose) and sugar is implicated in ABF formation, the issue of ingredient quality and ABF is a major concern to both soft drink and sugar manufacturers.

In the late 70's, 25% of US sugar production was consumed by soft drink manufacturers, and consequently, ABF was a major issue to the US industry. During this time researchers at the Sugar Processing Research Institute Inc. (SPRI, then the Cane Sugar Refining Research Project Inc.) conducted extensive research on the nature and cause of ABF in soft drinks manufactured from cane sugar. Clarke, Godshall, Roberts, and Carpenter are the authors of work produced by the SPRI during this period. Their published work is a major part of literature on this subject and is given coverage here. In the early 80's, high fructose corn syrup (HFCS) consumption by US soft drink manufacturers increased from 0.8 to over 3 million tons, with a consequent decrease in sucrose consumption to less than 300,000 tons (ca. 5% of production). In November 1984, major US soft drink companies announced that 100% HFCS would be used in their products (Moore *et al.*, 1991). ABF was no longer an issue to the US sugar industry and research in the US on this subject, previously considered essential, effectively ceased.

Both beet and cane sugar are reported to form ABF. The similarities and differences between ABF from cane and beet sugar are discussed. Where sugar is implicated in ABF formation, it is principally through the precipitation of polysaccharide impurities in sugar that are sparingly soluble in the acidic beverage. Therefore, the nature of polysaccharide

---

*Sugar Research Institute, Mackay, Australia

Quality of Fresh and Processed Foods, edited by Shahidi et al.
Kluwer Academic/Plenum Publishers, 2004.

impurities in sugar is considered. Chill haze in alcoholic beverages, caused by the interaction of polyphenols and proteins, also is relevant to ABF and is discussed here.

2.   THE SOFT DRINK MANUFACTURER'S PERSPECTIVE

From the perspective of the soft drink manufacturer, the appearance of ABF in soft drinks is a quality problem that is not easily managed. There is no easy, quick and reliable test for ABF in the sugar ingredient. When ABF is manifest in soft drinks the product is usually at the point of retail. While there is a strong link between ABF and sugar quality, ABF is a complex coacervate containing material from the sugar, water and most likely also the flavored syrup ingredients. Furthermore, the mechanism of ABF formation is still not well established. While to some extent water quality plays a role in ABF formation, nevertheless, no floc guarantees in sugar purchase contracts are often required by soft drink manufacturers. When ABF appears in soft drink, the sugar manufacturer is quickly blamed and compensation is sought.

Taylor (1998) describes the ingredients of soft drinks and briefly mentions an unsightly precipitate in soft drinks resulting from algal polysaccharides and polypeptides, and humic acids in the water supply. Taylor (1998) failed to acknowledge the contribution of sugar products to soft drink floc and this may be part of the parley between beverage and sugar manufacturers.

Taylor (1998), however, described the contents of soft drink syrup formulations. Food grade saponins are used to improve the foaming characteristics of cola and other formulations, and stabilizers, such as alginates, carrageen, vegetable gums, pectin, acacia, guar, tragacanth, xanthan and carboxymethylcellulose are used to improve mouthfeel, increase viscosity, and stabilize natural cloudiness (and in fruit drinks to disperse fruit solids). It is not surprising that it has been difficult to elucidate the chemistry of acid beverage floc formation and the contribution of sugar impurities to its cause, given that polysaccharides and saponins may be present in the syrup formulations as well as in the sugar as impurities (albeit at mg/kg quantities in both cases).

The section on carbonated beverages (Jones, 1978) in the Kirk-Othmer Encyclopaedia of Chemical Technology also lists and describes the ingredients of soft drinks. Poor quality sugar is noted to have extremely detrimental effects on beverage taste, odor and stability. Quality control is discussed and the effects of algae in the water supply mentioned to cause a sediment. No mention is made of ABF. However, in terms of quality control there is very little the beverage manufacturer can do about ABF, as the beverage would likely be on the retail shelf before the problem is manifest.

Hammond (1998) also states, 'The most common cause for floc formation in clear soft drinks is microbial growth.' He also attributes another cause of floc to algal polysaccharides in the water supply. Certainly, these possibilities should be eliminated before the sugar supply is implicated in floc formation.

3.   FLOC IN SOFT DRINKS FROM BEET SUGARS

As early as 1952, researchers at Spreckels Sugar Company, a Californian beet processor, had isolated floc from simulated acid beverages and found the floc to contain saponins and small amounts of high molecular weight materials (most likely pectin) (Eis et al., 1952). The saponins were determined to be the glycosides of the triterpene oleanolic acid. Subsequent to this work, Ridout et al., (1994) elucidated the structures of three beet saponins (all were glucuronic acid glycosides of oleanolic acid). Walker and Owens (1953) isolated a precipitate from acidified ABF positive beet sugar and found it to contain saponins, fats, carbohydrates, colloidal carbon particles from the manufacturing process and silica. Walker and Owens believed that saponins were primarily responsible for floc formation, and that the surfactant saponin carried fats into the refined sugar and the floc. Silica and colloidal carbon particles were believed to be swept up by the floc aggregate, rather than play active roles in the floc formation mechanism.

Clarke et al. (1997) were able to form ABF-like flocs in simulated beverages containing non-flocculating beet sugars by the addition of a methanolic extract of sugar beets or a beet saponin plus a protein. They proposed a mechanism based on interaction of negatively charged saponins or polysaccharides and positively charged proteins as an initial step in ABF formation. While the mechanism is prima facie believable, the evidence supporting the mechanism (viz., that a mixture of a methanolic extract of sugar beets or a beet saponin plus a protein forms floc) is by no means proof that ABF forms in this way in authentic soft drinks.

The ABF problems resulting from the use of beet sugar in soft drink manufacture have been essentially overcome by minimising the amount of saponin impurity in the sugar product. Since saponins are primarily in the leaves rather than the roots of beets, attention to topping (removal of the leafy head of the plant) during harvest significantly reduces saponins in beet processing streams. Diffusion at lower pH, carbonatation clarification, and ion exchange resins and activated carbon treatments all to some extent remove saponins from beet processing liquors.

4. FLOC IN SOFT DRINKS FROM CANE SUGAR

Stansbury and Hoffpauir (1959) isolated ABF from cane sugar sweetened soft drinks. Analysis of the floc revealed the presence of starch, lipids (wax), particles of activated carbon from refining, protein, and ash (principally silica). They concluded that ABF formation was initiated by adsorption of solutes onto the carbon particles, but later this was shown by Cohen et al. (1970) not to be the case.

Cohen et al. (1970) surveyed the impurity composition of raw and refines cane sugar, and investigated the nature of ABF in cane sugar. ABF was found to contain protein, inorganic material and polysaccharides. No correlation was found between floc and inorganic matter or polysaccharide concentrations. They also reported that all ABF forming sugars also developed a haze on addition of ethanol, but not all alcohol haze forming sugars would develop ABF.

Miki et al. (1975, 1980, 1984) isolated ABF from beverages made from Australian, Philippine, Cuban and South African raw sugars. All flocs contained polysaccharides (23.7 to 56.4%), silicates (24.8 to 43.2%) and protein (5.6 to 25.7%). The polysaccharide component contained glucose, mannose and galactose (and some arabinose, xylose and rhamnose). They postulated the existence of a galactoglucomannan, or a mixture of a galactomannan and a glucomannan from sugar in the ABF complex. Treatment of the polysaccharide component with starch and dextran hydrolysing enzymes reduced the glucose content, and after gel permeation chromatography a galactomannan was obtained. Miki (1984) proposed two possible structures for the galactomannan, viz., a backbone of $\alpha(1\text{-}6)$ and $\alpha(1\text{-}2)$ mannopyranosyl structure with single $\alpha(1\text{-}2)$ galactosyl side chains, or a backbone of $\alpha(1\text{-}6)$ mannopyranosyl structure with $\alpha(1\text{-}2)$ mannopyranosyl side chains terminating with either $\alpha(1\text{-}2)$ mannopyranosyl or $\alpha(1\text{-}2)$ galactopyranosyl units. The isolation of the galactomannan does not preclude the existence of a galactoglucomannan or a glucomannan in the floc complex.

Roberts and Carpenter (1974) isolated floc from simulated beverages and found it to contain silica (63.5%), fats or waxes (5.26%), starch (5.50%) and other organic compounds including protein and other polysaccharides. The polysaccharide fraction was further purified, hydrolysed and analysed. The polysaccharides consisted of arabinose (0.63%), rhamnose (0.48%), xylose (0.69%), mannose (1.21%), galactose (0.58%) and glucose (14.4% - all numbers based on % of original floc material). The amino acid composition of proteins in ABF also is reported.

Morel du Boil (1997) has reviewed the published work on ABF from cane sugar and identified types of floc (viz., alcohol haze in sweetened alcoholic beverages, precipitation of polysaccharides and proteins, silicate floc from water supply and microbial contamination). Morel du Boil's commentary with 48 references is a summary of the work of others, rather than a critical review. She subscribes to the mechanism proposed by Clarke et al. (1977) described below, and seems to disregard the work of Miki et al. (1975, 1980, 1984).

Dunsmore et al. (1978) investigated the removal of ABF forming constituents in refining processes, e.g., enzyme treatment, clarification and filtration. Clarification by carbonatation removed floc, but phosphatation (with and without flocculants) did not. Proteolytic enzymes were not suitable because pH and temperature optima restricted their use in refining. Syrups filtered (0.45 to 1.20 μm) at temperatures below 60 °C did not floc, and those filtered (0.45 μm) even at 80 °C did not floc. Filtering through celite and kieselguhr below 60 °C removed floc. Thereby, they were able to report practical solutions to prevention of floc formation, i.e., carbonatation clarification, and/or filtering at temperatures below 60 °C. However, they were unable to explain the reasons why these practical solutions appeared to work.

There is consensus among researchers that ABF in cane sugar sweetened soft drinks is a complex coacervate of silica, polysaccharides, protein, waxes and other organic compounds. Certainly, most of the polysaccharides, waxes and some of the proteins and silica are present as impurities in refined cane sugar (albeit at concentrations less than 100 mg/kg). Much of the research on ABF is concerned with the elucidation of polysaccharide structures in the floc coacervate. This is not surprising given that it is most likely that a sparingly soluble high molecular weight polysaccharide is pivotal to initial floc forming mechanisms. For this reason polysaccharide impurities in refined cane sugar are considered part of the scope of this review, and are summarised below.

5. POLYSACCHARIDES IMPURITIES IN REFINED CANE SUGAR

The polysaccharides impurities in cane sugar may be either indigenous to the cane plant or result from microbial activity in the field or the factory. The types of polysaccharides present in cane sugar and their sources have been well researched and are described in several texts (e.g., Gratius et al.,, 1995; Kitchen 1988). The polysaccharide impurities of cane sugar and their origins are summarised in Table 1.

The most well known and well researched polysaccharide impurity is dextran (e.g., Sutherland, 1960, Leonard and Richards 1969, Hidi et al., 1974, Charles, 1984). Dextran in sugar was first identified by Nicholson and Liliental (1959). Bruijn (1966) observed it in deteriorated cane and attributed its presence to Leuconostoc mesenteroides infection. Bruijn (1966) used globular proteins as standards in gel permeation chromatography and underestimated the molecular weight of dextran. Covacevich et al. (1977) elucidated the

Table 1. Summary of polysaccharide impurities in cane sugar.

Polysaccharide	Origin	Chemical nature
Starch	Cane plant	Amylose - linear α1-4 glucopyranosyl polymer; Amylopectin - α1-6 branched, α1-4 glucopyranosyl polymer
Dextran	Microbial infection	α1-4branched, α1-6 glucopyranosyl polymer
Cellulosans	Cane plant	Soluble fragments of cellulose (β1-4 glucopyranosyl polymer)
Hemicelluloses	Cane plant	Soluble fragments of xylans, mannans, galactans and arabinoglucuronoxylans
Indigenous sugarcane polysaccharide (ISP)	Cane plant	Polymer of arabinose, galactose and glucuronic acid
Sarkaran	Unknown, possibly microbial	linear α1-4 and α1-6 glucopyranosyl polymer - a pullulan
Pectin	Cane plant	α1-4 glacturonic acid polymer
Levans	Microbial infection	β2-6 fructofuranosyl polymer
Galactoglucomannan	Unknown	Either a galactoglucomannan or a mixture of a glucomannan and a galactomannan

structure of native dextran by methylation analysis and noted its similarity to that from *L. mesenteroides* NRRL B512-f.

Indigenous sugar cane polysaccharide (ISP) was first isolated by Roberts et al. (1964), and later identified as an arabinogalactan (Roberts *et al.*, 1976a). In subsequent work they reported the presence of glucuronic acid groups on the polysaccharide (Roberts *et al.*,1978a). Blake and Clarke (1984a) isolated ISP from Australian canes and compared it to samples of Roberts' Louisiana ISP. They found much lower levels of glucuronic acid in both the Australian and Louisiana ISP isolates, and disputed the existence of glucuronic acid in the ISP structure. To the best of our knowledge this remains an unresolved issue.

Sarkaran has been identified in raw sugar from stale (South African, Bruijn, 1966) and stand over (Australian, Blake and Clarke, 1984b) sugar canes. Bruijn (1966) noted the difference between the structure of the stale cane polysaccharide and pullulan, and therefore named it sarkaran. Blake and Clarke (1984b) obtained the sarkaran from stand-over cane, but were not able to obtain it from yeast or fungal cultures isolated from the same stand-over cane. However, neither Bruijn (1966) nor Blake and Clarke (1984b) seemed aware that Catley et al. (1966) had noted minor structural features of pullulans (viz., pullulan contains both maltotetrose and maltotriose subunits polymerised by α (1-6) linkages) that would encompass the sarkaran structure. Sarkaran is therefore a pullulan and most likely is from microbial sources.

6. TESTS FOR ACID BEVERAGE FLOC

Alcian blue or Basacryl orange dye tests are summarised by Clarke et al. (1997). These are sensitive to the presence of negatively charged polysaccharides. They were never good predictors of floc formation and the carcinogenic dyes are now banned substances.

Stansbury and Hoffpauir (1959) reported that floc tests used by the beet sugar industry did not work for cane sugars. They also investigated a test used by Pepsi Cola Company in the late 50's, that was based on measurement of turbidity after coagulation of negative charged colloids with a quaternary amine and found the procedure failed to differentiate ABF positive and ABF negative sugars. They were unable to make any marked improvement on the Coca Cola 10 day floc test.

Liuzzo and Wong (1982) developed a protein-dye-binding colorimetric method to measure trace amounts of protein in refined sugars. They compared the protein-dye-binding method to the Kjeldahl method and tested three floc forming sugars and three non-flocculating sugars. Floc forming sugars had protein concentrations between 0.3 and 0.4%, whereas non-flocculating sugars had protein concentrations between 0.004 and 0.006%. Liuzzo and Wong (1982) concluded that there was a clear difference in protein concentrations between ABF positive and ABF negative sugars. In this study on a limited number of samples, they did not describe or quantify the ABF from the three floc forming sugars. No follow up study on a larger sample of sugars has been reported, and to the best of our knowledge this protein-dye-binding method is not used by either soft drink or sugar manufacturers.

At this point in time there seems to be no published, reliable and rapid method to determine the floc propensity of sugars. The Coca Cola test based on observation of simulated acid beverages at high sucrose concentrations after 10 days is the only internationally recognised test for ABF. There are several variants of the Coca Cola test (see Clarke, 1997). In 1970, the International Commission for Uniform Methods of Sugar Analysis tentatively adopted the 10 day Coca Cola test (McGinnis, 1970). The Australian sugar industry uses a variant of this test which is based on semi-quantitative measurements at 7, 14, 21 and 28 days.

7. MECHANISM OF ABF FORMATION IN CANE SUGAR SWEETENED SOFT DRINKS

Roberts et al. (1976 b) investigated the effect of silica and polysaccharide interactions in ABF formation. While simulated beverages containing floc negative sugar, indigenous sugar cane polysaccharide (ISP) and silica formed a floc, those containing ISP and silica did not. They concluded that the sugar contained a factor that interacted with the ISP and down played the role of silica. In later work (Roberts and Godshall, 1978a, 1978b) colloidal ISP was found to carry a negative charge and floc was obtained in simulated beverages upon the addition of ISP and a protein fraction isolated from cane sugar.

Clarke et al. (1977, 1978) summarised the findings of the SPRI research efforts and proposed the a mechanism for ABF formation. This mechanism is shown in figure 1. In a

1997 paper on beet sugar ABF, Clarke et al. (1997) again outline the mechanism for ABF formation in cane sugar sweetened soft drinks (viz. ISP, a polysaccharide with glucuronic acid residues from plant cell walls, and protein from the plant or a processing additive form a coacervate that then interacts with neutral polysaccharides such as dextran and other compounds to form a floc). The assumption that the initial step is based on the interaction of ISP and protein is based on the ability to form a haze and then a floc when these compounds are mixed in a simulated beverage. While the mechanism is prima facie believable, the evidence supporting the mechanism is by no means proof that ABF forms in this way in authentic soft drinks. Clarke et al. (1997) also make brief mention of a regional specific floc; presumable they are referring to the work of Miki et al. (1975, 1980) who isolated a floc with a different polysaccharide composition.

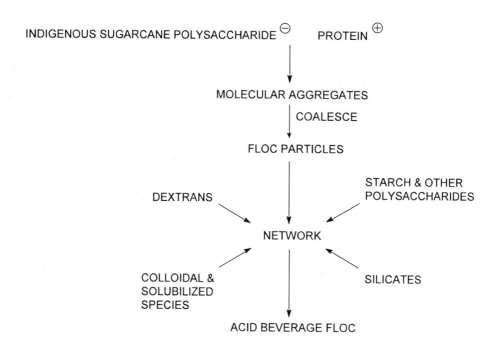

Figure 1. Mechanism of ABF formation proposed by Clarke et al. (1977)

Larsson et al. (1999) studied the effect of simple electrolytes (e.g., $MgCl_2$, Na_2SO_4, KBr) on the flocculation of nanosized silica particles (size 3 to 5 nm) and linear or branched cationic polymers. The linear polymer was polyacrylamide and the branched polymer was a diethylaminoethyl linked amylopectin. Their interest in these reactions was founded in the use of these types of polymers in the dewatering step of paper manufacture. The significance of this work to the mechanism of ABF formation is that it establishes that silica and charged polysaccharides (albeit positively charged) can form floc. It also is interesting that the stoichiometry of silica to polyelectrolyte, and the nature of the polyelectrolyte (branched or unbranched) affected the structure and size of the floc complex. This recent work is strong evidence for a more pivotal role of silica in ABF formation, and is supported by the work on chill haze in alcoholic beverages.

8.   CHILL HAZE IN ALCOHOLIC BEVERAGES

Many alcoholic beverages, and especially beers manufactured from malted barley and hops, may form a haze on chilling to temperatures of less than 5 °C. This haze, caused by the interaction of proteins and polyphenols, is well described in the literature (e.g., Siebert and Lynn 1999, Asano *et al.*, 1982, McMurrough *et al.*, 1999). Some literature has implications to ABF formation in non-alcoholic acid beverages. For example, the beer manufacturing industry uses silica to stabilise the product, and the interaction of silicates with proteins is certainly relevant to this review.

The mechanism that leads to chill haze involves the oxidation of polyphenols and subsequent interaction of these oxidised polyphenols with proteins at low temperatures (Fernyhough *et al.*, 1994). This chill haze disappears on heating. Oxidation of polyphenols leading to polymerisation may also produce an irreversible haze (O'Rourke, 1994). McMurrough et al. (1999) also attribute chill haze to colloidal instability of the protein-polyphenol complex, but note that hazes may still form by precipitation of polysaccharides or calcium oxalate.

It has been observed that the addition of silica hydrogel (a solution of colloidal silica (~30%) and water (~70%), with a narrow particle size distribution) is effective in decreasing the amounts of polyphenol sensitive proteins in beer (Fernyhough, 1994). The properties of the silica hydrogel particles such as the pore diameter and the specific surface area influence the removal of proteins from the beverage. Chill haze forming proteins are best removed by a silica hydrogel particles with a mean pore diameter in the range of 30-120 Å.

Silica surfaces bare residual valence electrons that react with water at acidic pH to form silanol (SiOH) groups. Polar organic molecules containing oxygen or nitrogen atoms (i.e., bearing a free pair of electrons) can hydrogen bond to the silanol groups. In this way proteins are adsorbed onto the silica surface. Proteins may also be adsorbed onto the silica surface by ionic bonding through quaternary ammonium ions, and by hydrophobic interactions. Fernyhough et al. (1994) have shown that proteins bind most strongly at their isoelectric point, and in some cases the pH range of binding may be very narrow. In beer, the adsorption of proteins onto silica hydrogel has been found to occur between pH 3 and pH 5. The adsorption capacity of the silica depends on size, surface area, porosity and density of silanol groups. In this pH range polyphenols and polysaccharides also were observed to bind to the silica hydrogel.

Studies by Siebert and Lynn (1999) have shown that silica gel can remove about 25% of haze active proteins in apple cider and about 85% of haze active proteins in beer. To explain this difference, they propose that the silica gel and polyphenols compete to bind haze active proteins. Apple cider contains a high concentration of polyphenols compared to protein, so there is little adsorption of proteins onto silica. Conversely, beer contains a low concentration of polyphenols compared to proteins, so more proteins are free from ployphenol interaction and can adsorb onto the silica surface.

The significance of this work to ABF in soft drinks is that it establishes the interaction of colloidal silicates with solutes by hydrogen bonding, and both ionic and hydrophobic interactions. The importance of stoichiometry of solutes in complex systems (involving more than one solute type) is also illustrated (i.e., the proportions of solutes must be optimal in order to achieve significant adsorption onto the silica surface). In fact, stoichiometric effects are also observed in solute-solute (i.e., protein-polyphenol) interactions. Siebert's model (1999) to explain the protein-polyphenol interaction proposes that if the amount of protein and polyphenol molecules are about the same, then there will be maximum haze development, and if the stoichiometry is not balanced there will be smaller particles and less haze.

9. CONCLUSIONS

While there is consensus among researchers that ABF in cane sugar sweetened soft drinks is a complex coacervate of silica, polysaccharides, protein, waxes and other organic compounds, the mechanism of ABF formation is still not completely elucidated. Certainly, most of the polysaccharides, waxes and some of the proteins and silica are present as impurities in refined cane sugar (albeit at concentrations less than 0.1%), so the implication of poor quality sugar ingredients in ABF formation is reasonable. The proposed mechanism of ISP and protein interaction as an initial step in ABF formation is based upon the observation that a mixture of ISP and protein in a simulated acid beverage forms floc. This is by no means proof that ABF forms in this way in authentic soft drinks, and in fact the presence of glucuronic acid groups (imparting a negative charge to ISP that is essential to the mechanism) remains in dispute. Literature on the related beverage quality issue (viz., chill haze in alcoholic beverages) establishes the interaction of colloidal silicates with solutes (such as proteins and polysaccharides) by hydrogen bonding, and both ionic and hydrophobic interactions. However, the role of silica in ABF formation is poorly understood, but may be pivotal in the initial mechanism of ABF flocculation.

10. REFERENCES

Ashurst, P. R. (Ed.), 1998, Chemistry and Technology of Soft Drinks and Fruit Juices, Sheffield Academic Press, Sheffield.

Asano, K., Shinagawa, K., and Hashimoto, N., 1982, Characterisation of Haze-Forming Proteins of Beer and Their Roles in Chill Haze Formation, *J. Amer. Soc. Brewing Chemists*, **40**:147-154.

Bruijn, J., 1966, Deterioration of sugar cane after harvesting, *Int. Sugar J.*, **68**:356-358.

Blake, J. D., and Clarke, M. L., 1984, Observations on the structure of ISP, *Int. Sugar J.*, **86**:295-299.

Blake, J. D., and Clarke, M. L., 1984, A water soluble polysaccharide from stand-over cane, *Int. Sugar J.*, **86**:276-279.

Catley, B. J., Robyt, J. F., and Whelan, W. J., 1966, A minor structural feature of pullulan, *Biochem J.*, **100**:5-6.

Charles, D. F., 1984, Polysaccharides in refined and raw sugar, *Int. Sugar J.*, **86**: 105-109.

Clarke M. A., Roberts, E. J., Godshall, M.A., and Carpenter, F.G., 1977, Beverage floc and cane sugar, *Proc. 16ᵗʰ ISSCT Congress*, 2587-2598.

Clarke M. A., Roberts, E. J., Godshall, M.A., and Carpenter, F.G., 1978, Beverage floc and cane sugar, *Int. Sugar J.*, **80**:197-201.

Clarke, M. A., Roberts, E. J., and Godshall, M. A., 1997, Acid beverage floc from beet sugars, Zuckerindustrie, **122**: 873-877.

Cohen, A., Dionisio, O. G., and Drescher, S. J., 1970, The isolation and characterisation of certain impurities responsible for quality problems in refined cane sugar, *Proc. 29th Ann. Meeting Sugar Ind. Technol.*, 123-165.

Covacevich, M. T., and Richards, G. N., 1977, Studies on dextrans isolated from raw sugar manufactured from deteriorated cane, *Int. Sugar J.*, **79**:3-9.

Dunsmore, A., Heal, M. J., Matic, M., and Runggas, F.M., 1978, A practical solution to the acid beverage floc problem, *Proc. 37ᵗʰ Ann. Meeting Sugar Ind. Technol.*, 514-527.

Eis, F. G., Clark, L. W., McGinnis, R. A., and Alston, P. W., 1952, Floc in carbonated beverages, *Ind. Eng. Chem.*, **44**:2844-2848.

Fernyhough, B., McKeown, I., and McMurrough, I., 1994, Beer Stabilisation with Silica Gel, *Brewers' Guardian*, **123**:44-50.

Gratius, I., Decloux, M., Dornier, M., and Cuvelier, G., 1995, The determination of polysaccharides in raw cane sugar syrups, *Int. Sugar J.*, **97**:296-300 & 339-343.

Hammond, D. A., 1998, Analysis of soft drinks and fruit juices, in Chemistry and Technology of Soft Drinks and Fruit Juices, Ed. Ashurst, P.R., Sheffield Academic Press, pp 166-196.

Hidi, P., Keniry, J. S., Mahoney, V.C., and Paton, N.H., 1974, Observations on the occurrence and nature of polysaccharides in sugar canes, *Proc. 15th ISSCT Congress*, 1255-1265.

Jones, M. B., 1978, Kirk-Othmer Encyclopedia of Chemical Technology, 3rd Ed. John Wiley & Sons, New York, Volume 4, pp. 710-725.

Kitchen, R. A., 1988, Polysaccharides of sugarcane and their effects on sugar manufacture, Chemistry and Processing of Sugarbeet and Sugarcane, Eds. Clarke, M. A. and Godshall, M. A., Elsevier Science Pub., Amsterdam, pp 208-235.

Larsson, A., Walldal, C., and Wall, S., 1999, Flocculation of cationic polymers and nanosized particles. *Colloids and Surfaces A: Physicochemical and Engineering Aspects*, **159**:65-76.

Leonard, G. J., and Richards, G. N., 1969, Polysaccharides as causal agents in production of elongated sucrose crystals from cane juice, *Int. Sugar J.*, **71**:263-265.

Liuzzo, J. A., and Wong , C. M., 1982, Detection of Floc-Producing Sugars by a Protein Dye-Binding Method, *J. Agric. Food Chem.*, **30**, 340-341.

McGinnis, R. A., 1970, Referees Report Subject 19 - Characteristics of white sugar, Report on the proceedings of the 15th session of ICUMSA, pp 200-213.

McMurrough, I., Madigan, D., Kelly, R., and O'Rourke, T., 1999, Haze Formation Shelf-Life Prediction for Lager Beer, *Food Technol.*, **53**:58-62.

Miki T., Saito, S., and Kamoda, M., 1975, Composition of polysaccharides in carbonated beverage floc, *Int. Sugar J.*, **77**:67-69.

Miki, T., Saito, S., Ito, H., and Kamoda, M., 1980, Composition of ploysaccharides in carbonated beverage floc from raw cane sugar, *Proc. 17th ISSCT Congress*, 2751-2762.

Miki, T., 1984, A galactomannan in carbonated beverage floc from raw cane sugar. *Carbohydr. Res.*, **129**:159-165.

Moore, W., and Buzzanell, P., 1991, Trends in U.S. Soft Drink Consumption - Demand Implications for Low-Calorie and Other Sweeteners, Sugar and Sweetener Outlook Report, USDA SSRV 16N3.

Morel du Boil, P. G., 1997, Refined sugar and floc formation, a survey of the literature, *Int. Sugar J.*, **99**:310-314.

Nicholson, R. I. and Liliental, B., 1959, Formation of a polysaccharide in sugarcane, *Aust. J. Biol. Sci.*, **12**:192-203.

O'Rourke, T., 1994, The requirements of beer stabilisation, *Brewers' Guardian*, **123**:30-33.

Ridout, C. L., Price, K. R., Parkin G., Dijoux, M. G., and Lavaud, C., 1994, Saponins from Sugar Beet and the Floc Problem, *J. Agric. Food Chem.*, **42**:279-282.

Roberts, E. J., and Carpenter, F. G., 1974, Composition of acid beverage floc, Proc. 1974 Tech. Session Cane Sugar Refining Research, 39-50.

Roberts, E. J., and Godshall, M. A., 1978a, Identification and estimation of glucuronic acid in indigenous polysaccharides of sugar cane, *Int. Sugar J.*, **80**:10-12.

Roberts, E. J., and Godshall, M. A., 1978b, The role of charged colloids in floc formation, *Int. Sugar J.*, **80**:105-109.

Roberts, E. J., Godshall, M. A., Carpenter, F. G., and Clarke M. A., 1976a, Composition of soluble indigenous polysaccharides from sugar cane, *Int. Sugar J.*, **78**:163-165.

Roberts, E. J., Godshall, M. A., Clarke M. A., and Carpenter, F. G., 1976b, Some observations on acid beverage floc, *Int. Sugar J.*, **78**:326-328.

Roberts, E. J., Jackson, J. F. and Vance, J. H., 1964, Progress in research on the soluble polysaccharides of sugarcane, *Proc. Tech. Sessions Cane Sugar refining Res.*, 76-84.

Siebert, K. J., and Lynn, P. Y., 1997, Mechanisms of Adsorbent Action in Beverage Stabilisation, *J. Agric. Food Chem.*, **45**:4275-4280.

Siebert, K. J., 1999, Protein-Polyphenol Haze in Beverages, *Food Technol.*, **53**:54-57.

Sutherland, G. K., 1960, An investigation of the polysaccharides present in sugar mill syrups, *Aust. J, Biol. Sci.*,**13**:300-306.

Stansbury, M. F., and Hoffpauir, C. L., 1959, Composition of "Floc" Formed in Acidified Sirups from Refined Granulated Cane Sugars, *J. Agric. Food Chem.*, **7**:353-358.

Taylor, R. B., 1998, Ingredients, in Chemistry and Technology of Soft Drinks and Fruit Juices, Ed. Ashurst, P.R., Sheffield Academic Press, pp 16-54.

Walker, Jr., H. G., and Owens, H. S., 1953, Beet Sugars: Acid insoluble constituents in selected samples, *J. Agric. Food Chem.*, **6**:450-453.

INFLUENCE OF DNA ON VOLATILE GENERATION FROM MAILLARD REACTION OF CYSTEINE AND RIBOSE

Yong Chen,[1] Chee-Kok Chin,[2] and Chi-Tang Ho[1]

1. INTRODUCTION

A number of studies utilizing various analytical techniques in recent years have revealed some very important aroma compounds imparting meat flavor perception (Mottram and Madruga, 1994; Hofmann and Schieberle, 1995; MacLeod, 1986). Out of more than 1000 volatiles identified from cooked meats, a few have been characterized as meaty flavor impact compounds. In general, thiol-substituted furans and thiophenes and related disulfides possess strong meat-like and/or roast aromas with low odor threshold values. For instances, 2-methyl-3-furanthiol and its disulfide bis-(2-methyl-3-furyl)-disulfide were identified as major contributors to the meaty aroma of cooked beef, chicken, and pork (Gasser and Grosch, 1988; 1990) while 2-furfurylthiol was identified as a contributor to roasty and coffee-like notes (MacLeod, 1998).

It is well known that the Maillard reaction between reducing sugars and amino acids is one of the crucial routes for the formation of these compounds during cooking. Studies have shown that cysteine and pentoses are important precursors participating in the thermal reaction to generate meaty aroma compounds. In meats ribose is one of the major sugars, and it originates principally from ribonucleotides, in particular, adenosine triphosphate (Mottram, 1998). Postmortem effects result in a large amount of nucleotides from degradation of deoxyribonucleic acid (DNA) and ribonucleic acid (RNA).

Besides serving as the source for pentose sugar, nucleotides have been found to act synergistically with glutamic acid or monosodium glutamate (MSG) to enhance meaty, brothy, and MSG-like taste while suppressing sulfurous notes (MacLeod, 1998). Two nucleotides, inosine-5'-monophosphate (IMP) and guanosine-5'-monophosphate (GMP),

[1] Department of Food Science and [2]Department of Plant Sciences, Cook College Rutgers University, New Brunswick, NJ 08901-8520

Quality of Fresh and Processed Foods, edited by Shahidi et al.
Kluwer Academic/Plenum Publishers, 2004.

which accumulate in post-slaughter muscle as a result of the hydrolysis of inosine triphosphate and guanosine triphosphate, respectively (Mottram and Madruga, 1994), are taste active. They both have been used as flavor enhancers in savory foods and are believed to contribute to the "umami" taste (Maga, 1983).

Thermally, these nucleotides are not stable. Their contents can be decreased by half within approximately 30 min at 121 °C in highly acidic conditions (Maga, 1998). The thermal degradability of these nucleotides makes them potential modifiers of thermally generated flavors. However, previous studies related to the influence of nucleic acids on meaty flavor generation have primarily focused on IMP. In a meat model system, IMP enhanced the formation of thiols and novel disulfides containing furan moieties (Mottram and Madruga, 1994). Volatiles generated from the reaction of both alliin-IMP and deoxyalliin-IMP elicited a pungent garlic flavor with roasted notes, caused by sulfur-containing compounds plus a number of pyrazines and thiazoles, including methylpyrazine, ethylpyrazine, 2,5-dimethylpyrazine, 2-propylthiazole, and 2-ethyl-4-propylthiazole (Yu *et al.*, 1994).

The level of IMP was 0.106-0.443% in beef and 0.075-0.122% in chicken; chicken appeared to contain a somewhat higher amount of AMP, CMP, UMP than beef (Maga, 1998). It was reported that the normal concentration of DNA in longisimus muscle was about 1091.9 µg/g muscle (Koohmaraie *et al.*, 1995). The high level of DNA in meats potentially has some effects on meat flavor production during cooking. However, information regarding the overall effect of DNA on meaty flavor is virtually unavailable. The influence of DNA on the formation of volatile compounds at both pH 5 and 8.5 using a Maillard reaction model system of ribose and cysteine was examined in this study.

2. MATERIALS AND METHODS

2.1 Materials

D-ribose, tridecane and anhydrous sodium sulfate were purchased from Aldrich Chemical Company (Milwaukee, WI). DNA and L-cysteine were bought from Sigma Chemical Co. (St. Louis, MO). Methylene chloride used was of HPLC grade and purchased from Fisher Scientific (Fair Lawn, NJ).

2.2 Thermal Reactions

The powdered DNA was made from Herring sperm and had a fishy odor. For each reaction involving DNA, 3 g of DNA powder was weighted and thermally dissolved in distilled water. The solution was extracted with 120 ml (40 x 3) of hexane to remove possible lipid residues. L-cysteine and D-ribose (0.01 mol) with or without 3 g of DNA were dissolved in 100 ml of distilled water. The addition of DNA into the solution of ribose and cysteine lowered the pH by about 1 unit. The pH of solutions was adjusted to 5.0 or 8.5 with sodium hydroxide prior to thermal reaction in 150 ml Hoke stainless steel cylinders (Hoke Inc., Clifton, NJ), which were heated at 180 °C in an oven for 2 h. The reactions were immediately stopped by cooling under a stream of cold water. Other thermal reactions of DNA, between DNA and ribose or cysteine were all conducted in a similar manner and under the same conditions.

2.3 Liquid/liquid Extraction of Volatile/Semivolatile Compounds

After cooling, the brown reaction mixture was mixed with 0.5 ml of a solution of internal standard (tridecane, 1 mg/ml) and extracted with methylene chloride (50 ml x 3 times). The extract was dehydrated by anhydrous sodium sulfate and concentrated under a nitrogen flow to 10 ml in a flask and then transferred to a Kuderna-Danish concentrator and further concentrated to 1-1.5 ml before subjecting to further analysis.

2.4 GC/Mass Spectrometry Analysis

The concentrated isolates from different reaction mixtures were analyzed by GC/mass spectrometry (GC/MS), using a Hewlett-Packard 6890 GC equipped with a fused silica capillary column (60 m x 0.25 mm i.d.; 1 μm thickness, DB-1) coupled to a Hewlett-Packard 5973 series mass selective detector. Mass spectra were obtained by electron ionization at 70 eV and a source temperature of 250 °C.

2.5 Identification of the Volatiles

The identification of volatile compounds was based on GC/MS analysis. The compounds from the isolate were identified by comparing the mass spectral data with those of authentic compounds available in the Wiley 275 library or previous publications (Hofmann and Schieberle, 1995; Mottram and Whitfield, 1995a,b; Werkhoff *et al.*, 1993). Quantification of the volatile/semivolatile compounds from the liquid phases was based on using tridecane as an internal standard.

3. RESULTS AND DISCUSSION

The effect of DNA on volatile generation was investigated using Maillard reaction solutions of ribose and cysteine with or without DNA at both starting pH of 5 and 8.5, which were heated at roasting temperature of 180 °C for 2 h. As indicated by Table 1, pH changed 2-3 units after the reactions. Changes of pH are common in the reaction systems in the absence of buffering agents (Madruga and Mottram, 1995).

Organoleptically, the odor elicited from the reaction of ribose and cysteine under acidic conditions was sulfurous and meaty, but under basic conditions, the odor had additional roasty and burnt notes, and the sulfurous, meaty aroma became less obtrusive. The presence of DNA in the reaction systems imparted some sugary, fruity, and sweet notes in addition to the aroma characteristics perceived in the absence of DNA.

Compounds identified from the liquid phases of the reactions between ribose and cysteine with or without DNA are listed on Table 2. The data provided clear evidence that DNA can act as a major source for pentose. The addition of DNA increased the levels of furans, especially 2-furfurylalcohol, furfural and difurylmethane. In the reaction systems without DNA, 2-furfurylalcohol was formed in small quantities but addition of DNA increased its concentration by many folds. Believed to have a fruity and sugary aroma, 2-furfurylalcohol has been identified as a major component from baked sweet potato (Sun *et al.*, 1993) and rice cake (Buttery *et al.*, 1999). It can also be generated in large amounts from the reaction between cysteine and glucose (Shibamoto and Yeo,

Table 1. Final pH, final appearance and flavor description of the thermal reaction mixtures of ribose and cysteine with/without DNA

Solutions of Model Systems	Initial pH	Final pH	Aroma Characteristic
DNA	5	4.27	Fruity, sugary
	8.5	5.55	Slightly burnt
DNA + Ribose	8	3.61	Sugary
	8.5	4.25	Sugary, fruity
Ribose + Cysteine	5	3.56	Sulfurous, meaty
	8.5	5.25	Sulfurous, burnt, nutty
Cysteine + DNA	5	5.31	Sulfurous
	8.5	7.08	Sulfurous, burnt
Ribose + Cysteine + DNA	5	3.67	Sulfurous, meaty
	8.5	5.42	Nutty, burnt

1994) and found to have strong inhibitory ability toward hexanal oxidation in commercial beer (Wei *et al*., 2001). Furfural serves as an important precursor for the formation of other furanoids and heterocyclic compounds such as thiophenes and pyrroles. When H_2S or NH_3 are present, replacement of the oxygen on the furan ring with sulfur or nitrogen results in the formation of the corresponding thiophenes or pyrrole derivatives.

Consistent with previous research reports (Mottram and Madruga, 1994; Shu *et al*., 1985; Tressl *et al*., 1989), the results indicated that more sulfur compounds were produced at lower pH. The meaty flavor impact compound 2-methyl-3-furanthiol and its oxidized dimer bis (2-methyl-3-furyl) disulfide were formed in a much larger amount under acidic conditions and DNA appeared to inhibit their formation. These two compounds have the potency to dictate the characteristics flavor of meat. It has been shown that compounds containing a 2-methyl-3-furanyl group could originate from the reaction of hydrogen sulfide with 4-hydroxy-5-methyl-3(2H)-furanone (HMF) (Whitfield *et al*., 1993), which may derive from pentoses in the Maillard reaction or from the dephosphorylation of ribose phosphate (Madruga and Mottram, 1995; Grosch and Zeilr-Hilgart, 1992). It is well known that formation of HMF from pentose sugar is favored under more acidic conditions. A significant amount of 2-furfurylthiol was generated from reaction of ribose with cysteine at pH 5, much less formed at pH 8.5. This is in accordance with previous studies showing that 2-furfurylthiol formation was greatly affected by pH and lowering the pH would significantly increase its formation (Parliament and Stahl, 1994). The formation of 2-furfurylthiol probably resulted from the reaction of H_2S with the ribose breakdown product furfural (Whitfield *et al*., 1993). Although DNA increased the amount of furfural, it did not show the same effect on 2-furfurylthiol formation. Instead, it greatly reduced the quantity of 2-furfurylthiol and its corresponding dimer, *bis* (2-furfuryl) disulfide. Possessing a characteristic coffee-like

Table 2. Volatile Compounds Generated from the Liquid Phases of the Thermal Reactions of Ribose and Cysteine, with/without DNA at pH 5 or pH 8.5

Compounds	Amount (mg / g ribose)			
	RC 5[a]	RCD 5[b]	RC 8.5[c]	RCD 8.5[d]
Furanoids				
Furfural	0.172	0.487	0.011	0.023
2/3-Furfuryl alcohol	0.023	1.471	0.027	5.021
2-Acetyl-5-methylfuran		0.013		
Difurylmethane	0.006	0.247		0.049
2-(2-Furanylmethyl)-5-methylfuran		0.007		
1-(2-Furyl)-2-propanone	0.016	0.003		
2,2'-(1,2-Ethenediyl)*bis*furan	0.020			
5-Methyl-2(5H)-furanone		0.022		0.003
Thiophenoids				
2-Methylthiophene	0.468	0.132	0.423	0.237
3-Methylthiophene	0.078	0.022	0.074	0.039
2, 3-Dimethylthiophene	0.122	0.054	0.106	0.021
2-Acetylthiophene			0.011	0.009
2-Ethylthiophene	0.035	0.002	0.081	0.007
2-Propylthiophene	0.009		0.010	
2-Acetyl-3/5-methylthiophene	0.011	0.005		
Dimethylformylthiophene	0.058	0.017	0.028	0.016
2/5-Methyl-thieno[2,3-b]thiophene	0.016	0.006		
3-Thiophenecarboxaldehyde		0.045		0.023
2-Hydroxymethylthiophene		0.135		0.005
3/5-Methyl-2-thiophenecarboxaldehyde		0.009	0.048	0.082
2-Formyl-3-methylthiophene		0.007	0.028	0.081
2-Formyl-2,3-dihydrothiophene				0.195
2/3-Thiophenemethanol	0.051	0.133	0.044	0.113
Thieno[3,2-b]thiophene	0.148	0.047	0.090	0.051
Dihydrothienothiophene			0.044	0.033
Methyldihydrothienothiophene	0.071	0.015	0.127	0.022
5-Methylthieno[2, 3-d]thiophene	0.125	0.036	0.036	0.007
2-Thiophenethiol	0.157		0.095	
2-Methyl-3-thiophenethiol	0.096	0.026	0.170	0.045
2-Thienylmethanol	0.164	0.065	0.070	0.117
Dihydro-3-(2H)-thiophenone	0.007		0.007	0.008
Dihydro-2-methyl-3(2H)-thiophenone	0.064	0.026	0.161	0.149
5-Methyl-2(5H)-thiophenone	0.014	0.017		
1-(2-Thienyl)-1-propanone			0.015	
Other sulfur-containing compounds				
2-Methyl-3-furanthiol	0.276	0.017	0.075	
bis(2-Methyl-3-furyl)disulfide	0.178	0.009	0.033	
2-Furfurylthiol	0.834	0.321	0.488	0.276
bis(2-Furfuryl)disulfide	0.420	0.205	0.198	0.124
3-Mercapto-2-pentanone	0.124	0.086		

	a	b	c	d
Furfuryl sulfide	0.020	0.012		0.015
2-[(Methylthio)methyl]furan	0.011	0.015		
2-Methyl-3-(methylthio)-butane			0.097	
2, 3-Dihydro-6-methylthieno[2,3c]furan			0.071	0.065
1,2-Dithian-4-one			0.015	
3-Methyl-1,2,4-trithiane	0.004		0.012	0.013
1,3,5-Trithiolane				0.013
3, 5-Dimethyl-1, 2, 4-trithiolane	0.110	0.134	0.223	0.127
Pyrazines				
Pyrazine			0.076	0.101
Methylpyrazine			0.205	0.088
2,5 or 1,6-Dimethylpyrazine				0.021
Ethylpyrazine				0.043
2-(n-Propyl)-pyrazine				0.002
Dihydro-cyclopentapyrazine			0.028	0.043
5H-5-Methyl-6,7-dihydrocyclopentapyrazine			0.010	0.016
Thiazoles				
Thiazole	0.040	0.074	0.160	0.234
Isothiazole	0.020	0.012	0.004	0.003
2-Acetylthiazole			0.067	0.112
5-Ethyl-2, 4-dimethylthiazole	0.035	0.016	0.041	0.016
2, 4, 5-Trimethylthiazole			0.015	0.009
2-Propionylthiazole				0.007

[a] Ribose/cysteine reaction system at pH 5; [b] Ribose/cysteine/DNA reaction system at pH 5; [c] Ribose/cysteine reaction system at pH 8.5; [d] Ribose/cysteine/DNA reaction system at pH 8.5.

aroma, 2-furfurylthiol has been identified from the reaction between cysteine and inosine 5'-monophosphate (Zhang and Ho, 1991).

The thiophenoids were the most numerous compounds identified. Among them, 2-methylthiophene was the most abundant, followed by thieno[3,2-b]thiophene and 5-methylthieno[2,3-d] thiophene. In general, addition of DNA inhibited the formation of these thiophenoids with a few exceptions. Compounds such as 2-methylthiophene, 2,3-dimethylthiophene, 2-ethylthiophene, 2-propylthiophene, dihydro-2-methyl-3(2H)-thiophenone and a couple of thienothiophenes all exhibited at least 2 to 3 folds reduction in quantity when DNA was present under both pH conditions. 2-Thiophenethiol and the meaty flavor compound 2-methyl-3-thiophenethiol, which have antioxidant capacity, were also reduced. However, the data also showed that the contents of several thiophenes such as thiophenemethanol, 3-thiophenecarboxaldehyde, 2-hydroxymethylthiophene, and 2-formyl-3-methylthiophene were increased as a result of DNA addition. The increase of thiophenemethanol in the presence of DNA is because DNA increases the concentration of furfuryl alcohol, which could react with H_2S to form thiophenemethanol by a single replacement. Thiophenoids are commonly identified from cooked meats with odor threshold in the ppb range. The alkyl thiophenes have aromas reminiscent of roasted onions. So far none of the thiophenes identified in meat appears to be directly responsible for the meaty characteristics (Mottram, 1994), but may be considered as important contributors to the meaty aroma.

Pyrazine formation in model systems depends largely on pH value (Mottram, 1994; Leahy and Reineccius, 1989). The condensation of amino acid with dicarbonyl compounds known as the Strecker degradation or the condensation between ammonia and dicarbonyls are known possible pathways for pyrazine formation. Since these condensations are favored at higher pH due to the higher availability of unprotonated amino groups in less acidic solutions, pyrazines usually are formed at higher pH environment. In this study, pyrazines were only identified when the pH was increased and the presence of DNA enhanced their formation to a certain extent. One possible explanation is that DNA acted as a nitrogen source necessary for pyrazine formation.

Formation of thiazoles was also pH sensitive. Except that isothiazole was favored at more acidic condition, other thiazoles were either more abundant or only detected when the pH was increased. DNA increased the concentration of thiazole and 2-acetylthiazole but decreased that of 5-ethyl-2,4-dimethylthiazole. Found in roast beef, grilled pork and fried chicken, 5-ethyl-2,4-dimethylthiazole has a nutty, roast, meaty, liver-like flavor and possesses a low odor threshold value of 2 ppb (Mottram, 1994). Thiazole compounds are usually alkyl substituented and found in roast, grilled or fried products. 2-Acetylthiazole is the only thiazole containing another functional group and commonly detected in meat volatiles (Forss, 1972). As indicated by Vernin and Parkanyi (1982) the most likely route for thiazole formation is the reaction between hydrogen sulfide, ammonia and carbonyl compounds, which are the breakdown products of cysteine and ribose in this system. Addition of DNA into the system possibly increased the availability of ammonia as a result of thermal degradation of DNA.

In order to obtain a better understanding about the effect of DNA on volatile generation in a system of ribose and cysteine, thermal reactions were conducted in several other solutions containing only DNA, or DNA and ribose, or DNA and cysteine. They were heated under the same conditions without pH adjustment, or with pH adjusted to both starting pH 5 and 8.5. The pH of DNA solution was 1.74 before adjustment. Heating this solution generated only a few volatile compounds with a large amount of 4-oxo-pentanoic acid methyl ester, amounting to over 95% of the total area on the GC chromatogram. The other volatiles included cyclohexene and 5-methyl-2(5H)-furanone in small quantities (data not shown). However, thermal treatments of the pH adjusted DNA solutions had much less 4-oxo-pentanoic acid methyl ester but more other volatile compounds such as 2-furfuryl alcohol, 2,3-dihydro-1H-inden-1-one, cyclopentanedione, and a couple of bisfurans (Table 3). Being the most abundant in solutions of both pH, 2-furfuryl alcohol appeared to be generated favorably under more basic conditions. Condensation and degradation of 2-furfuryl alcohol likely gave rise to the variety of furan-containing compounds identified. Comparing the volatiles from heating DNA alone to those from heating DNA and ribose together (Table 3), one noteworthy point is the furfural formation. Thermal degradation of DNA did not generate identifiable amounts of furfural, but when ribose was added, a large amount was produced. This quantitative difference is possibly due to the fact that DNA contains deoxyribose instead of ribose that is specific to the formation of furfural. In other words, thermal degradation of deoxyribose produces 2-furfuryl alcohol whereas thermal degradation of ribose generates furfural. In the systems containing both ribose and cysteine without DNA, both furfural and furfuryl alcohol can be generated (Table 2). In this case, formation of furfuryl alcohol can be derived from reduction of furfural as shown in Figure 1. On the

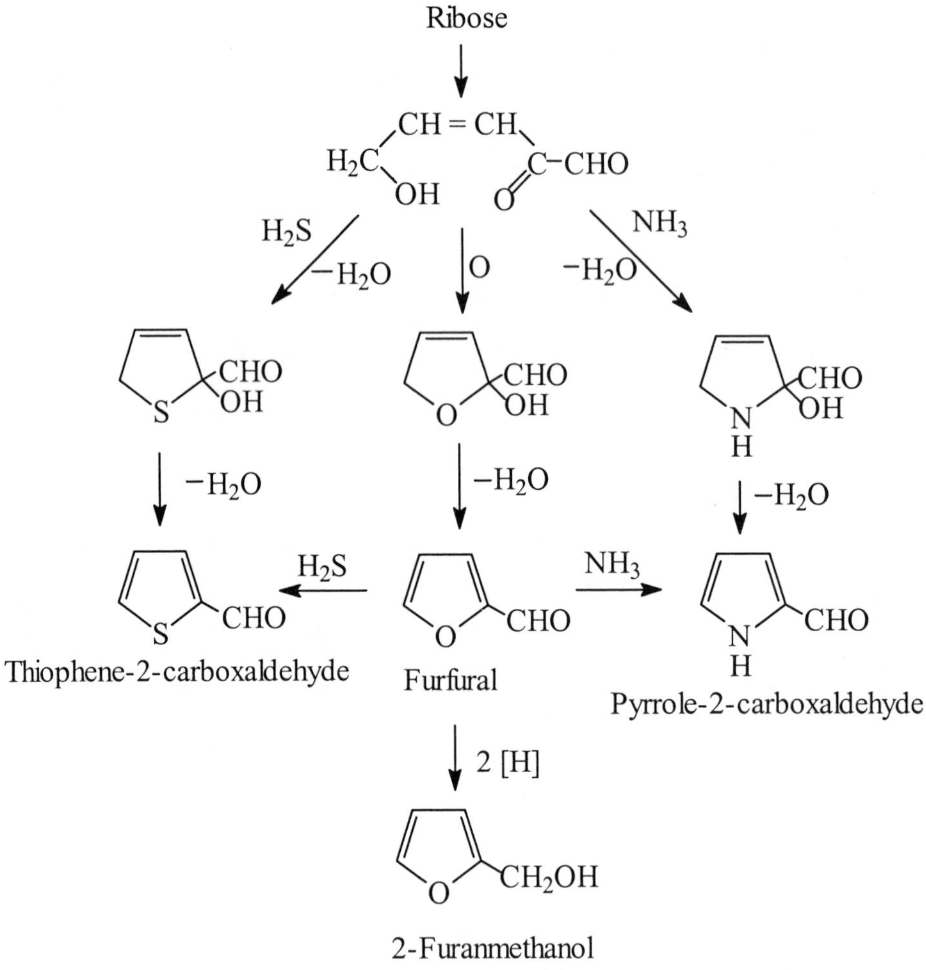

Figure 1. Formation of some related furan, thiophene and pyrrole derivatives from Maillard intermediates (MacLeod and Ames, 1988).

Table 3. Volatile compounds generated from the thermal reactions of DNA and between DNA and ribose at both starting pH of 5 and 8.5.

Compounds	Amount (mg) (3g DNA)		Amount (mg) (1.5 g Ribose + 3 g DNA)	
	pH 5	pH 8.5	pH 5	pH 8.5
Furfural			23.396	13.309
2-Furfuryl alcohol	1.245	9.532	1.83	3.223
1-(2-Furanyl)-ethanone	0.080			0.315
5-Methyl-2(5H)-furanone	0.141		0.285	0.075
5-Methyl-2(3H)-furanone	0.090		0.233	
1-(2-Furanyl)-1-propanone				0.012
2-(2-Furanylmethyl)-5-methylfuran				0.038
5-Methyl-2-furancarboxaldehyde	0.006		0.021	0.058
2, 2'-Methylenebisfuran	0.086	0.062	0.299	0.185
2, 2'-[Oxybis(methylene)]*bis*furan	0.008	0.015	0.038	
2, 5-*bis*(2-Furanylmethyl)-furan	0.010			
1-(2-Furanylmethyl)-1H-pyrrole		0.016		
1H-Pyrrole-2-carboxaldehyde				0.026
1,2-Cyclopentanedione	0.180	1.013	0.142	0.494
Pentanoic acid, 4-oxo-methyl ester	0.039	0.012	0.138	0.049
2-Hydroxy-3-methyl-2-cyclopenten-1-one		0.034		0.15
3-Ethyl-2-hydroxy-2-cyclopenten-1-one		0.038		0.092
2-Cyclopentene-1, 4-dione	0.041		0.021	0.026
4-Methoxy-2-cyclopenten-1-one	0.028		0.031	0.023
1, 2-Cyclohexanedione	0.052			0.130
2, 3-Dihydro-1H-inden-1-one	0.351	0.056	0.427	0.387
2 or 3 or 5-Methyl-1H-indole		0.015		
1H-Indole-4-carboxaldehyde	0.007			
Kinetin	0.009		0.031	

other hand, the formation of furfuryl alcohol from heating DNA other than from furfural conversion suggests a different route, which likely follows hydrolysis of DNA and dehydration of the deoxyribose as illustrated in Figure 2. One interesting compound identified from heating DNA alone or DNA and ribose together is kinetin, which can be derived from the condensation of 2-furfuryl alcohol or furfural with adenine by losing a molecule of H_2O as shown in Figure 3. Its formation appear to be favored under more acidic conditions and facilitated by ribose as shown by the data. Formation of kinetin suggests that DNA bases such as adenine were relatively stable against heat treatment in this experiment. Kinetin is a well-known plant hormone demonstrating a wide variety of biological activities including regulation of gene expression, stimulation of calcium flux, the cell cycle and as an anti-stress and anti-aging compound (Barciszewski *et al.*, 2000).

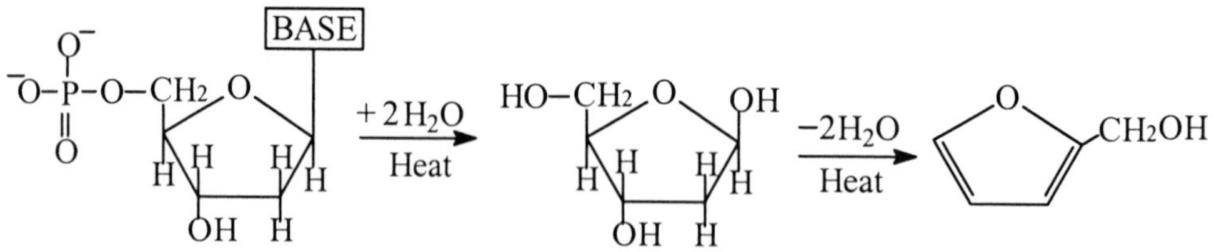

Figure 2. Formation of 2-furfuryl alcohol from DNA

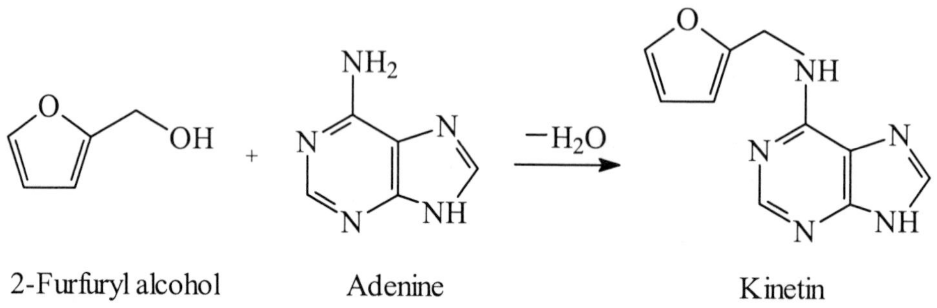

2-Furfuryl alcohol   Adenine   Kinetin

Figure 3. Formation of kinetin from 2-furfuryl alcohol and adenine.

Recent *in vitro* and *in vivo* studies revealing its biological effects made it even more scientifically interesting and commercially attractive as an ingredient for cosmetic products (Barciszewski *et al.*, 2000). Data obtained also indicated that the formation of 2-furfuryl alcohol was preferred under less acidic conditions, while furfural formation was favored under more acidic environments, in which ribose reacts with ammonia via 1,2-enolization to form 3-deoxysone that produces furfural via cyclization. Among volatiles listed in Table 3, there were several nitrogen containing compounds including a couple of indoles and pyrroles. 1-(2-Furanylmethyl)-1H-pyrrole was found from heating DNA alone at higher pH. Shown in Figure 1, if pyrrole formation results from the reaction of ammonia with pentose degradation product as proposed by MacLeod and Ames (1988), the formation of 1-(2-furanylmethyl)-1H-pyrrole would indicate that DNA can produce ammonia under thermal conditions, since condensation between 2-furfurylalcohol and pyrrole would give rise to the formation of 1-(2-furanylmethyl)-1H-pyrrole. However, the small amount of these nitrogen-containing compounds suggest that DNA was not a major nitrogen source for providing ammonia.

The reactions between DNA and cysteine generated some well-known sulfur containing compounds, including 2-furfurylthiol, 2-acetyl-3/5-methylthiophene, 2-methyl-thieno[2,3-b]thiophene and two 3, 5-dimethyl-1, 2, 4-trithiolane isomers (Table 4). A large amount of 2-thiophenemethanol was produced likely as the result of a single oxygen atom replacement by sulfur on the furan ring of 2-furfuryl alcohol. Again, furfural was not identified without ribose present in the reaction system. The large amount of 2-furfurylthiol suggest that its formation could arise from condensation not only between furfural and H_2S but also between furfuryl alcohol and H_2S. This indicates that furfuryl alcohol from DNA could compete with furfural from ribose for H_2S derived from cysteine.   Therefore, the increase of thiophenemethanol and the decrease of furfurylthiol resulting from the addition of DNA into the mixture of ribose and cysteine were a result of competition for H_2S. On the other hand, it also suggested it was easier for H_2S to replace the oxygen atom on the furan ring than to react with the carbonyl group of furfural or the hydroxy residue of furfuryl alcohol. In addition, reaction between DNA and cysteine produced several pyrrole-containing compounds, more than those produced by heating DNA alone. Since addition of cycteine could increase the availability of ammonia after thermal degradation, the number and concentration of pyrrole compounds also increased. In this reaction system, no pyrazines were identified. Instead, there were a couple of pyridines produced. This observation further indicates that deoxyribose from DNA is not acting like ribose, which would otherwise follow the Strecker degradation pathway to react with cysteine to produce dicarbonyls and aminoketones for the formation of pyrazine at higher pH. Data in Table 3 indicated that heating DNA itself, thermal reactions between DNA and ribose did not generate any pyrazine either. These results suggest that formation of pyrazine necessitate the participation of both ribose and cysteine.

In conclusion, the results demonstrated that DNA had complex effects on volatile formation from the Maillard reaction model system. DNA acted as a source to provide deoxyribose for the formation of 2-furfuryl alcohol and other degradation products. Deoxyribose and ribose likely follow different pathways to produce mainly 2-furfuryl alcohol and furfural respectively. They compete with each other for ammonia or sulfur precursors available in the system. DNA exerted a quenching effect on the quantities of some heterocyclic sulfur containing compounds such as 2-methyl-3-furanthiol, 2-furfurylthiol and their respective disulfides. Although DNA also exhibited some facilitating ability on the formation of certain nitrogen-containing compounds including pyrazines and thiazoles, the bases that DNA contained were relatively stable and should not be regarded as major nitrogen sources.

Table 4. Volatile compounds identified from the reaction between cysteine and DNA

Compounds	Amount (mg) (1.212 g Cysteine + 3 g DNA)	
	pH 5	pH 8.5
Furans		
2-Furfuryl alcohol	2.584	5.064
2, 2'-Methylenebisfuran	0.026	
5-Methyl-2(3H)-furanone	0.018	
2, 5-Dimethylfuran	0.006	
2-Acetyl-5-methylfuran	0.105	
Ketones		
2-Cyclopentene-1, 4-dione	0.006	
Cyclopentanone	0.034	0.077
2-Ethylcyclopentanone	0.007	0.014
2-Methyl-2-cyclopenten-1-one		0.034

3-Methyl-2-cyclopenten-1-one	0.007	0.014
3-Methyl-2-cyclohexen-1-one		0.032
4-(1-Methylethyl)-2-cyclohexen-1-one	0.017	
Thiophenes		
2/3-Methylthiophene	0.064	0.080
Dihydro-3(2H)-thiophenone	0.006	0.021
Dihydro-2-methyl-3(2H)-thiophenone	0.016	0.020
2-Thiophenemethanol	0.652	0.887
2-Acetyl-3-methylthiophene		0.012
2-Acetyl-5-methylthiophene	0.035	0.097
Thieno[3, 2-b]thiophene		0.014
2-Methyl-thieno[2, 3-b]thiophene	0.017	0.117
Pyridines		
2, 6-Dimethylpyridine	0.010	0.009
3-Ethylpyridine		0.026
Thieno[3, 2-c]pyridine		0.255
Pyrroles		
3-Methyl-1H-pyrrole		0.084
2/3-Ethyl-1H-pyrrole		0.038
2-Methyl-4-ethyl-1H-pyrrole	0.027	0.052
3, 4-Diethyl-2-methyl-1H-pyrrole	0.026	
1H-Pyrrole-2-carboxaldehyde		0.450
Thiazole		
2-Acetylthiazole	0.016	0.132
2-Methylthiazole	0.009	
4, 5-Dihydro-2-methylthiazole	0.009	
5-Ethyl-2-methylthiazole	0.013	
1, 2-Benzisothiazole	0.071	
Others		
Pentanoic acid, 4-oxo-methyl ester	0.023	
2-Furfurylthiol	0.255	0.026
2 or 3 or 5-Methyl-1H-indole		0.020
Furfuryl sulfide	0.013	
3, 5-Dimethyl-1, 2, 4-trithiolane	0.183	0.187
3, 5-Dimethyl-1, 2, 4-trithiolane	0.186	0.157
1, 2, 5-Trithiepane		0.018
1, 2-Ethanedithiol	0.013	

4. REFERENCES

Barciszewski, J., Siboska, G., Rattan, S. I. S., and Clark, B. F. C., 2000, Occurrence, biosynthesis and properties of kinetin (N6-furfuryladenine), *Plant Growth Regulation,* **32**:257-265.

Buttery, R. G., Orts, W. J., Takeoka, G. R., and Nam, Y. 1999, Volatile flavor components of rice cakes, *J. Agric. Food Chem,,* **47**:4353-4356.

Forss, D. A., 1972, Odor and flavor compounds from lipids, *Prog. Chem. Fats Other Lipids*, **13**:181-258.

Gasser, U., Grosch, W., 1988, Identification of volatile flavor compounds with high aroma values from booked beef, *Z. Lebensm. Unters. Forsch.*, **186**:489-494.

Gasser, U., and Grosch, W., 1990, Primary odorants of chicken broth-a comparative-study with meat broths from cow and ox, *Z. Lebensm. Unters. Forsch.*, **190**:3-8.

Grosch, W., and Zeiler-Hilgart, G., 1992, Formation of meatlike flavor compounds, in: *Flavor Precursors: Thermal and Enzymatic Conversions*, R. Teranishi, G. R. Takeoka, M. Güntert, eds., ACS Symposium Series 490, American Chemical Society, Washington, DC, pp. 183-192.

Hofmann, T., Schieberle, P., Evaluation of the key odorants in a thermally treated solution of ribose and cysteine by aroma extract dilution techniques, *J. Agric. Food Chem.*, **1995**, 43:2187-2194.

Koohmaraie, M., Shackelford, S. D., Wheeler, T. L., Lonergan, S. M., Doumit, M. E., 1995, A muscle hypertrophy condition in lamb (Callipyge): Characterization of effects on muscle growth and meaty quality traits, *J. Animal Sci.*, 73:3596-3607.

Leahy, M. M., Reineccius, G. A., 1989, Kinetics of formation of alkylpyrazines: effect of type of amino acid and type of sugar, in: *Flavor Chemistry: Trends and Developmenmt*, R. Teranishi, R. G. Buttery, F. Shahidi, eds., ACS Symposium Series 388; American Chemical Society, Washington, DC, pp. 76-91.

MacLeod, G., 1986, The scientific and technological basis of meat flavors, in: *Developments in Food Flavors*, G. C. Birch, M. G. Lindley, eds., Elsevier, London, pp. 191-223.

MacLeod, G., Ames, J. M., 1988, Soy flavor and its improvement, *CRC Crit. Rev. Food Sci. Nutr.*, 27:219-400.

Macleod, G., 1998, The flavor of beef, in: *Flavor of Meat, Meat Products and Seafoods*, F. Shahidi, ed., Blackie Academic & Professional, New York, pp. 27-60.

Madruga, M. S., and Mottram, D. S., 1995, The effect of pH on the formation of Maillard-derived aroma volatile using a cooked meat system, *J. Sci. Food Agric.*, 68:305-310.

Maga, J. A., 1983, Flavor potentiators, *Crit. Rev. Food Sci. Nutr.*, *18*:231-312.

Maga, J. A., 1998, Umami flavor of meat, in: *Flavor of Meat, Meat Products and Seafoods*, F. Shahidi, ed., Blackie Academic & Professional, New York, pp. 197-216.

Mottram, D., 1994, Meat flavor, in: *Understanding Natural Flavours*, J. R. Piggott, A. Paterson, eds., Blackie Academic & Professional, New York, pp. 141-163.

Mottram, D. S., Madruga, M. S. 1994, Important sulfur-containing compounds in foods, in: *Sulfur Compounds in Foods*, C. J. Mussinan, M. Keelan, eds., American Chemical Society, Washington, DC, pp. 180-187.

Mottram, D. S., Whitfield, F. B., 1995a, Volatile compounds from the reaction of cysteine, ribose, and phospholipid in low-moisture systems, *J. Agric. Food Chem.*, 43:984-988.

Mottram, D. S., Whitfield, F. B., 1995b, Maillard-lipid interaction in nonaqueous systems: volatiles from the reaction of cysteine and ribose with phosphatidylcholine, *J. Agric. Food Chem.*, *43*:1302-1306.

Mottram, D. S., 1998, Flavor formation in meat and meat products: a review, *Food Chem.*, *62*:415-424.

Parliament, T. H., Stahl, H. D., 1994, Generation of furfuryl mercaptan in cysteine-pentose model systems in relation to roasted coffee, in: *Sulfur Compounds in Foods*, C. J. Mussinan, M. E. Keelan, eds., ACS Symposium Series No. 564, American Chemical Society, Washington, DC, pp. 161-170.

Shibamoto, T., Yeo, H., 1994, Flavor in the cysteine-glucose model system prepared in microwave and conventional ovens, in: *Thermally Generated Flavors: Maillard, Microwave, and Extrusion Processes*, T. H. Parliment, M. J. Morello, R. J. McGorrin, eds., ACS Symposium Series No. 543, American Chemical Society, Washington, DC, pp. 457-465.

Shu, C. K., Hagedorn, M. L., Mookherjee, B. D., Ho, C.T., 1985, pH effect on the volatile components in the thermal degradation of cysteine, *J. Agric Food Chem.*, *33*:442-446.

Sun, J. B., Severson, R. F., Kays, S. J., 1993, Quantitative technique for measuring volatile components of baked sweet potatoes, *Hortscience*, *28*:1110-1113.

Tressl, R., Kelak, B., Martin, N., Kersten, E., 1989, Formation of amino acid specific Maillard products and their contribution to thermally generated aromas, in: *Thermal Generation of Aromas*, T. H. Parliament, R. J. McCorrin, C. T. Ho, eds., ACS Symposium Series 409, American Chemical Society, Washington, DC, pp. 156-171.

Vernin, G., Parkanyi, C., 1982, Mechanisms of formation of heterocyclic compounds in Maillard and pyrolysis reactions, in: *Chemistry of Heterocyclic Compounds in Flavors and Aromas*, G. Vernin, ed., Halsted Press, New York, pp. 151-207.

Wei, A., Mura, K., Shibamoto, T., 2001, Antioxidative activity of volatile chemicals extracted from beer, *J. Agric. Food Chem.*, 49:4097-40101.

Werkhoff, P., Brüning, J., Emberger, R., Güntert, M., Hopp, R., 1993, Flavor chemistry of meat volatiles. New results on flavor components from beef, pork, and chicken, in: *Recent Developments in Flavor and Fragrance Chemistry*, R. Hopp, K. Mori, eds., VCH, Weinheim, pp. 183-213.

Whitfield, F. B., Mottram, D. S., Shaw, K. J., 1993, The formation of some novel thiols and disulphides in model systems containing 4-hydroxy-5-methyl-3(2H)-furanone, in: *Progress in Flavor Precursor Studies,* P. Schreier, P. Winterhalter, eds., Allured Publ. Co., Carol Stream, IL, pp. 395-400.

Yu, T. H., Wu, C. M., Ho, C.T., 1994, Volatile compounds generated from thermal interactions of inosine-5'-monophosphate and alliin or deoxyalliin, in: *Sulfur Compounds in Foods,* C. J. Mussinan, M. E. Keelan, eds., ACS Symposium Series No. 564, American Chemical Society, Washington, DC, pp. 188-198.

Zhang, Y., Ho, C.T., 1991, Formation of meat like aroma compounds from thermal reaction of inosine 5'-monophosphate with cysteine and glutathione, *J. Agric. Food Chem.,* 39:1145-1148.

INDEX